D1560571

Study Guide to

CALCULUS

Gilbert Strang

Jennifer Carmody

Correlated with the text:

CALCULUS by Gilbert Strang
First Printing 1991, Revised Printing 1992

WELLESLEY-CAMBRIDGE PRESS
Box 812060 Wellesley MA 02181

Study Guide to Calculus
Gilbert Strang and Jennifer Carmody
ISBN 0-9614088-4-7
to accompany **CALCULUS** by Gilbert Strang

Typeset by M&N Toscano, Somerville MA
Printed by Glidway Corporation, Canton MA

Other texts from Wellesley-Cambridge Press

Introduction to Applied Mathematics
Gilbert Strang, ISBN 0-9614088-0-4

An Analysis of the Finite Element Method
Gilbert Strang and George Fix, ISBN 0-13-032946-0

Introduction to Linear Algebra
Gilbert Strang, ISBN 0-9614088-5-5, First Printing 1993.

Wellesley-Cambridge Press
Box 812060
Wellesley MA 02181 USA
(617) 431-8488

STUDY GUIDE TO CALCULUS

This Student Study Guide accompanies the textbook ***Calculus*** by Gilbert Strang. It is correlated section by section with the essential points of the text. The Guide contains four components which experience has shown are most helpful:

1. Model problems with comments and complete solutions.
2. Extra drill problems included with exercises for chapter review.
3. Read-through questions from the text with the blanks filled in.
4. Solutions to selected even-numbered problems in each section.

The Guide can be obtained from bookstores or by writing directly to Wellesley–Cambridge Press. Our address is Box 812060, Wellesley MA 02181 (prepayment by check). The press also publishes two other basic textbooks by Gilbert Strang: ***Introduction to Linear Algebra*** and ***Introduction to Applied Mathematics***. This Study Guide shares the same goal, to teach mathematics in an active and purposeful way.

1.1 Velocity and Distance

The first step in calculus (in my opinion) is to begin working with *functions*. The functions can be described by formulas or by graphs. The first graphs to study are straight lines, and the functions that go with them are linear: $y = mx + b$ or (using other letters) $f = vt + C$. It takes practice to connect the formula $y - 4 = 7(x - 2)$ to the straight line with slope 7 going through the point (2,4). You need to be able to rewrite that formula as $y = 7x - 10$. You also need to be able to find the formula from the graph. Here is an example: *Find the equations for these straight lines.*

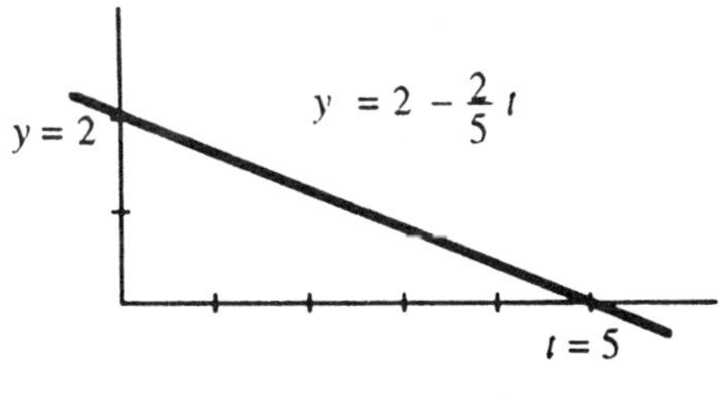

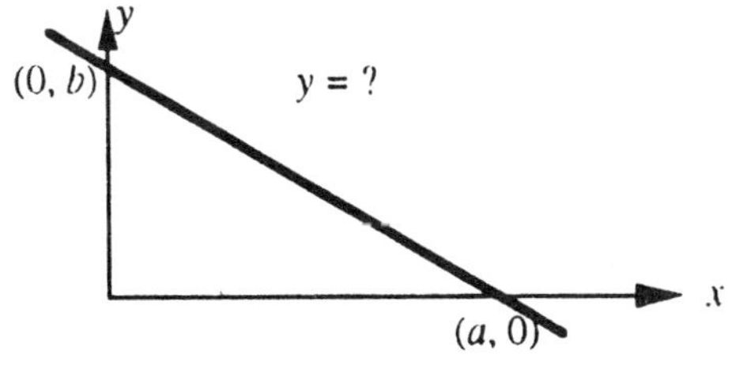

The first line goes down 2 and across 5. The slope is $\frac{-2}{5} = -.4$. The minus sign is because y decreases as t increases – the slope of the graph is negative. The "starting point" is $y = 2$ when $t = 0$. The equation is $y = 2 - .4t$. Check that this gives $y = 0$ when $t = 5$, so the formula correctly predicts that second point.

Problem 1: *Find the equation for the second line.* Comment: The points $(0, b)$ and $(a, 0)$ are given by letters not numbers. This is typical of mathematics, which is not so much about numbers as most people think – it is really about patterns. Sometimes numbers help, other times they get in the way. Many exercises in mathematics books are really asking you to find the pattern by solving a general problem (with letters) instead of a special problem (with numbers). But don't forget: You can always substitute 5 and 2 for the letters a and b.

Solution to Problem 1: The slope is $-\frac{b}{a}$. The equation is $y = b - \frac{b}{a}x$. Compare with $y = 2 - \frac{2}{5}t$.

Pairs of functions. Calculus deals with two functions at once. We have to understand both functions, and *how they are related.* In writing the book I asked myself: Where do we find two functions? What example can we start with? The example should be familiar and it should be real – not just made up. The best examples build on what we already know. The first pair of functions is distance $f(t)$ and velocity $v(t)$.

Comment: This example is taken from *life*, not from physics. Don't be put off by symbols like $f(t)$ and $v(t)$. The velocity is certainly connected to the distance traveled. This connection is clearest when the velocity is constant. If you go 120 miles in 2 hours at steady speed, then $v = \frac{120 \text{ miles}}{2 \text{ hours}} = 60$ miles per hour. That example converts to straight line graphs: The distance graph goes up with slope 60. Starting from $f(0) = 0$, the equation for the line is $f(t) = 60t$.

Most people see how $f = 60t$ leads to $v = 60$. Now look at the connection between $f(t)$ and $v(t)$ in the next graphs.

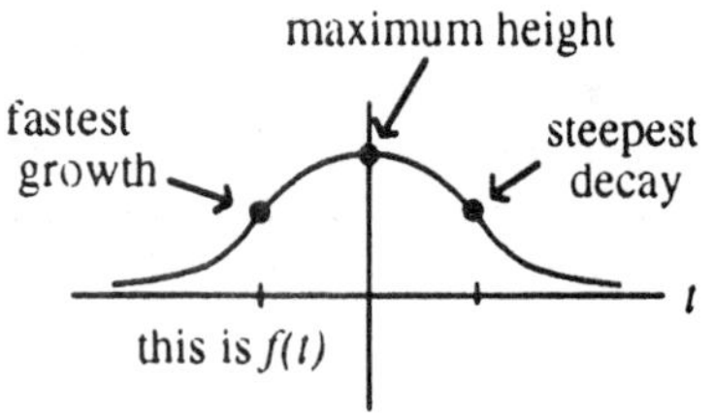

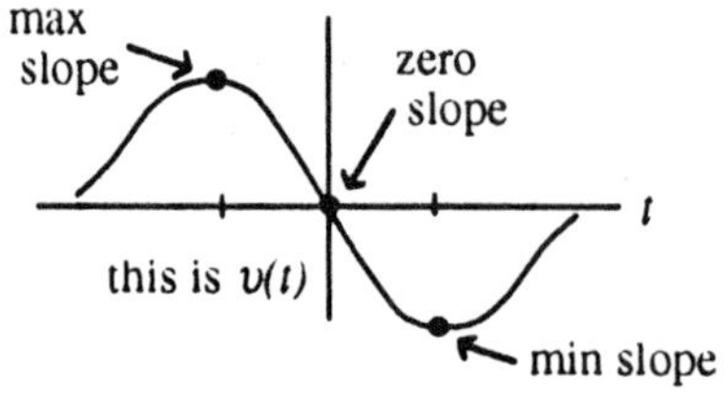

What do you see? The slope of f is zero at the center. Thus $v = 0$ where f is a maximum. The maximum of v is where the graph of f is steepest (upwards). Then v is largest (positive). When the graph of f flattens out, v drops back toward zero.

With formulas I could specify these functions exactly. The distance might be $f(t) = \frac{1}{1+t^2}$. Then Chapter 2 will find $\frac{-2t}{(1+t^2)^2}$ for the velocity $v(t)$. Very often calculus is swept up by formulas, and the ideas get lost. You need to know the rules for computing $v(t)$, and exams ask for them, but it is not right for calculus to turn into pure manipulations. Our goal from the start is to *see the ideas.*

Again comes the question: Where to go after $f = 60t$ and $v = 60$? The next step is to allow two velocities (two slopes). The velocity jumps. The distance graph switches from one straight line to another straight line – with a different slope. It takes practice to write down the formulas and draw the graphs. Here are six questions about straight line graphs that change slopes. The figure shows $f(t)$, the questions are about $v(t)$.

Answers are given for the first graph. Test yourself on the second graph.

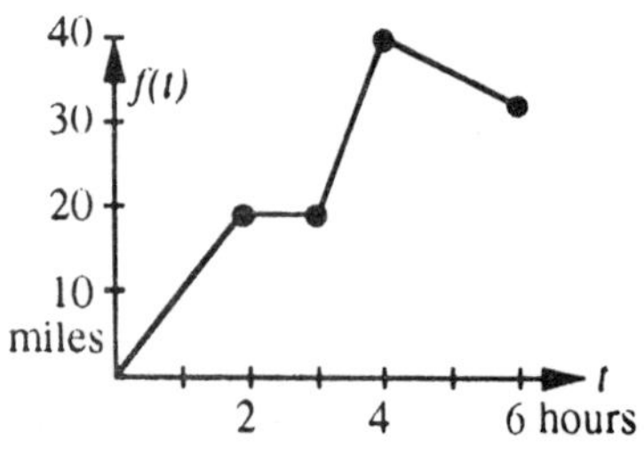

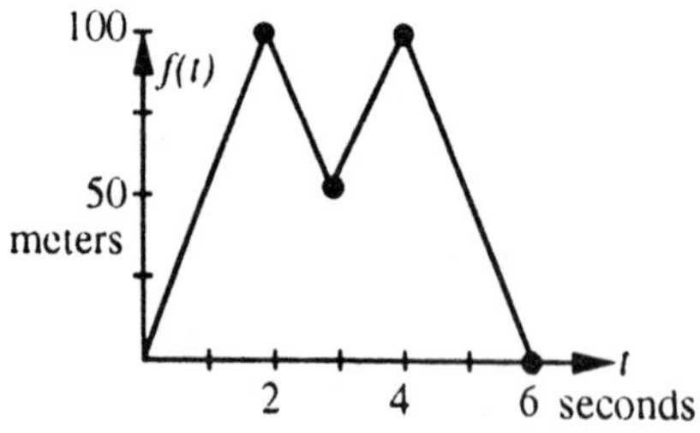

2 During the first two hours, the car travels ______ miles. The velocity during this time is ______ miles per hour. The slope of the distance graph from $t = 0$ to $t = 2$ is ______.
Answer: Miles traveled = 20, velocity = 10 miles per hour, slope = $\frac{20}{2} = 10$.

3 From $t = 2$ to $t = 3$ the car travels ______ miles. The velocity during this time is ______ miles per hour. The slope of this segment is ______. Answer: 0,0,0.

4 From $t = 3$ to $t = 4$ the car travels ______ miles. The velocity during this time is ______ mph. The slope of the f-graph from $t = 3$ to $t = 4$ is ______. Answer: $20, 20, \frac{40-20}{4-3} = 20$.

5 From $t = 4$ to $t = 6$, the car travels *backwards* ______ miles. The change in f is ______ miles. The velocity is ______ miles per hour. The slope of this segment of the f-graph is ______.
Answer: Backwards 10 miles, change in f is -10, velocity -5 miles per hour, slope $\frac{30-40}{6-4} = -5$.

6 *Draw graphs of $v(t)$ from those two graphs of $f(t)$.*

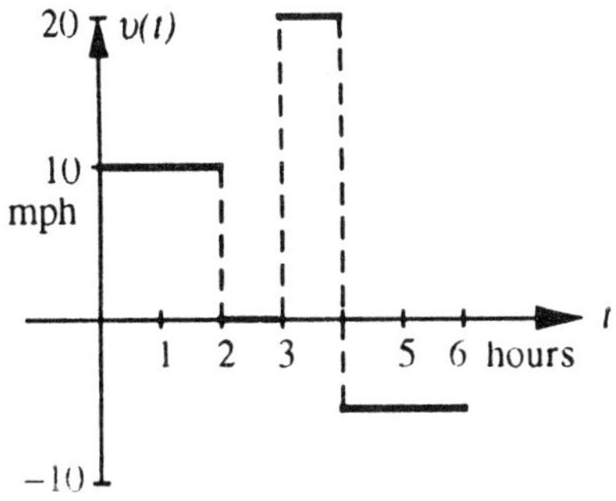

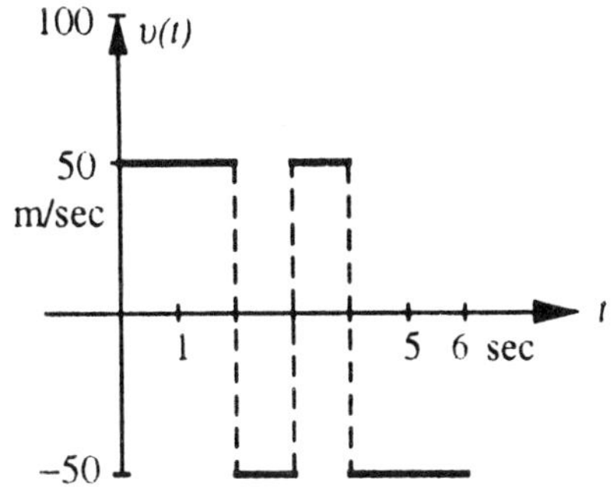

7 Give a 4-part formula for $f(t)$. Each piece is a line segment described by a linear function $f(t) = vt + C$. The first piece is just $f(t) = vt$ because it starts from $f(0) = 0$.

- Use the form $vt + C$ if you know the slope v and the height C when $t = 0$. With different letters this form is $y = mx + b$. Use $f(t) = f(3) + v(t - 3)$ if you know v and you also know the distance $f(3)$ at the particular time $t = 3$. Section 2.3 on *tangent lines* uses both forms. The second form becomes $y = f(a) + f'(a)(x - a)$ and the slope is denoted by $f'(a)$.

From $t = 3$ to $t = 4$, our slope is 20 and the starting point is (3,20). The formula for this segment is $f(t) = 20 + 20(t - 3), 3 \le t \le 4$. This simplifies to $f(t) = 20t - 40$. The complete solution is:

$$f(t) = \begin{cases} 10t + 0, & 0 \le t \le 2 \\ 20, & 2 \le t \le 3 \\ 20 + 20(t-3), & 3 \le t \le 4 \quad \text{or} \quad 20t - 40 \\ 40 - 5(t-4), & 4 \le t \le 6 \quad \text{or} \quad -5t + 60 \end{cases}$$

IMPORTANT If we are given $v(t)$ we can discover $f(t)$. But we do need to know the distance at one particular time like $t = 0$. The table of f's below was started from $f(0) = 0$. Questions **8–11** allow you to fill in the missing parts of the table. Then the piecewise constant $v(t)$ produces a piecewise linear $f(t)$.

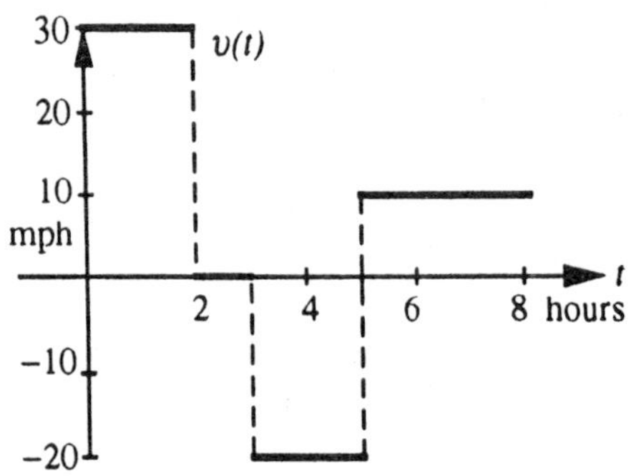

t	$f(t)$
0	0
2	60
3	$60 + 0$
5	$60 + 0 +$ ______
8	$60 + 0 + (-40)$
	$+$ ______ $=$ ______

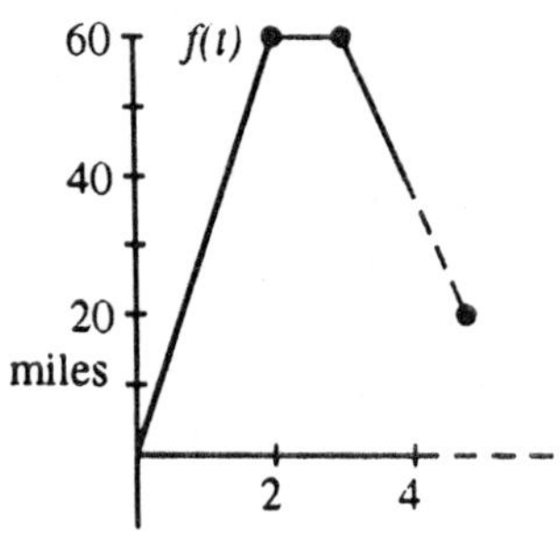

8 From $t = 0$ to $t = 2$, the velocity is ______ mph. The distance traveled during this time is ______ miles. The area of this rectangular portion of the graph is ______ $\times$ ______ $=$ ______. Answer: 30, 60, $30 \times 2 = 60$.

9 From $t = 2$ to $t = 3$, the velocity is ______ miles per hour. The "area" of this portion of the graph (a rectangle with no height) is ______ $\times$ ______ $=$ ______. Answer: $0, 1 \times 0 = 0$.

10 From $t = 3$ to $t = 6$, the velocity is ______ mph. During this time the vehicle travels *backward* ______ miles. The area of this rectangular portion is ______ $\times$ ______ $=$ ______. Keep in mind that area below the horizontal axis counts as *negative*. Answer: $-20, 40, -20 \times 2 = -40$.

11 From $t = 5$ to $t = 8$ the velocity is ______ mph. During this time the vehicle travels forward ______ miles. The area of this rectangular portion is ______ $\times$ ______ $=$ ______. Answer: 10, 30, $10 \times 3 = 30$.

12 Using the information in **8-11**, finish the table and complete the distance graph out to time $t = 8$. Connect the points $(t, f(t))$ in the table with straight line segments.

13 What is the distance traveled from $t = 0$ to $t = 4$? From $t = 0$ to $t = 7$?
Answer: The points (4, 40) and (7, 40) can be read directly from the graph. The distance in each case is 40 miles.

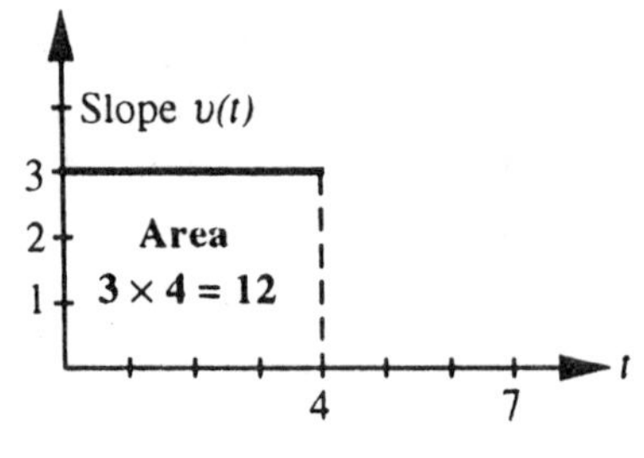

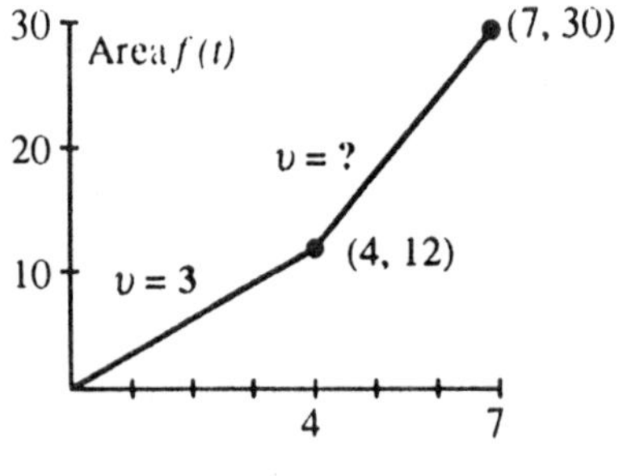

14 This is Problem 1.1.32 in the text. Suppose $v = 3$ up to time $t = 4$. What constant velocity will lead to $f(7) = 30$ if $f(0) = 0$? Give graphs and formulas for $v(t)$ and $f(t)$.

- We know that $f = 12$ when $t = 4$ because distance = rate $\times$ time = $3 \times 4 = 12$. Note that the slope of the f-graph from $t = 0$ to $t = 4$ is 3, the velocity. Since we want $f(7) = 30$, we must go $30 - 12 = 18$ more miles. The slope of the second piece of the f-graph should be $\frac{30-12}{7-4} = \frac{18}{3} = 6$. The desired velocity is $v = 6$. The equation of this segment is $f(t) = 12 + 6(t - 4)$. At the breakpoint $t = 4$, the velocity $v(4)$ is *not* defined, but the distance $f(4)$ *is* defined. The symbol in $v(t)$ is $<$ while the symbol in $f(t)$ is $\leq$.

$$v(t) = \begin{cases} 3, & 0 < t < 4 \\ 6, & 4 < t < 7 \end{cases} \qquad f(t) = \begin{cases} 3t + 0, & 0 \le t \le 4 \\ 12 + 6(t-4), & 4 \le t \le 7. \end{cases}$$

Here are solutions to the read-through questions and selected even-numbered exercises for Section 1.1.

Starting from $f(0) = 0$ at constant velocity v, the distance function is $f(t) = \mathbf{vt}$. When $f(t) = 55t$ the velocity is $v = \mathbf{55}$. When $f(t) = 55t + 1000$ the velocity is still **55** and the starting value is $f(0) = \mathbf{1000}$. In each case v is the **slope** of the graph of f. When $\mathbf{v(t)}$ is negative, the graph of $\mathbf{f(t)}$ goes downward. In that case area in the v–graph counts as **negative**.

Forward motion from $f(0) = 0$ to $f(2) = 10$ has $v = \mathbf{5}$. Then backward motion to $f(4) = 0$ has $v = \mathbf{-5}$. The distance function is $f(t) = 5t$ for $0 \le t \le 2$ and then $f(t)$ equals $\mathbf{5(4-t)}$ (not $-5t$). The slopes are **5** and **−5**. The distance $f(3) = \mathbf{5}$. The area under the v–graph up to time 1.5 is **7.5**. The domain of f is the time interval $\mathbf{0 \le t \le 4}$, and the range is the distance interval $\mathbf{0 \le f \le 10}$. The range of $v(t)$ is only **5 and −5**.

The value of $f(t) = 3t + 1$ at $t = 2$ is $f(2) = \mathbf{7}$. The value 19 equals $f(\mathbf{6})$. The difference $f(4) - f(1) = \mathbf{9}$. That is the change in distance, when $4 - 1$ is the change in **time**. The ratio of those changes equals **3**, which is the **slope** of the graph. The formula for $f(t) + 2$ is $3t + 3$ whereas $f(t+2)$ equals $\mathbf{3t+7}$. Those functions have the same **slope** as f : the graph of $f(t) + 2$ is shifted **up** and $f(t+2)$ is shifted **to the left**. The formula for $f(5t)$ is $\mathbf{15t+1}$. The formula for $5f(t)$ is $\mathbf{15t + 5}$. The slope has jumped from 3 to **15**.

The set of inputs to a function is its **domain**. The set of outputs is its **range**. The functions $f(t) = 7 + 3(t-2)$ and $f(t) = vt + C$ are **linear.** Their graphs are **straight lines** with slopes equal to **3** and **v**. They are the same function, if $v = \mathbf{3}$ and $C = \mathbf{1}$.

10 $v(t)$ is negative-zero-positive; $v(t)$ is above 55 then equal to 55; $v(t)$ increases in jumps; $v(t)$ is zero then positive. All with corresponding $f(t)$.

26 The function increases by 2 in one time unit so the slope (velocity) is 2; $f(t) = \mathbf{2t + C}$ with constant $C = f(0)$.

36 At $t = 0$ the reading was $.061 + 10(.015) = \mathbf{.211}$. A drop of $.061 - .04 = .021$ would take $\mathbf{.021/.015}$ hours. This was the Exxon Valdez accident in Alaska.

1.2 Calculus Without Limits (page 14)

This section was rewritten for the second printing of the book, in order to bring out the central ideas. One idea starts with numbers like $f = 3, 5, 9$. Their differences are 2 and 4. The sum of those differences is $2 + 4 = 6$. This equals $9 - 3$, or $f_{\text{last}} - f_{\text{first}}$. (It is like a card trick, where you think of numbers and only tell me the differences. Then I tell you the last number minus the first number.) The middle number 5 cancels out:

$$(5-3) + (9-5) = 9 - 3 \quad \text{or} \quad (f_1 - f_0) + (f_2 - f_1) = f_2 - f_0.$$

Taking sums is the "inverse" *of taking differences.* This is an early pointer to calculus.

- The "*derivative*" of $f(t)$ is the slope $v(t)$. This is like differences.
- The "*integral*" of $v(t)$ is the area $f(t)$. This is like sums.

Teaching question: *Why mention these ideas so early?* They are not proved. They can't even be explained in complete detail. Here is my answer: You the student should know where the course is going. I believe in seeing an idea several times, instead of "catch it once or lose it forever". Here was a case where a few numbers give a preview of the Fundamental Theorem of Calculus.

As the idea unfolds and develops, so does the notation. At first you have specific numbers 3,5,9. Then you have general numbers f_0, f_1, f_2. Then you have a piecewise linear function. Then you have a general function $f(t)$.

Similarly: At first you have differences $5-3$ and $9-5$. Then you have $v_1 = f_1 - f_0$ and $v_j = f_j - f_{j-1}$. Then you have a piecewise constant function. Then you have the derivative $v(t) = \frac{df}{dt}$.

The *equation for a straight line* is another idea that unfolds and develops. It was already in some of the problems for Section 1.1. Now Section 1.2 brings it out again, especially the income tax example. The tax in the second income bracket is

$$f(x) = \text{ tax on } \$20,350 + (\text{ tax rate } .28)(\text{income over } \$20,350).$$

This is so typical of what comes later: $f(x) = f(a) + (\text{slope})\ (x-a)$. Do those letters x and a make the equation less clear? For many students I believe they do, at least at first. The number 20,350 is definite where the letter a is vague. The number .28 is specific where the slope $f'(a)$ will be general. A teacher is always caught by this dilemma. Mathematics expresses the pattern by symbols, but it is understood first for numbers.

The best way is to see the same idea several times, which we do for straight lines. Section 2.3 finds the tangent line to a curve. Section 3.1 uses the slope for a linear approximation. Section 3.7 explains Newton's method for solving $f(x) = 0$ – replace the curve by its tangent line. Will you see that the underlying idea is the same? I hope so, because it is the key idea of differential calculus: *Lines stay close to curves* (at least for a while).

1 Suppose $v = 3$ for $t < T$ and $v = 1$ for $t > T$. If $f(0) = 0$, find formulas for $f(t)$.

- The equation of the first segment is $f(t) = 0 + 3t$. This continues to $t = T$ when $f = 3T$. Using the fact that $(T, 3T)$ starts the second segment and $v = 1$ is the slope, we find $f(t) = 3T + 1(t-T)$ for $T \le t$.

2 Suppose f_0, f_1, f_2, f_3 are the distances 0, 8, 12, 14 at times $t = 0, 1, 2, 3$. Find velocities v_1, v_2, v_3 and a formula that fits v_j. Graph $v(t)$ and $f(t)$ with constant and linear pieces.

- The differences are $v_1 = 8, v_2 = 4$, and $v_3 = 2$. The velocities are halved at each step: $v_1 = \frac{1}{2}(16), v_2 = \frac{1}{4}(16), v_3 = \frac{1}{8}(16)$. The formula $v_j = (\frac{1}{2})^j(16)$ or $v_j = 16/2^j$ fits all velocities.

3 Using the differences $v_1, v_2, \cdots, v_j$ and the starting value f_0, give a formula for f_j.

- $f_j = f_0 + v_1 + v_2 + \cdots + v_j$. ***You should notice this formula.***

4 Suppose the tax rates are increased by 2 percentage points to .17, .30, and .33. At the same time, a tax credit of \$500 is allowed. If your original income was x, does your tax $f(x)$ go up or down after the changes? What if the \$500 is only a deduction from income?

- The answer depends on x. At $x = \$10,000$ the rate increase costs only 2% of \$10,000, which is \$200. You win. At $x = \$25,000$ the 2% rate increase balances the \$500 credit.

  If the \$500 is *deducted from income* of \$10,000, then you pay 17% of \$9,500 instead of 15% of \$10,000. This time you lose. To find the income at which deduction balances the 2% rate increase, solve $.15x = .17(x - 500)$ to find $x = \$4,250$.

5 (Algebra) If the distances are $f_j = j^2$, show that the velocities are $v_j = 2j - 1$.

- Since the difference v_j is $f_j - f_{j-1}$, you have to subtract $(j-1)^2$ from j^2 :

$$j^2 - (j-1)^2 = j^2 - (j^2 - 2j + 1) = 2j - 1.$$

Please notice the summary at the end of Section 1.2, and then go over these read-through problems. Here are the blanks filled in, followed by even-numbered solutions.

Start with the numbers $f = 1, 6, 2, 5$. Their differences are $v = \mathbf{5, -4, 3}$. The sum of those differences is **4**. This is equal to f_{last} minus $\mathbf{f_{first}}$. The numbers 6 and 2 have no effect on this answer because in $(6-1) + (2-6) + (5-2)$ the numbers 6 and 2 **cancel**. The slope of the line between $f(0) = 1$ and $f(1) = 6$ is **5**. The equation of that line is $f(t) = \mathbf{1 + 5t}$.

With distances 1, 5, 25 at unit times, the velocities are **4 and 20**. These are the **slopes** of the f-graph. The slope of the tax graph is the tax **rate**. If $f(t)$ is the postage cost for t ounces or t grams, the slope is the **cost** per **ounce (or per gram)**. For distances 0, 1, 4, 9 the velocities are **1, 3, 5**. The sum of the first j odd numbers is $f_j = \mathbf{j^2}$. Then f_{10} is **100** and the velocity v_{10} is **19**.

The piecewise linear sine has slopes **1, 0, −1, −1, 0, 1**. Those form a piecewise **constant** cosine. Both functions have **period** equal to 6, which means that $f(t+6) = \mathbf{f(t)}$ for every t. The velocities $v = 1, 2, 4, 8, \ldots$ have $v_j = \mathbf{2^{j-1}}$. In that case $f_0 = 1$ and $f_j = \mathbf{2^j}$. The sum of 1, 2, 4, 8, 16 is **31**. The difference $2^j - 2^{j-1}$ equals $\mathbf{2^{j-1}}$. After a burst of speed V to time T, the distance is **VT**. If $f(T) = 1$ and V increases, the burst lasts only to $T = \mathbf{1/V}$. When V approaches infinity, $f(t)$ approaches a **step** function. The velocities approach a **delta** function, which is concentrated at $t = 0$ but has area **1** under its graph. The slope of a step function is **zero or infinity.**

8 $f(t) = \mathbf{1 + 10t}$ for $0 \le t \le \frac{1}{10}$, $f(t) = \mathbf{2}$ for $t \ge \frac{1}{10}$

10 $f(3) = \mathbf{12}$; $g(f(3)) = g(12) = \mathbf{25}$; $g(f(t)) = g(4t) = \mathbf{8t + 1}$. When t is changed to $4t$, distance increases **four** times as fast and the velocity is multiplied by 4.

28 The second difference $f_{j+1} - 2f_j + f_{j-1}$ equals $(f_{j+1} - f_j) - (f_j - f_{j-1}) = v_{j+1} - v_j = 4$.

32 The period of $v + w$ is **30**, the smallest multiple of both 6 and 10. Then v completes five cycles and w completes three. An example for functions is $v = \sin \frac{\pi x}{3}$ and $w = \sin \frac{\pi x}{5}$.

42 The ratios $\frac{\Delta f}{\Delta t}$ are $\frac{e-1}{1} = 1.718$ and $\frac{e^{.1}-1}{.1} = 1.052$ and $\frac{e^{.01}-1}{.01} = 1.005$. They are approaching 1.

1.3 The Velocity at an Instant (page 21)

Worked examples 1-4 bring out the key ideas of Section 1.3. You are computing the slope of a curve - genuine calculus! To compute this slope at a particular time $t = 1$, find the change in distance $f(1+h) - f(1)$. To compute the velocity at a general time t, find the change $f(t+h) - f(t)$. In both cases, divide by h for average slope. Let $h \to 0$ for instantaneous velocity = slope at point.

1 For $f(t) = t^2 - t$ find the *average* speed between (a) $t = 1$ and $t = 2$ (b) $t = 0.9$ and $t = 1.0$ (c) $t = 1$ and $t = 1 + h$. (d) Use part (c) as $h \to 0$ to find the *instantaneous* speed at $t = 1$. (e) What is the *average* speed from t to $t + h$? (f) What is the formula for $v(t)$?

- (a) $f(2) = 2^2 - 2 = 2$ and $f(1) = 1^2 - 1 = 0$. Average speed $= \frac{2-0}{2-1} = 2$
- (b) $f(0.9) = 0.81 - 0.9 = -0.09$. Average speed is $\frac{f(1) - f(0.9)}{1 - (0.9)} = \frac{.09}{0.1} = 0.9$

- (c) $\frac{f(1+h)-f(1)}{(1+h)-(1)} = \frac{[(1+h)^2-(1+h)]-[1^2-1]}{(1+h)-1} = \frac{[(1+2h+h^2)-(1+h)]-0}{h} = \frac{h^2+h}{h} = h+1.$
- (d) Let h get extremely small. Then $h+1$ is near 1. The speed at $t=1$ is $v=1$.
- (e) $\frac{f(t+h)-f(t)}{(t+h)-t} = \frac{[(t+h)^2-(t+h)]-[t^2-t]}{h} = \frac{t^2+2ht+h^2-t-h-t^2+t}{h}.$
After cancelling this is

$$\frac{h^2+2ht-h}{h} = h+2t-1 = \textit{average speed}.$$

- (f) Let h go to zero. The average $h+2t-1$ goes to $0+2t-1$. Thus $v(t)=2t-1$. *Check* $v=1$ at $t=1$.

2 If $v(t)=6-2t$, find a formula for $f(t)$ and graph it.

- $f(t)$ is the area under the v-graph from 0 to t. For small t, this is the area of a *trapezoid* with heights 6 and $6-2t$. The base is t. Since area = average height times base, $f(t) = \frac{1}{2}(6+6-2t)t = (6-t)t = 6t-t^2$. At $t=3$ we reach $f(3)=18-9=9$. This is the area of the triangle.

  Does this area formula $f(t)=6t-t^2$ continue to apply after $t=3$, when some area is negative? *Yes it does.* The triangle beyond $t=3$ has base $t-3$ and height (or depth) $6-2t$. Its area is $\frac{1}{2}bh = \frac{1}{2}(t-3)(6-2t) = -9+6t-t^2$. After adding the positive area 9 of the first triangle, we still have $6t-t^2$.

  The negative height automatically gave negative area. The graph is a parabola that turns down at $t=3$, when new area is negative.

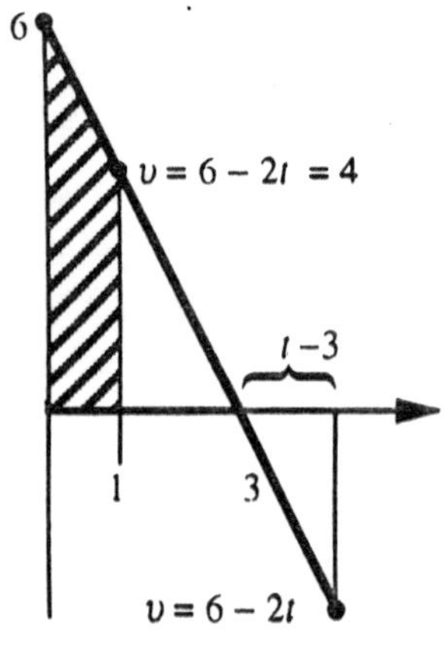

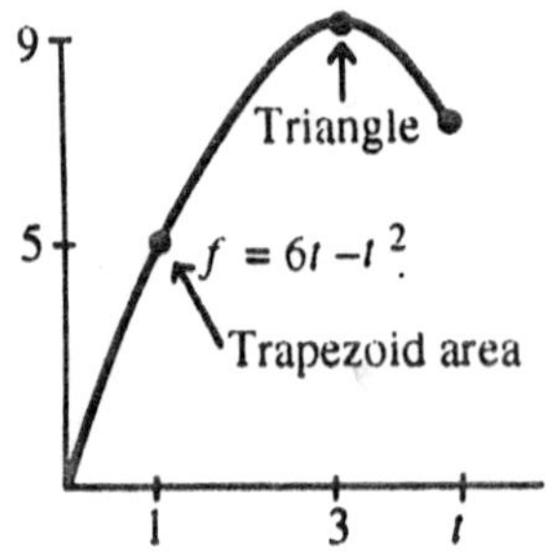

3 If $f(t)=4t^2$, find the slope at time t. Start with the average from t to $t+h$.

- Step 1: The average slope is

$$\frac{f(t+h)-f(t)}{(t+h)-t} = \frac{4(t+h)^2-4t^2}{h} = \frac{(4t^2+8th+4h^2)-4t^2}{h}$$

  This simplifies to $\frac{8th+4h^2}{h} = \mathbf{8t+4h}$.

- Step 2: Decide what happens to the average slope as h approaches zero. The average $8t + 4h$ approaches the slope at a point (the instantaneous velocity) which is $8t$.

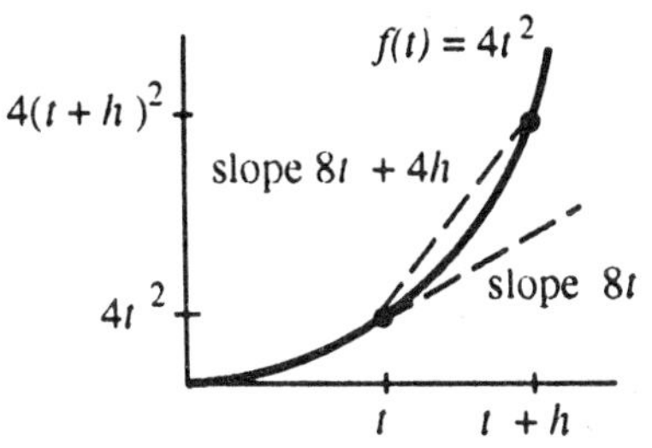

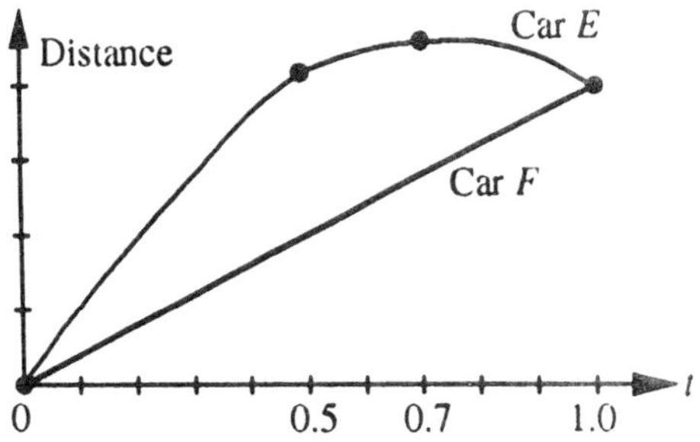

4 The graph shows distances for two cars. E and F start and end at the same place (why?) but their trips are different. Car F travels at a steady velocity. Describe what is happening to car E at these times:

(a) $t = 0$ to $t = .5$ (b) $t = .5$ (c) $t = .5$ to $t = .7$ (d) $t = .7$ (e) $t = .7$ to $t = 1$.

- (a) Car E covers more ground than F. It is going faster than F.
- (b) Since the slope of the graph *at this instant* is the same for E and F, their speed is the same.
- (c) E gains less ground than F, so E is going slower than F.
- (d) *At this instant E has speed zero.* The distance graph is level. *Maximum distance, zero speed.*
- (e) Car E loses ground, the slope of the distance graph is negative: E is going backward.

Here are the read-throughs and selected solutions to even-numbered problems.

Between the distances $f(2) = 100$ and $f(6) = 200$, the average velocity is $\frac{100}{4} = \mathbf{25}$. If $f(t) = \frac{1}{4}t^2$ then $f(6) = \mathbf{9}$ and $f(8) = \mathbf{16}$. The average velocity in between is **3.5**. The instantaneous velocities at $t = 6$ and $t = 8$ are **3** and **4**.

The average velocity is computed from $f(t)$ and $f(t+h)$ by $v_{\text{ave}} = \mathbf{\frac{1}{h}(f(t+h) - f(t))}$. If $f(t) = t^2$ then $v_{\text{ave}} = \mathbf{2t + h}$. From $t = 1$ to $t = 1.1$ the average is **2.1**. The instantaneous velocity is the **limit** of v_{ave}. If the distance is $f(t) = \frac{1}{2}at^2$ then the velocity is $v(t) = \mathbf{at}$ and the acceleration is **a**.

On the graph of $f(t)$, the average velocity between A and B is the slope of **the secant line**. The velocity at A is found by **letting B approach A.** The velocity at B is found by **letting A approach B.** When the velocity is positive, the distance is **increasing.** When the velocity is increasing, the car is **accelerating.**

2 (c) $\frac{\frac{1}{2}a(t^2+2th+h^2)-\frac{1}{2}at^2}{h} = at + \frac{1}{2}ah$. The limit at $h = 0$ is $v = at :=$ (acceleration) $\times$ (time). (e) $\frac{6-6}{h} = 0$ (limit is 0); (f) the limit is $v(t) = 2t$ (and $f(t) = t^2$ gives $\frac{(t+h)^2-t^2}{h} = 2t + h$).

14 True (the slope is $\frac{\Delta f}{\Delta t}$); False (the curve is partly steeper and partly flatter than the secant line which gives the average slope); True (because $\Delta f = \Delta F$); False (V could be larger in between).

18 The graph is a parabola $f(t) = \frac{1}{2}t^2$ out to $f = 2$ at $t = 2$. After that the slope of f stays constant at 2.

20 Area to $t = 1$ is $\frac{1}{2}$; to $t = 2$ is $\frac{3}{2}$; to $t = 3$ is 2; to $t = 4$ is $\frac{3}{2}$; to $t = 5$ is $\frac{1}{2}$; area from $t = 0$ to $t = 6$ is zero. The graph of $f(t)$ through these points is parabola-line-parabola (symmetric)-line-parabola to zero.

26 $f(t) = t - t^2$ has $v(t) = 1 - 2t$ and $f(3t) = 3t - 9t^2$. The slope of $f(3t)$ is $3 - 18t$. This is $\mathbf{3v(3t)}$.

28 To find $f(t)$ multiply the time t by the average velocity. This is because $v_{\text{ave}} = \frac{f(t)-f(0)}{t-0} = \frac{f(t)}{t}$.

1.4 Circular Motion (page 28)

This section is important but optional (if that combination is possible). It is important because it introduces sines and cosines, and it also introduces the idea of a "parameter" t. No other example could do this better. These are *periodic* functions, which repeat every time you go around the circle. For any class studying physics, circular motion is essential. For calculus the section is optional, because we return in Section 2.4 to compute the derivatives of $\sin t$ and $\cos t$ from $\Delta f/\Delta t$.

It is certainly possible to study *circular motion* and omit up-and-down *harmonic motion*. In the circular case the ball is at $x = \cos t, y = \sin t$. In the up-and-down case the shadow is still at height $y = \sin t$, but now $x = 0$. My idea was that this up-and-down motion is made easy by its connection to circular motion. When we know the velocity of the ball on the circle, we know the derivatives of $\sin t$ and $\cos t$.

The standard motion has speed 1. The velocity vector is tangent to the unit circle of radius 1. The derivative of $\sin t$ is $\cos t$, and the derivative of $\cos t$ is $-\sin t$. Every student is going to learn those rules – by seeing the circle, you see what they mean. Calculus concentrates so much on a few particular functions (amazingly few), and sines and cosines are on that short list. I am not in favor of "late trigonometry," because seeing a function several times is the way to understand it.

Conclusion: This section *introduces* $\sin t$ and $\cos t$. *Don't study them to death.* Just begin to see how they work.

Questions **1-4** refer to a ball going counterclockwise around the unit circle, centered at (0,0) with radius 1. The speed is 1 (say 1 meter/sec to have units). Assume the ball starts at angle zero. Then $x = \cos t$ and $y = \sin t$.

1 How long does it take for 10 revolutions? This means 10 times around the circle.

- The distance around a circle is $2\pi r$, which is 2π meters since $r = 1$. One revolution at speed 1 m/sec takes 2π seconds. So 10 revolutions take 20π seconds.

2 At time $t = \pi$, where is the ball?

- The ball is halfway around the circle at $x = -1, y = 0$.

3 Where is the ball (approximately) at $t = 14$ seconds?

- $\frac{14}{2\pi} \approx 2.23$, so the ball has made about $2\frac{1}{4}$ revolutions. Its position after that $\frac{1}{4}$ revolution is near the top of the circle where $x = 0, y = 1$. More accurately $y = \sin 14 = .99$.

4 Suppose that string is cut at $t = \frac{5\pi}{3}$. When and where would the ball hit the x axis?

- You need to know some trigonometry, which is shown in the figure. The ball is at $B = (\cos\frac{5\pi}{3}, \sin\frac{5\pi}{3}) = (\frac{1}{2}, -\frac{\sqrt{3}}{2})$. When the string is cut, the ball travels along BP at one meter per second. The tangent of the angle $POB = \frac{\pi}{3}$ is $\frac{PB}{OB} = \frac{\sqrt{3}}{1}$. Therefore the ball goes from B to P in $\sqrt{3}$ seconds. The time of arrival is $\frac{5\pi}{3} + \sqrt{3}$.
  The right triangle has $OB^2 + BP^2 = 1^2 + (\sqrt{3})^2 = 4$, so the hypotenuse is $OP = 2$.

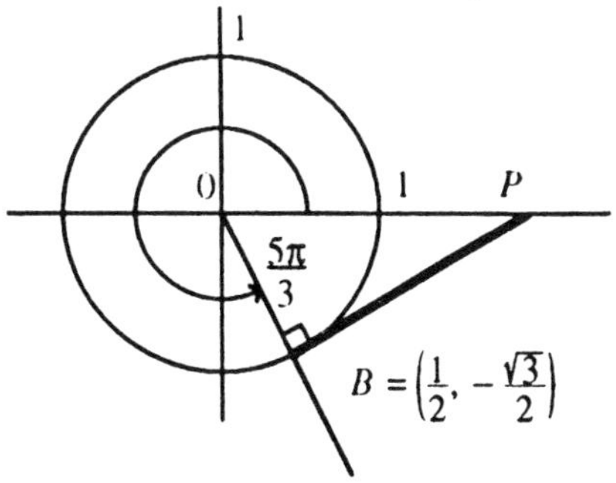

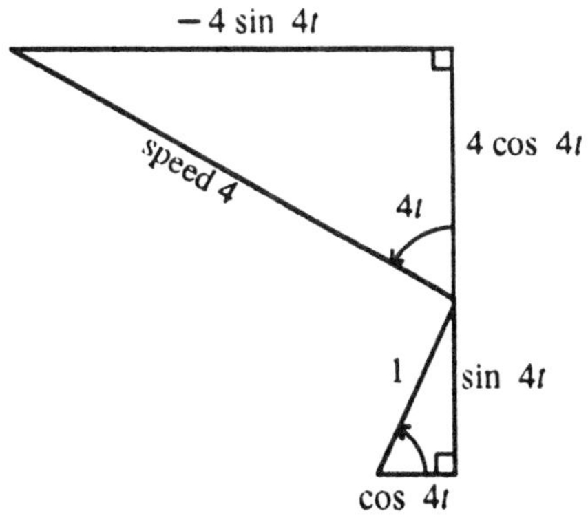

5 Another ball travels the unit circle counterclockwise with speed 4 m/sec starting at angle 0.

(a) What angle does it reach at time t? Answer: On the unit circle, a distance of 1 on the circumference corresponds to a central angle of 1 radian. Since the speed is 4 m/sec the central angle is increasing at 4 radians/sec. *At time t the angle is* $4t$ *radians.*

(b) What are the ball's x and y coordinates at time t? Answer: $(x, y) = (\cos 4t, \sin 4t)$.

(c) What are the ball's x and y velocities at time t? Answer: The velocity (the tangent vector) is drawn with length 4 because the speed is 4. The triangle inside the circle is similar to the triangle outside the circle. The sides of the larger triangle are four times longer. Thus the vertical velocity is $4\cos 4t$ and the horizontal velocity is $-4\sin 4t$.

- Summary: Motion around the unit circle with speed k (counterclockwise) gives $x = \cos kt$ and $y = \sin kt$. The velocity is $-k\sin kt$ (horizontal) and $k\cos kt$ (vertical).

6 What are the components of velocity if a ball travels at 1 radian per second around a circle of radius R feet (starting at angle 0)?

- The ball's position at time t is $(R\cos t, R\sin t)$. Since the ball travels R feet along the circumference for each radian, the speed is R ft/sec. The horizontal velocity is $-R\sin t$ and the vertical velocity is $R\cos t$.

Motion around a circle of radius R feet at k radians per second gives $x = R\cos kt$ and $y = R\sin kt$. The speed is kR ft/sec. The horizontal velocity is $-Rk\sin kt$. The vertical velocity is $Rk\cos kt$.

Here are read-throughs and selected solutions.

A ball at angle t on the unit circle has coordinates $x = \mathbf{cos\ t}$ and $y = \mathbf{sin\ t}$. It completes a full circle at $t = \mathbf{2\pi}$. Its speed is **1**. Its velocity points in the direction of the **tangent**, which is **perpendicular** to the radius coming out from the center. The upward velocity is **cos t** and the horizontal velocity is $-$ **sin t**.

A mass going up and down level with the ball has height $f(t) = \mathbf{sin\ t}$. This is called simple **harmonic** motion. The velocity is $v(t) = \mathbf{cos\ t}$. When $t = \pi/2$ the height is $f = \mathbf{1}$ and the velocity is $v = \mathbf{0}$. If a speeded-up mass reaches $f = \sin 2t$ at time t, its velocity is $v = \mathbf{2\cos 2t}$. A shadow traveling *under* the ball has $f = \cos t$ and $v = \mathbf{-sin\ t}$. When f is distance = area = integral, v is **velocity = slope = derivative.**

14 The ball goes halfway around the circle in time π. For the mass to fall a distance 2 in time π we need $2 = \frac{1}{2}a\pi^2$ so $a = 4/\pi^2$.

16 The area is $f(t) = \sin t$, and $\sin\frac{\pi}{6} - \sin 0 = \frac{1}{2}$.

18 The area is still $f(t) = \sin t$, and $\sin\frac{3\pi}{2} - \sin\frac{\pi}{2} = -1 - 1 = -2$.

20 The radius is 2 and time is speeded up by 3 so the speed is 6. There is a minus sign because the cosine starts downward (ball moving to left).

26 Counterclockwise with radius 3 starting at (3,0) with speed 12.

1.5 Review of Trigonometry (page 33)

Right triangles show the usual picture of trigonometry. The sine and cosine and tangent are ratios of the sides. The cosecant and secant and cotangent are the same ratios turned upside down.

A circle shows a better picture of trigonometry. The angle goes all the way to 360° and beyond. Of course we change 360° to 2π radians. Then distance around the circle is $r\theta$ = radius times angle. The derivative of $\sin t$ is $\cos t$, when angles are in radians. Otherwise we would have factors $\frac{2\pi}{360}$ which nobody wants. This section is helpful as a quick reference to the laws of trigonometry. *Don't forget:*

$$\sin 2t = 2\sin t\cos t \quad \text{and} \quad \cos 2t = 2\cos^2 t - 1 = 1 - 2\sin^2 t.$$

Three more to remember: $\mathbf{sin^2}\theta + \mathbf{cos^2}\theta = \mathbf{1}$ and $\mathbf{1} + \mathbf{tan^2}\theta = \mathbf{sec^2}\theta$ and $\mathbf{1} + \mathbf{cot^2}\theta = \mathbf{csc^2}\theta$. Those all come from $x^2 + y^2 = r^2$, when you divide by r^2 then x^2 then y^2.

The distance from (2,5) to (3,3) is $\sqrt{1^2 + 2^2} = \sqrt{5}$. The book proves the addition formulas $\cos(s - t) = \mathbf{cos\ s\ cos\ t + sin\ s\ sin\ t}$ and after sign change $\cos(s + t) = \mathbf{cos\ s\ cos\ t - sin\ s\ sin\ t}$.

When $s = t$ you get $2t$, the "double-angle": $\cos 2t = \mathbf{cos^2 t - sin^2 t}$ or $\mathbf{2\cos^2 t - 1}$. Therefore $\frac{1}{2}(1 + \cos 2t) = \mathbf{cos^2 t}$, a formula needed in calculus.

1 (Problem 1.5.8) Find the distance from (1,0) to (0,1) along (a) a straight line (b) a quarter-circle and (c) a semicircle centered at $(\frac{1}{2}, \frac{1}{2})$.

- (a) Straight distance $\sqrt{1^2+1^2} = \sqrt{2}$. (b) The central angle is $\frac{\pi}{2}$, so the distance is $r\theta = \frac{\pi}{2}$. (c) The central angle is π and the radius is $\frac{\sqrt{2}}{2}$, half the length in part (a). Distance $\frac{\sqrt{2}}{2}\pi$.

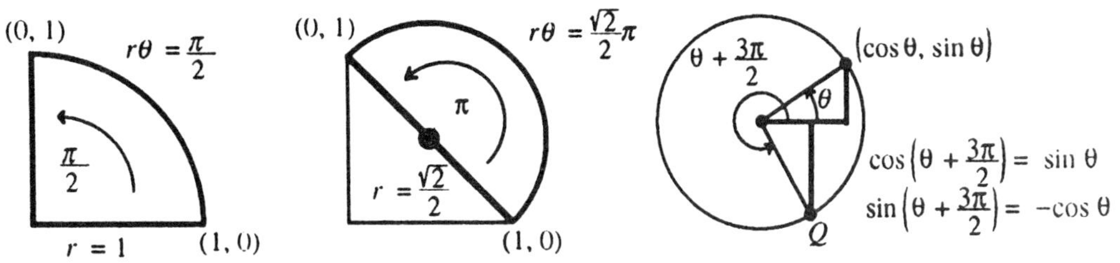

2 Simplify $\sin(\theta + \frac{3\pi}{2})$. Use formula (a), then use the circle.

- $\sin\theta\cos\frac{3\pi}{2} + \cos\theta\sin\frac{3\pi}{2}$ equals $(\sin\theta)(0) + (\cos\theta)(-1) = -\cos\theta$.
- Graphical solution: $\theta + \frac{3\pi}{2}$ is $\frac{3}{4}$ of a rotation ahead of θ. From the picture you see that $\sin(\theta + \frac{3\pi}{2})$ is the y coordinate of Q. This is $-\cos\theta$.

3 Find every angle θ that satisfies $\cos\theta = -1$.

- We know that $\cos\pi = -1$. Since the cosine repeats when θ is increased or decreased by 2π, the answer is $\theta = \pi \pm n(2\pi)$ for $n = 0, 1, 2, \cdots$. This gives the set $\theta = \pm\pi, \pm 3\pi, \pm 5\pi, \cdots$.

Read-throughs and selected even-numbered solutions :

Starting with a **right** triangle, the six basic functions are the **ratios** of the sides. Two ratios (the cosine x/r and the **sine** y/r) are below 1. Two ratios (the secant r/x and the **cosecant r/y**) are above 1. Two ratios (the **tangent** and the **cotangent**) can take any value. The six functions are defined for all angles θ, by changing from a triangle to a **circle**.

The angle θ is measured in **radians**. A full circle is $\theta = 2\pi$, when the distance around is $2\pi r$. The distance to angle θ is θr. All six functions have period 2π. Going clockwise changes the sign of θ and **sin** θ and **tan** θ. Since $\cos(-\theta) = \cos\theta$, the cosine is **unchanged (or even)**.

4 $\cos 2(\theta + \pi)$ is the same as $\cos(2\theta + 2\pi)$ which is $\cos 2\theta$. Since $\cos^2\theta = \frac{1}{2} + \frac{1}{2}\cos 2\theta$, this also has period π.

14 $\sin 3t = \sin(2t+t) = \sin 2t \cos t + \cos 2t \sin t$. This equals $(2\sin t\cos t)\cos t + (\cos^2 t - \sin^2 t)\sin t$ or $3\sin t - 4\sin^3 t$.

26 $\sin\theta = \theta$ at $\theta = \mathbf{0}$ **and never again**. Reason: The right side θ has slope 1 and the left side has slope $\cos\theta < 1$. (So the graphs of $\sin\theta$ and θ can't meet a second time.)

30 $A\sin(x+\phi)$ equals $A\sin x\cos\phi + A\cos x\sin\phi$. That expression must match with $a\sin x + b\cos x$. Thus $a = A\cos\phi$ and $b = A\sin\phi$. Then $a^2 + b^2 = A^2\cos^2\phi + A^2\sin^2\phi = A^2$. Therefore $\mathbf{A} = \sqrt{\mathbf{a^2} + \mathbf{b^2}}$ and $\tan\phi = \frac{A\sin\phi}{A\cos\phi} = \frac{b}{a}$.

34 The amplitude and period of $2\sin\pi x$ are both **2**.

1.6 A Thousand Points of Light (page 34)

These two pages are to explain the figures on the back cover. They are not for studying. If **Mathematica** is available, or any other way to draw these point-graphs, there are many that you could experiment with. Try $y = \sin cn$ for different choices of c and different ranges of n. The graphs are surprising. The **Mathematica** command is ListPlot [Table[Sin [cn], {n,N}]]. Put in c and N.

1.7 Computing in Calculus (page 36)

Every year brings progress in three key directions. All three are essential, as computer labs and computer assignments become an accepted part of calculus courses. Without any claim to completeness, here are notes on recent activity.

1 ***Graphing Calculators*** The TI-85 and HP-48 have extended the range of the TI-81 and HP-28. For subjects up to and including calculus, the TI-81 (least powerful of the four) is still a good choice. The more powerful HP's can be computing tools also in advanced courses. Clemson University is a leader in teaching with HP's, and calculators will soon be allowed on College Board exams. You may be familiar with them before starting calculus.

I find that quick calculations in class bring out the meaning of symbols. An example is already in the exercises for Section 1.2 – the computation of $(2^{.1} - 1)/.1$ and $(2^{.01} - 1)/.01$. The expression $[f(x + \Delta x) - f(x)]/\Delta x$ and also the function $f(x) = 2^x$ can be encountered early and actively.

2 ***Calculus Laboratories*** These are becoming widespread. The goal is to add a valuable new experience, but to avoid overburdening students or faculty. The start-up effort is considerable. In the long run, it will pay off.

DERIVE is one popular choice of software. At Duke University it is combined with MathCAD. Other universities have selected MAPLE (for example RPI), and they find it increasingly accessible and convenient. A third strong choice is **Mathematica** (at Georgia Tech). At this writing **Mathematica** is not offered in an inexpensive student version. MATLAB has just taken that step – this is the premier code for matrix calculations (with graphics).

The rate of progress with software needs to extend to laboratory manuals. It is still difficult to identify outstanding sources of problems – which should go beyond a translation of old calculus questions into computer programs. At its best, a calculus lab can be a welcome addition, provided *subtraction* is also allowed – all of us are less bound by the syllabus than we think.

3 ***Projects and Activities*** Projects take 1-2 weeks, activities take 1–2 days. An essential part of each project is a report. Work in groups is demonstrated as very successful – the problem of equal work for equal grade is not severe in actual practice, and the advantages are great.

Ithaca College and New Mexico State have been pioneers with projects. My favorite in-class activity is this ***Water Tank Problem***, which can come early to emphasize slopes and graphs. It was created by Peter Taylor, adapted at Ithaca College, and is slightly extended here. Teachers and students are invited to work through it together, on the board and *in the class*.

At time $t = 0$, water begins to flow into an empty tank at the rate of 40 liters/minute. This flow rate is held constant for two minutes. From $t = 2$ to $t = 4$, the flow rate is gradually reduced to 5 liters/minute. This rate is constant for the final two minutes. At $t = 6$ the tank contains 120 liters.

A Draw a graph of the volume $V(t)$ of water in the tank for $0 \le t \le 6$.

B What is the average rate of flow into the tank over the entire six-minute period? Show how this can be interpreted on your graph.

C Starting at time $t = 2$, water is pumped out at the constant rate of 15 liters/minute. Now $V(t)$ above represents the total amount of water that has flowed ***in***, and we let $W(t)$ represent the total amount pumped ***out***. Plot $W(t)$ on the same axes. Show how to interpret the volume of water in the tank at any time.

D (*This is calculus*) Find on your graph the point at which the water level in the tank is a ***maximum***. What is the instantaneous flow rate from the hose into the tank at this time?

E Graph the ***rates*** of flow in and out, for $0 \le t \le 6$. Mark the point where the water level is a maximum. Extra question: Can the flow rate into the tank be linear from $t = 2$ to $t = 4$?

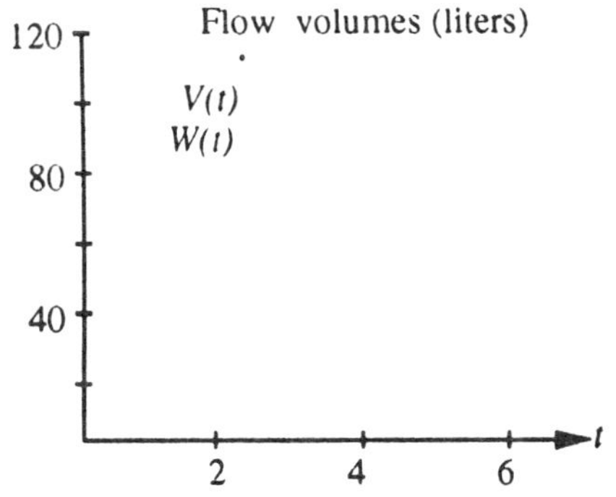

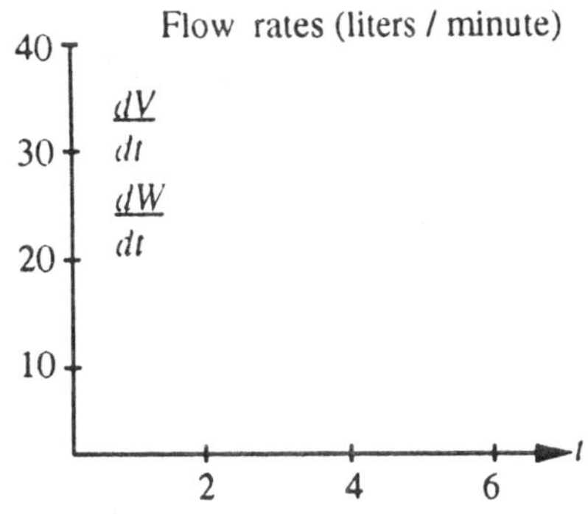

1 Chapter Review Problems

Graph Problems These questions are asking for sketches by hand, not for works of art. It is surprising how much of calculus shows up in these pictures.

G1 Draw $f(x) = \sin x$ between $x = 0$ and $x = 4\pi$. Mark all maximum points by M and all minimum points by m. Mark the points where the slope is steepest (going up or down) by S.
Under that graph draw $v(x) = \cos x$ between $x = 0$ and $x = 4\pi$. *Line it up correctly.* Repeat the letters M, m, S down on the cosine graph. Notice that M and m are at points where the cosine is ______, and S is at points where the cosine is ______.

G2 Draw $y = 2 - x^2$ between $x = -1$ and $x = 1$. Mark the maximum point by M and the steepest points by S. Directly under that graph draw $v = -2x$. Move the letters M and S down to this slope graph. When y has a maximum its slope is ______. When y is steepest its slope has a ______.

G3 Draw a smooth curve that goes up then down then up again. A specific choice would be the graph of $y = x^3 - x$ from $x = -2$ to $x = 2$. Mark the maximum, minimum, and steepest points by M, m, S. Under your curve draw a graph of its slope. (The slope of $x^3 - x$ is $3x^2 - 1$.) If the graphs are lined up, then M and m are at points where the slope is zero.
New question: What is different about the M point and the m point? How can you tell a maximum from a minimum, looking at the "zero crossing" on the slope graph?
Answer: At a maximum of $y(x)$, its slope crosses from ______ to ______.

G4 **The triangles of trigonometry.** Draw a right triangle with sides marked $\cos\theta, \sin\theta$, and 1. Enlarge it to a triangle with sides marked $1, \tan\theta$, and $\sec\theta$. (You multiplied the sides of the first triangle by $\sec\theta$, which is greater than 1.) From the second triangle read off the identity $1+\tan^2\theta =$ ______. Now draw a third triangle with sides $\cot\theta$, 1, and $\csc\theta$. You have multiplied the first triangle by $\csc\theta$. This is $1/\sin\theta$ and again it is greater than 1. From this third triangle read off the identity $1+\cot^2\theta =$ ______.
Question: Is the third triangle always larger than the second triangle? *Answer: No.* To get the third triangle from the second, multiply by $\frac{\cos\theta}{\sin\theta}$. This is larger than 1 when θ is larger than ______. The triangles have $0 < \theta < 90°$. At $\theta = 45°$, which two triangles are the same?

Computing Problems Use a computer or calculator. You do not need a supercomputer.

C1 Find the tax on incomes of $x_1 = \$1,000,000$ and $x_2 = \$2,000,000$. Tax rates for a single person are in Section 1.2. How large is the tax difference $f(x_2) - f(x_1)$?

C2 What income x would leave you with \$1,000,000 after paying tax in full?

C3 For the function $f(x) = x^2 + x^3 + x^4$ compute $f(1.1), f(1.01)$, and $f(1.001)$. Subtract $f(1)$ from each of those and divide by .1, .01, and .001.

C4 Starting from $f_0 = 0$ with differences $v = 1, \frac{1}{2}, \frac{1}{3}, \frac{1}{4}, \frac{1}{5}, \cdots$ compute $f_{100}, f_{200}, f_{400}, f_{800}$.

Review Problems

R1 Under what condition on y_2 will the line through $(1,5)$ and $(4, y_2)$ have a positive slope?

R2 Under what condition on x_2 will the line through $(1,5)$ and $(x_2, 3)$ have a positive slope?

R3 Find the slope of the line through (x_1, y_1) and (x_2, y_2). When is the slope infinite?

R4 Explain the basic idea of "*Calculus without limits*" in Section 1.2.

R5 Explain how the average velocity $= \frac{\text{change in } f}{\text{change in } t}$ leads to an instantaneous velocity.

R6 For $f(t) = t^3$ find the average velocity between $t = 0$ and $t = h$. What is the instantaneous velocity at $t = 0$?

R7 For $f(t) = t^3$ find the average velocity between $t = 1$ and $t = 1 + h$. As $h \to 0$ find $v(1)$.

Drill Problems

D1 Find the equation of the line through $x = 0, y = 3$ with slope 5.

D2 Find the equation of the line through $(0,3)$ and $(2,7)$.

D3 Find the distance function $f(t)$ if $f(0) = 3$ and $f(2) = 7$ and the velocity is constant.

D4 If $f(0) = 6$ and $f(10) = 26$ find the average velocity between $t = 0$ and $t = 10$. *True or false:* The actual velocity is above and below that average for equal times (5 above and 5 below).

D5 For $f(t) = 7t - 5$ write down $f(t+h)$, subtract $f(t)$, and divide by h.

D6 For $f(t) = 7t^2 - 5$ write down $f(t+h)$, subtract $f(t)$, and divide by h.

D7 Find the differences v for $f = 1, 2, 4, 2, 1$. The sum of differences is ______.

D8 Find f_3 starting from $f_0 = 10$ if the differences v are $2, -6, 7$.

D9 Express $\cos 2t$ in terms of $\cos^2 t$ and then in terms of $\sin^2 t$.

CHAPTER 2 DERIVATIVES

2.1 The Derivative of a Function (page 49)

In this section you are mainly concerned with learning the meaning of the derivative, and also the notation. The list of functions with known derivatives includes $f(t) =$ constant, Vt, $\frac{1}{2}at^2$, and $1/t$. Those functions have $f'(t) = 0$, V, at, and $-1/t^2$. We also establish the "square rule", that the derivative of $(f(t))^2$ is $2f(t)f'(t)$. Soon you will see other quick techniques for finding derivatives. But learn the basics first.

The derivative is the slope of the tangent line. Mathematicians often say the "slope" of a function when they mean the "derivative." Questions 1–7 refer to this figure:

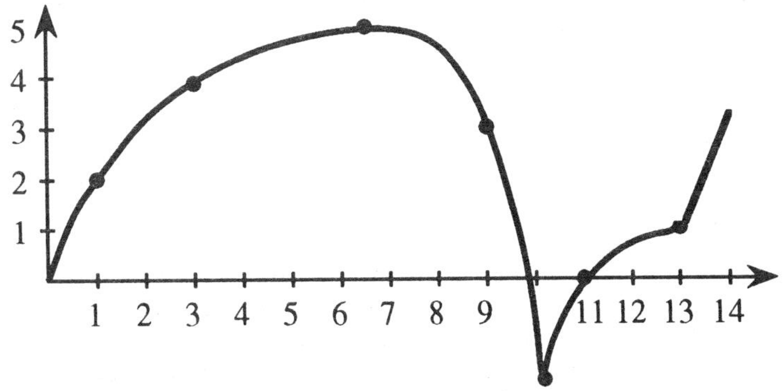

1. If the slope of the tangent line at (1,2) is $\frac{3}{2}$, this means that $f'(\_\_) = \_\_\_\_$.
   - It means that $f'(1) = \frac{3}{2}$. The first coordinate of (1,2) goes into $f'(\_\_)$.
2. If $f'(3) = \frac{1}{2}$, then the slope of the tangent line through (___, ___) is $\frac{1}{2}$.
   - The point is (3,4).
3. The graph indicates that $f'(\_\_) = 0$. This zero derivative means ____ .
   - The graph has $f'(6\frac{1}{2}) = 0$. The tangent is horizontal at $x = 6\frac{1}{2}$. Later we learn: This happens at a maximum or minimum.
4. $\frac{dy}{dx}$ is negative when ___ $< x <$ ___ . In this interval the function is ____ .
   - The slope is negative when $6\frac{1}{2} < x < 10$. The function is decreasing (left to right).
5. The derivative $\frac{dy}{dx}$ is not defined when $x =$ ____ and ____ .
   - $\frac{dy}{dx}$ is undefined at $x = 10$ because the tangent line is vertical. You might say "infinite slope." Also there is no derivative at $x = 13$. The graph has a *corner.* The slope is different on the left side and right side of the corner.
6. $f(x)$ is (positive or negative?) at (9,3) while $f'(x)$ is (positive or negative?).
   - $f(9) = 3$ is positive. The graph is above the y axis. The slope $f'(9)$ is negative.
7. $f(10.5)$ is (positive or negative?) while $f'(10.5)$ is ____ .
   - At $(10.5, -1)$ the graph is below the y axis and rising. $f(x)$ is negative and $f'(x)$ is positive.

The derivative is the rate of change. When the function is the distance $f(t)$, its derivative is (instantaneous) velocity. When the function is the velocity $v(t)$, its derivative is (instantaneous) acceleration.

8. Suppose a vehicle travels according to the rule $f(t) = 3t^3 - t$. Find its velocity at time $t = 4$.

- Even though you have in mind a special time $t = 4$, you cannot substitute $t = 4$ until the end. (You don't want $f(t) = 44$. A constant function has velocity zero! What you will use is $f(4) = 44$.) First find the average velocity $\frac{\Delta f}{\Delta t}$ using Δt or h:

$$\frac{\Delta f}{\Delta t} = \frac{f(t+\Delta t) - f(t)}{\Delta t} = \frac{3(t^2 + 2t\,\Delta t + (\Delta t)^2) - (t + \Delta t) - 3t^2 + t}{\Delta t}.$$

Notice that $3t^2$ cancels with $-3t^2$ and also t cancels $-t$. This produces

$$\frac{\Delta f}{\Delta t} = \frac{6t\Delta t + 3(\Delta t)^2 - \Delta t}{\Delta t} = 6t + 3\Delta t - 1.$$

The division removed Δt from the denominator. *Now let* Δt *go to* 0. The limit is $f'(t) = 6t - 1 = v(t)$. At $t = 4$, the velocity is $v(4) = 6 \times 4 - 1 = 23$. To find acceleration, go back to the formula for velocity (*before* you plugged in $t = 4$). Take the derivative of $v(t) = 6t - 1$:

$$\frac{\Delta v}{\Delta t} = \frac{v(t+\Delta t) - v(t)}{\Delta t} = \frac{[6(t+\Delta t) - 1] - (6t - 1)}{\Delta t} = \frac{6\Delta t}{\Delta t} = 6.$$

As $\Delta t \to 0$, the average $\frac{\Delta v}{\Delta t}$ stays at 6. So the acceleration is $\frac{dv}{dt} = 6$. The graph of $v(t) = 6t - 1$ is a line with constant slope 6.

- A simple but important point is that if the graph is shifted up or down, its slope does not change. Another way to say this is that if $g(x) = f(x) + c$, where c is a constant number, then $g'(x) = f'(x)$. Question: *Does* $f'(x)$ *change if the graph is shifted right or left?* Yes it does. The slope graph shifts too. The slope of $f(x) + 1$ is $f'(x)$, but the slope of $f(x+1)$ is $f'(x+1)$.

9. Find functions that have the same derivative as $f(x) = \frac{1}{x}$.

- $g(x) = \frac{1}{x} - 5$ and $h(x) = \frac{1}{x} + 2$ and any $\frac{1}{x} + C$. No other possibilities!

Read-throughs and selected even-numbered solutions :

The derivative is the **limit** of $\Delta f/\Delta t$ as Δt approaches **zero.** Here Δf equals $\mathbf{f(t+\Delta\, t) - f(t)}$. The step Δt can be positive or **negative.** The derivative is written v or **df/dt** or $\mathbf{f'(t)}$. If $f(x) = 2x + 3$ and $\Delta x = 4$ then $\Delta f = \mathbf{8}$. If $\Delta x = -1$ then $\Delta f = \mathbf{-2}$. If $\Delta x = 0$ then $\Delta f = \mathbf{0}$. The slope is not 0/0 but $df/dx = \mathbf{2}$.

The derivative does not exist where $f(t)$ has a **corner** and $v(t)$ has a **jump.** For $f(t) = 1/t$ the derivative is $\mathbf{-1/t^2}$. The slope of $y = 4/x$ is $dy/dx = \mathbf{-4/x^2}$. A decreasing function has a **negative** derivative. The **independent** variable is t or x and the **dependent** variable is f or y. The slope of y^2 **(is not)** $(dy/dx)^2$. The slope of $(u(x))^2$ is $\mathbf{2u(x)\ du/dx}$ by the square rule. The slope of $(2x+3)^2$ is $\mathbf{2(2x+3)2 = 8x + 12}$.

2 (a) $\frac{\Delta f}{h} = \frac{2hx+h^2}{h}$ becomes $\mathbf{2x}$ at $h = 0$ (b) $\frac{(x+5h)^2 - x^2}{5h} = \frac{10hx + 25h^2}{5h} = 2x + 5h$ becomes $\mathbf{2x}$ at $h = 0$ (c) $\frac{(x+h)^2 - (x-h)^2}{2h} = \frac{4xh}{2h} = \mathbf{2x}$ *always* (d) $\frac{(x+1)^2 - x^2}{h} = \frac{2x+1}{h} \to \infty$ as $h \to 0$ **4** $x^2 + 1, x^2 + 10, x^2 - 100$

6 The line and parabola have slopes 1 and $2x$. So the touching point must have $x = \frac{1}{2}$. There $y = \frac{1}{2}$ for the line, $y = (\frac{1}{2})^2 + c$ for the parabola so $c = \mathbf{\frac{1}{4}}$.

22 The graph of $f(t)$ has slope -2 until it reaches $t = 2$ where $f(2)$ equals -1; after that it has slope zero.

36 (a) **False** First draw a curve that stays below $y = x$ but comes upward steeply for negative x. Then create a formula like $\mathbf{y = -x^2 - 10}$. (b) **False** $f(x)$ could be any constant, for example $f(x) = 10$. Note what is *true*: If $\frac{df}{dx} \le 1$ and $f(x) \le x$ at some point then $f(x) \le x$ everywhere **beyond that point.**

2.2 Powers and Polynomials (page 56)

The derivatives of x^5 and $x^5+x^{-\sqrt{2}}$ and $3x^5$ come from **2B, 2C, 2D**. Practice until $5x^4$ and $5x^4-\sqrt{2}x^{-\sqrt{2}-1}$ and $15x^4$ are automatic. It's not the end of the world if the binomial formula escapes you, but you have to find derivatives of powers and polynomials. Note that the derivative of x^n is nx^{n-1} for *all* numbers n (fractions, negative numbers, $\cdots$). So far this has only been proved when n is a positive integer. *Find* $\frac{dy}{dx}$ *in Problems* 1–6:

1. $y=4x+2$. This line has slope $\frac{dy}{dx}=4(1)+0=4$. Use **2B** (for x) and **2D** (for $4x$) and **2C** (for $4x+2$).
2. $y=\frac{1}{x^2}=x^{-2}$. This has $n=-2$ so $\frac{dy}{dx}=-2x^{-3}=\frac{-2}{x^3}$.
3. $y=5x^{-\frac{2}{5}}+10x^{\frac{1}{5}}$. Fractional powers give $\frac{dy}{dx}=5(-\frac{2}{5})x^{-\frac{7}{5}}+10(\frac{1}{5})x^{-\frac{4}{5}}=-2x^{-\frac{7}{5}}+2x^{-\frac{4}{5}}$. *Why* $-\frac{7}{5}$?
4. $y=4\sqrt{x}=4x^{\frac{1}{2}}$. Here $n=\frac{1}{2}$. Then $\frac{dy}{dx}=4(\frac{1}{2})x^{-\frac{1}{2}}=\frac{2}{\sqrt{x}}$.
5. $y=(3x-5)^2$. The answer is *not* $\frac{dy}{dx}=2(3x-5)$! Use the square rule to get $2(3x-5)(3)=18x-30$. *The extra* 3 *comes from the derivative of* $3x-5$. You can expand $(3x-5)^2$ and take the derivative of each term:

$$y=(3x-5)(3x-5)=9x^2-15x-15x+25 \quad \text{so} \quad \frac{dy}{dx}=9(2x)-15-15=18x-30.$$

6. $y=4x^{\sqrt{3}}$. Don't let $\sqrt{3}$ throw you! Just forge ahead to find $\frac{dy}{dx}=4\sqrt{3}x^{\sqrt{3}-1}$.
7. Find the second derivative of $y=4x^3-2x+\frac{6}{x}$.
   - The first derivative is $\frac{dy}{dx}=12x^2-2-6x^{-2}$. Then the second derivative is $24x+0+12x^{-3}$.
8. Find a function that has $\frac{dy}{dx}=-y^2$. This is a *differential equation*, with y on both sides.
   - Solving means finding $y(x)$ from information about $\frac{dy}{dx}$. At this point we don't know the derivatives of too many functions. We can guess $y=cx^n$ and work backward. Then $\frac{dy}{dx}=cnx^{n-1}$ and $-y^2=-c^2x^{2n}$. If we can choose c and n so that $cnx^{n-1}=-c^2x^{2n}$, we have solved the differential equation. Match the powers $n-1=2n$ to get $n=-1$. Now we want cn to equal $-c^2$, so $c=1$. The answer is $y=x^{-1}$. Check that $\frac{dy}{dx}=-y^2$. Both sides equal $-x^{-2}$.

Read-throughs and selected even-numbered solutions :

The derivative of $f=x^4$ is $f'=\mathbf{4x^3}$. That comes from expanding $(x+h)^4$ into the five terms $\mathbf{x^4+4x^3h+6x^2h^2+4xh^3+h^4}$. Subtracting x^4 and dividing by h leaves the four terms, $\mathbf{4x^3+6x^2h+4xh^2+h^3}$. This is $\Delta f/h$, and its limit is $\mathbf{4x^3}$.

The derivative of $f=x^n$ is $f'=\mathbf{nx^{n-1}}$. Now $(x+h)^n$ comes from the **binomial** theorem. The terms to look for are $x^{n-1}h$, containing only one **h**. There are **n** of those terms, so $(x+h)^n=x^n+\mathbf{nx^{n-1}h}+\cdots$. After subtracting $\mathbf{x^n}$ and dividing by h, the limit of $\Delta f/h$ is $\mathbf{nx^{n-1}}$. The coefficient of $x^{n-j}h^j$, not needed here, is "n choose j" = **n!/j!(n-j)!**, where $n!$ means $\mathbf{n(n-1)\cdots(1)}$.

The derivative of x^{-2} is $\mathbf{-2x^{-3}}$. The derivative of $x^{1/2}$ is $\mathbf{\frac{1}{2}x^{-1/2}}$. The derivative of $3x+(1/x)$ is $\mathbf{3-1/x^2}$, which uses the following rules: the derivative of $3f(x)$ is $\mathbf{3f'(x)}$ and the derivative of $f(x)+g(x)$ is $\mathbf{f'(x)+g'(x)}$.

Integral calculus recovers **y** from dy/dx. If $dy/dx = x^4$ then $y(x) = \mathbf{x^5/5 + C}$.

2 $f(x) = \frac{1}{7}x^7$ (or $\frac{1}{7}x^7 + C$)

8 $\frac{df}{dx} = \frac{1}{n!}(nx^{n-1}) = \frac{x^{n-1}}{(n-1)!}$. Note the step $\frac{n}{n(n-1)\cdots(1)} = \frac{1}{(n-1)\cdots(1)} = \frac{1}{(n-1)!}$

10 $f'(x) = \frac{2}{3}(\frac{3}{2}x^{1/2}) + \frac{2}{5}(\frac{5}{2}x^{3/2}) = x^{1/2} + x^{3/2}$.

14 The slope of $x + \frac{1}{x}$ is $1 - \frac{1}{x^2}$ which is zero at $x = 1$. At that point the graph of $x + \frac{1}{x}$ levels off. (The function reaches its minimum, which is 2. For any other positive x, the combination $x + \frac{1}{x}$ is larger than 2.)

22 If $y = \frac{1}{\sqrt{x}}$ then $\Delta y = \frac{1}{\sqrt{x+h}} - \frac{1}{\sqrt{x}} = \frac{\sqrt{x}-\sqrt{x+h}}{\sqrt{x+h}\sqrt{x}}$ = (multiply top and bottom by $\sqrt{x} + \sqrt{x+h}$) = $\frac{x-(x+h)}{\sqrt{x+h}\sqrt{x}(\sqrt{x}+\sqrt{x+h})}$. Cancel $x - x$ in the numerator and divide by h : $\frac{\Delta y}{h} = \frac{-1}{\sqrt{x+h}\sqrt{x}(\sqrt{x}+\sqrt{x+h})}$. Now let $h \to 0$ to find $\frac{dy}{dx} = \frac{-1}{2x^{3/2}} = -\frac{1}{2}x^{-3/2}$ (which is nx^{n-1}).

38 If $y = y_0 + cx$ then $E(x) = \frac{dy/dx}{y/x} = \frac{c}{\frac{y_0}{x}+c}$ which approaches **1** as $x \to \infty$.

42 $y = x^n$ has $E = \frac{dy/dx}{y/x} = \mathbf{n}$. The revenue $xy = x^{n+1}$ has $E = \mathbf{n+1}$.

44 Marginal propensity to save is $\frac{dS}{dI}$. Elasticity is not needed because S and I have the same **units**. Applied to the whole economy this is **macroeconomics**.

2.3 The Slope and the Tangent Line (page 63)

Questions 1–5 refer to the curve $y = x^3 - 2x^2$. The derivative is $\frac{dy}{dx} = 3x^2 - 4x$.

1. Find the slope at $x = 3$.

   - Substitute $x = 3$ in $\frac{dy}{dx}$ (not in y!). The slope is $27 - 12 = 15$.

2. Find the equation of the tangent line at $x = 3$.

   - From question 1, the slope at $x = 3$ is 15. The function itself is $y = 27 - 18 = 9$. Use the point-slope form to get the equation of the tangent line: $y - 9 = 15(x - 3)$. You can rewrite this as $y = 15x - 36$.

3. Find the equation of the normal line at $x = 3$. This line is perpendicular to the curve.

   - We know that the slope of the curve at (3,9) is 15. Therefore the slope of the normal is $-\frac{1}{15}$. *When two lines are perpendicular, the second slope is* $\frac{-1}{\text{first slope}}$. The normal line is $y - 9 = -\frac{1}{15}(x - 3)$.

4. Find the secant line from (0,0) to (3,9). A *secant* connects two points on the curve.

   - The slope of the secant is $\frac{9-0}{3-0} = 3$. The secant line is $y - 0 = 3(x - 0)$, or $y = 3x$.

5. Where does the curve $y = x^3 - 2x^2$ have a horizontal tangent line? This means $\frac{dy}{dx} = 0$.

   - The slope is $\frac{dy}{dx} = 3x^2 - 4x = x(3x - 4)$. Then $\frac{dy}{dx} = 0$ when $x = 0$ or $x = \frac{4}{3}$. Horizontal (flat) tangent lines are found at (0,0) and $(\frac{4}{3}, \frac{-32}{27})$.

6. (This is Problem 2.3.13) At $x = a$ compute (a) the equation of the tangent line to the curve $y = \frac{1}{x}$ and (b) the points where that line crosses the axes. (c) The triangle between the tangent line and the axes always has area ______ .

   - (a) $y = x^{-1}$ has slope $\frac{dy}{dx} = -x^{-2} = -\frac{1}{x^2}$. At $(a, \frac{1}{a})$ this slope is $-\frac{1}{a^2}$. The tangent line has equation $y - \frac{1}{a} = -\frac{1}{a^2}(x - a)$.

- (b) This line crosses the x axis when $y = 0$ so $0 - \frac{1}{a} = -\frac{1}{a^2}(x - a)$. Multiply by $-a^2$ to find $a = x - a$ and $x = 2a$. The line crosses the y axis when $x = 0 : y - \frac{1}{a} = -\frac{1}{a^2}(0 - a)$ which gives $y = \frac{2}{a}$.
- (c) The area of the triangle is $\frac{1}{2}$(base $\times$ height) $= \frac{1}{2}(2a)(\frac{2}{a})$. This area is always 2.

4. Turn off your calculator and use the methods of this section to estimate $\sqrt[3]{66} = 66^{1/3}$.

- Let $f(x) = x^{1/3}$. We know that $64^{1/3} = 4$. (64 is chosen because it is the closest perfect cube to 66.) The plan is to find the tangent line through the point $x = 64$, $y = 4$. Use this tangent line to approximate the cube root function:

$$x^{1/3} \text{ has derivative } \frac{1}{3}x^{-2/3} \text{ and } f'(64) = \frac{1}{3}(64)^{-2/3} = \frac{1}{3}(4^{-2}) = \frac{1}{3(16)} = \frac{1}{48}.$$

The tangent line is $y - 4 = \frac{1}{48}(x - 64)$. If $x = 66$ then $y = 4 + \frac{1}{48}(2) = 4\frac{1}{24}$. The cube root curve goes up approximately $\frac{1}{24}$ to $66^{1/3} \approx 4\frac{1}{24}$.

Read-throughs and selected even-numbered solutions :

A straight line is determined by **2** points, or one point and the **slope.** The slope of the tangent line equals the slope of the **curve.** The point-slope form of the tangent equation is $y - f(a) = \mathbf{f'(a)(x - a)}$.

The tangent line to $y = x^3 + x$ at $x = 1$ has slope **4.** Its equation is $\mathbf{y - 2 = 4(x - 1)}$. It crosses the y axis at $\mathbf{y = -2}$ and the x axis at $\mathbf{x = \frac{1}{2}}$. The normal line at this point (1,2) has slope $-\frac{1}{4}$. Its equation is $y - 2 = -\frac{1}{4}\mathbf{(x-1)}$. The secant line from (1,2) to (2, **10**) has slope **8.** Its equation is $y - 2 = \mathbf{8(x-1)}$.

The point $(c, f(c))$ is on the line $y - f(a) = m(x - a)$ provided $m = \mathbf{\frac{f(c)-f(a)}{c-a}}$. As c approaches a, the slope m approaches $\mathbf{f'(a)}$. The secant line approaches the **tangent** line.

2 $y = x^2 + x$ has $\frac{dy}{dx} = 2x + 1 = 3$ at $x = 1, y = 2$. The tangent line is $\mathbf{y - 2 = 3(x - 1)}$ or $y = 3x - 1$. The normal line is $\mathbf{y - 2 = -\frac{1}{3}(x - 1)}$ or $y = -\frac{x}{3} + \frac{7}{3}$. The secant line is $\mathbf{y - 2 = m(x - 1)}$ with $m = \frac{(1+h)^2+(1+h)-2}{(1+h)-1} = 3 + h$.

8 $(x - 1)(x - 2)$ is zero at $x = 1$ and $x = 2$. If this is the slope (it is $x^2 - 3x + 2$) then the function can be $\mathbf{\frac{1}{3}x^3 - \frac{3}{2}x^2 + 2x}$. We can add any $Cx + D$ to this answer, and the slopes at $x = 1$ and 2 are still equal. $y = x^4 - 2x^2$ has $\frac{dy}{dx} = 4x^3 - 4x$. At $\mathbf{x = 1}$ **and** $\mathbf{x = -1}$ the slopes are zero and the y's are equal. The tangent line (horizontal) is the same.

18 Tangency requires $4x = cx^2$ and also (slopes) $4 = 2cx$ at the same x. The second equation gives $x = \frac{2}{c}$ and then the first is $\frac{8}{c} = \frac{4}{c}$ which has no solution.

30 The tangent line is $y - f(a) = f'(a)(x - a)$. This goes through $y = g(b)$ at $x = b$ if $\mathbf{g(b) - f(a) = f'(a)(b - a)}$. The slopes are the same if $\mathbf{g'(b) = f'(a)}$.

46 To *just* pass the baton, the runners reach the same point at the same time ($vt = -8 + 6t - \frac{1}{2}t^2$) and with the same speed ($v = 6 - t$). Then $(6 - t)t = -8 + 6t - \frac{1}{2}t^2$ and $\frac{1}{2}t^2 - 8 = 0$. Then $\mathbf{t = 4}$ and $\mathbf{v = 2}$.

2.4 The Derivative of the Sine and Cosine (page 70)

This section proves that $\frac{\sin h}{h} \to 1$ as $h \to 0$. In other words $\lim_{h\to 0} \frac{\sin h}{h} = 1$. The separate limits of $\sin h$ and h lead to $\frac{0}{0}$ which is undefined. The key to differential calculus is that the ratio approaches a definite limit.

Once that limit is established, algebraic substitution lets us conclude that $\lim_{5h\to 0} \frac{\sin 5h}{5h} = 1$. Similarly $\frac{\sin(-2z)}{(-2z)}$ approaches 1. The limit of $\frac{\sin\Box}{\Box}$ is 1 as long as both boxes are the same and approach zero. However $\frac{\sin 4h}{h}$ approaches 4. We show this by multiplying by $\frac{4}{4}$ to get $\frac{4\sin 4h}{4h}$. The limit is 4 times the limit of $\frac{\sin 4h}{4h}$, or $4 \times 1 = 4$.

You can factor out 4 or any constant from inside the limit. Questions 1–3 use this trick.

1. Find $\lim_{\theta\to 0} \frac{\theta}{\sin 6\theta}$.

   - Multiply and divide by 6, to get $\frac{1}{6} \cdot \frac{6\theta}{\sin 6\theta}$. As $\theta \to 0$ the limit is $\frac{1}{6} \times 1 = \frac{1}{6}$.

2. Find $\lim_{h\to 0} \frac{\sin 4h}{\sin 2h}$.

   - Write $\frac{\sin 4h}{\sin 2h}$ as $\frac{\sin 4h}{4h} \cdot \frac{2h}{\sin 2h} \cdot 2$. (Check this to satisfy yourself that we have just multiplied by 1 in a good way.) As $h \to 0$, we have $\frac{\sin 4h}{4h} \to 1$ and $\frac{2h}{\sin 2h} \to 1$. The limit of $\frac{\sin 4h}{\sin 2h}$ is $1 \cdot 1 \cdot 2 = 2$.

3. Find $\lim_{z\to 0} \frac{\tan z}{z}$.

   - The ratio is $\frac{\tan z}{z} = \frac{\sin z}{z\cos z}$. We know that $\frac{\sin z}{z} \to 1$ and $\cos z \to 1$ as $z \to 0$. Divide to find $\frac{\tan z}{z} \to 1$. What does this mean about the graph of $y = \tan z$? It means that the slope of $\tan z$ at $z = 0$ is $\frac{dy}{dz} = 1$. Reason: $\frac{\tan z}{z}$ is the average slope. Its limit is the exact slope $\frac{dy}{dz} = 1$.

4. Find $\frac{dy}{dx}$ for $y = \sin 2x$, first at $x = 0$ and then at every x.

   - The slope at $x = 0$ is $\lim \frac{\sin 2x}{x}$. (Maybe I should say $\lim \frac{\sin 2h}{h}$ – the same.) This is $2\lim \frac{\sin 2x}{2x} = 2$.
   - The slope at any x is $\lim \frac{\sin 2(x+h) - \sin 2x}{h}$. We need a formula for $\sin 2(x + h)$. Equation (9) on page 32 gives the addition formula $\sin 2x \cos 2h + \cos 2x \sin 2h$. Group the $\sin 2x$ terms separately from the $\cos 2x$ term to get $\frac{\sin 2x(\cos 2h - 1) + \cos 2x \sin 2h}{h}$. Now take limits of those two parts:

$$\begin{aligned}\frac{dy}{dx} &= (\sin 2x)\lim_{h\to 0}\left(\frac{\cos 2h - 1}{h}\right) + (\cos 2x)\lim_{h\to 0}\left(\frac{\sin 2h}{h}\right)\\ &= (\sin 2x)(0) + (\cos 2x)(2) = 2\cos 2x.\end{aligned}$$

The same methods show that $y = \sin(nx)$ has $\frac{dy}{dx} = n\cos(nx)$. Similarly $y = \cos(nx)$ has $\frac{dy}{dx} = -n\sin(nx)$. These are important. *Notice the extra factor n – later it comes from the chain rule.*

5. Write the equation of the tangent line to $f(x) = \cos x$ at $x = \frac{\pi}{4}$. Where does this line cross the y axis?

   - At $x = \frac{\pi}{4}$ (which is 45°) the cosine is $y = \frac{\sqrt{2}}{2}$ and the slope is $y' = -\sin\frac{\pi}{4} = -\frac{\sqrt{2}}{2}$. The tangent line is $y - \frac{\sqrt{2}}{2} = -\frac{\sqrt{2}}{2}(x - \frac{\pi}{4})$. Set $x = 0$ to find the y-intercept $\frac{\sqrt{2}}{2} + \frac{\sqrt{2}\pi}{8} \approx 1.26$.

Read-throughs and selected even-numbered solutions :

The derivative of $y = \sin x$ is $y' = \mathbf{\cos x}$. The second derivative (the **derivative** of the derivative) is $y'' = \mathbf{-\sin x}$. The fourth derivative is $y'''' = \mathbf{\sin x}$. Thus $y = \sin x$ satisfies the differential equations $y'' = \mathbf{-y}$

and $y'''' = \mathbf{y}$. So does $y = \cos x$, whose second derivative is $-\mathbf{\cos x}$.

All these derivatives come from one basic limit: $(\sin h)/h$ approaches **1**. The sine of .01 radians is very close to **.01**. So is the **tangent** of .01. The cosine of .01 is not .99, because $1 - \cos h$ is much **smaller** than h. The ratio $(1 - \cos h)/h^2$ approaches $\mathbf{\frac{1}{2}}$. Therefore $\cos h$ is close to $1 - \frac{1}{2}h^2$ and $\cos .01 \approx$ **.99995**. We can replace h by x.

The differential equation $y'' = -y$ leads to **oscillation.** When y is positive, y'' is **negative.** Therefore y' is **decreasing.** Eventually y goes below zero and y'' becomes **positive.** Then y' is **increasing.** Examples of oscillation in real life are **springs** and **heartbeats.**

4 $\tan h = 1.01h$ at $h = 0$ and $h = \pm.17$; $\tan h = h$ at $h = 0$.

10 (a) $\frac{1-\cos h}{h^2} = \frac{1-\cos^2 h}{(1+\cos h)h^2} = \frac{1}{1+\cos h}(\frac{\sin h}{h})^2 \to \mathbf{\frac{1}{2}}$ (b) $\frac{1-\cos^2 h}{h^2} = (\frac{\sin h}{h})^2 \to \mathbf{1}$ (c) $\frac{1-\cos^2 h}{\sin^2 h} = \mathbf{1}$
(d) $\frac{1-\cos 2h}{h} = 2\frac{1-\cos 2h}{2h} \to 2(0) = \mathbf{0}$.

24 The maximum of $y = \sin x + \sqrt{3}\cos x$ is at $\mathbf{x} = \mathbf{\frac{\pi}{6}}$ (or 30°) where $\mathbf{y} = \mathbf{\frac{1}{2}} + \mathbf{\sqrt{3}\frac{\sqrt{3}}{2}} = \mathbf{2}$. The slope at that point is $\cos x - \sqrt{3}\sin x = \frac{\sqrt{3}}{2} - \frac{\sqrt{3}}{2} = \mathbf{0}$. Note that y is the same as $2\cos x$ shifted to the right by $\frac{\pi}{6}$.

26 (a) **False** (use the square rule) (b) **True** (because $\cos(-x) = \cos x$) (c) **False** for $y = x^2$ (happens to be true for $y = \sin x$) (d) **True** (y'' = slope of y' = positive when y' increases)

2.5 The Product and Quotient and Power Rules (page 77)

You have to learn the rules that are boxed on page 76. Many people memorize the product rule $uv' + vu'$ this way: *The derivative of a product is the first times the derivative of the second plus the second times the derivative of the first.* The derivative of a quotient is : "The bottom times the derivative of the top minus the top times the derivative of the bottom, all divided by the bottom squared." I chant these to myself as I use them. (This is from Jennifer Carmody. Professor Strang just mumbles them.) Questions 1–6 ask for $\frac{dy}{dx}$ from the rules for derivatives.

1. $y = (4x^3 - 2x + 7)^5$.

   - This has the form u^5 where $u = 4x^3 - 2x + 7$. Use the power rule $5u^4\frac{du}{dx}$ noting that $\frac{du}{dx} = 12x^2 - 2$. Then $y' = 5(4x^3 - 2x + 7)^4(12x^2 - 2)$.

2. $y = \sqrt{\cos x + \sin x}$.

   - Here $y = u^{1/2}$ where $u = \cos x + \sin x$ and $u' = -\sin x + \cos x$. Use the power rule with $n = \frac{1}{2}$:

$$y' = \frac{1}{2}u^{-\frac{1}{2}}\frac{du}{dx} = \frac{1}{2}(\cos x + \sin x)^{-\frac{1}{2}}(-\sin x + \cos x).$$

3. $y = (4x - 7)^3(2x + 3)^9$.

   - Use the product rule for $y = uv$, with $u = (4x - 7)^3$ and $v = (2x + 3)^9$. But we need the power rule to find $u' = 3(4x - 7)^2(4)$ and $v' = 9(2x + 3)^8(2)$. *Where did the 4 and 2 come from?*

Putting it all together, y' is

$$\begin{aligned} uv' + vu' &= (4x-7)^3(9)(2x+3)^8(2) + (2x+3)^9(3)(4x-7)^2(4) \\ &= 18(4x-7)^3(2x+3)^8 + 12(2x+3)^9(4x-7)^2. \end{aligned}$$

4. Verify that $(\sec x)' = \sec x \tan x$. Since $\sec x = \frac{1}{\cos x}$, we can use the reciprocal rule.

- *This is the quotient rule with $u = 1$ on top.* The bottom is $v = \cos x$. We want $\frac{-v'}{v^2}$:

$$(\sec x)' = (\frac{1}{\cos x})' = \frac{-(-\sin x)}{\cos^2 x} = \frac{1}{\cos x} \cdot \frac{\sin x}{\cos x} = \sec x \tan x.$$

5. $y = \frac{\sin x}{1+x^2}$.

- Use the quotient rule with $u = \sin x$ and $v = 1 + x^2$. Find $u' = \cos x$ and $v' = 2x$:

$$y' = \frac{vu' - uv'}{v^2} = \frac{(\text{bottom})(\text{top})' - (\text{top})(\text{bottom})'}{(\text{bottom})^2} = \frac{(1+x^2)\cos x - \sin x(2x)}{(1+x^2)^2}.$$

6. $y = \cos x(1 + \tan x)(1 + \sin^2 x)$. Use the triple product rule from Example 5 on page 72.

- The three factors are $u = \cos x$ with $u' = -\sin x$; $v = 1 + \tan x$ with $v' = \sec^2 x$; $w = 1 + \sin^2 x$ with $w' = 2 \sin x \cos x$ (by the power rule). The triple product rule is $uvw' + uv'w + u'vw$:

$$(uvw)' = \cos x(1 + \tan x)(2 \sin x \cos x) + \cos x(\sec^2 x)(1 + \sin^2 x) - \sin x(1 + \tan x)(1 + \sin^2 x).$$

Read-throughs and selected even-numbered solutions :

The derivatives of $\sin x \cos x$ and $1/\cos x$ and $\sin x/\cos x$ and $\tan^3 x$ come from the **product** rule, **reciprocal** rule, **quotient** rule, and **power** rule. The product of $\sin x$ times $\cos x$ has $(uv)' = uv' + \mathbf{u'v} = \mathbf{\cos^2 x - \sin^2 x}$. The derivative of $1/v$ is $\mathbf{-v'/v^2}$, so the slope of sec x is $\mathbf{\sin x/\cos^2 x}$. The derivative of u/v is $\mathbf{(vu' - uv')/v^2}$ so the slope of $\tan x$ is $\mathbf{(\cos^2 x + \sin^2 x)/\cos^2 x = \sec^2 x}$. The derivative of $\tan^3 x$ is $\mathbf{3 \tan^2 x \sec^2 x}$. The slope of x^n is $\mathbf{nx^{n-1}}$ and the slope of $(u(x))^n$ is $\mathbf{nu^{n-1} du/dx}$. With $n = -1$ the derivative of $(\cos x)^{-1}$ is $\mathbf{-1(\cos x)^{-2}(-\sin x)}$, which agrees with the rule for $\sec x$.

Even simpler is the rule of **linearity,** which applies to $au(x) + bv(x)$. The derivative is $\mathbf{au'(x) + bv'(x)}$. The slope of $3 \sin x + 4 \cos x$ is $\mathbf{3 \cos x - 4 \sin x}$. The derivative of $(3 \sin x + 4 \cos x)^2$ is $\mathbf{2(3 \sin x + 4 \cos x)(3 \cos x - 4 \sin x)}$. The derivative of $\mathbf{\sin^4 x}$ is $4 \sin^3 x \cos x$.

2 $\frac{dy}{dx} = (x^2+1)(2x) + (x^2-1)(2x) = \mathbf{4x^3}$ **4** $\frac{-2x}{(1+x^2)^2} + \frac{-(-\cos x)}{(1-\sin x)^2}$.

6 $(x-1)^2 2(x-2) + (x-2)^2 2(x-1) = 2(x-1)(x-2)(x-1+x-2) = \mathbf{2(x-1)(x-2)(2x-3)}$.

8 $x^{1/2}(1+\cos x) + (x + \sin x)\frac{1}{2}x^{-1/2}$ or $\frac{3}{2}x^{1/2} + x^{1/2}\cos x + \frac{1}{2}x^{-1/2}\sin x$

10 $\frac{(x^2-1)2x - (x^2+1)2x}{(x^2-1)^2} + \frac{\cos x(\cos x) - \sin x(-\sin x)}{\cos^2 x} = \frac{-4x}{(x^2-1)^2} + \frac{1}{\cos^2 x}$.

12 $x^{3/2}(3\sin^2 x \cos x) + \frac{3}{2}x^{1/2}\sin^3 x + \frac{3}{2}(\sin x)^{1/2}\cos x$

14 $\sqrt{x}(\sqrt{x}+1)\frac{1}{2}x^{-1/2} + \sqrt{x}(\sqrt{x}+2)\frac{1}{2}x^{-1/2} + (\sqrt{x}+1)(\sqrt{x}+2)\frac{1}{2}x^{-1/2} = (3x + 6\sqrt{x} + 2)\frac{1}{2}x^{-1/2}$ (or other form).

16 $10(x-6)^9 + 10\sin^9 x \cos x$.

18 $\csc^2 x - \cot^2 x = \frac{1}{\sin^2 x} - \frac{\cos^2 x}{\sin^2 x} = \frac{\sin^2 x}{\sin^2 x} = 1$ so the derivative is **zero.**

20 $\frac{(\sin x + \cos x)(\cos x + \sin x) - (\sin x - \cos x)(\cos x - \sin x)}{(\sin x + \cos x)^2} = \frac{2\sin^2 x + 2\cos^2 x}{(\sin x + \cos x)^2} = \frac{2}{(\sin x + \cos x)^2}$

22 $\frac{x\cos x}{\sin x}$ has derivative $\frac{\sin x(-x\sin x+\cos x)-x\cos x(\cos x)}{\sin^2 x} = \frac{-x+\sin x\cos x}{\sin^2 x}$ (or other form).
24 $[u(x)]^2(2v(x)\frac{dv}{dx}) + [v(x)]^2(2u(x)\frac{du}{dx})$
26 $x\cos x + \sin x - \sin x = \mathbf{x\cos x}$ (we now have a function with derivative $x\cos x$).
34 (a) $y = \frac{1}{4}x^4$ (b) $y = -\frac{1}{2}x^{-2}$ (c) $y = -\frac{2}{5}(1-x)^{5/2}$ (This one is more difficult.) (d) $y = -\frac{1}{3}\cos^3 x$

2.6 Limits (page 84)

Limits are not seen in algebra. They are special to calculus. You do use algebra to simplify an expression beforehand. But that final gasp of "taking the limit" needs a definition, which involves epsilon (ϵ) and delta (δ).

The idea that $3x+2$ approaches 5 as x approaches 1 is pretty clear. We pin this down (and make it look difficult) by following through on the epsilon-delta definition

$3x+2$ is near 5 (as near as we want) when x is near 1
$(3x+2)-5$ is small (as small as we want) when $x-1$ is small
$|(3x+2)-5| < \epsilon$ (for any fixed $\epsilon > 0$) when $0 < |x-1| < \delta$ (δ depends on ϵ).

This example wants to achieve $|3x-3| < \epsilon$. This will be true if $|x-1| < \frac{1}{3}\epsilon$. So choose δ to be $\frac{1}{3}\epsilon$. By making that particular choice of δ we can say: If $|x-1| < \delta$ then $|(3x+2)-5| < \epsilon$. The number ϵ can be as small as we like. Therefore $L = 5$ is the correct limit.

A small point. Are we saying that "$f(x)$ comes *closer* to L as x comes closer to 1"? No! That is true in this example, but not for all limits. It gives the idea but it is not exactly right. Invent the function $y = x\sin\frac{1}{x}$. Since the sine stays below 1, we have $|y| < \epsilon$ if $|x| < \epsilon$. (This is extra confusing because we can choose $\delta = \epsilon$. The limit of $y = x\sin\frac{1}{x}$ is $L = 0$ as $x \to 0$). The point is that this function actually *hits* zero many times, and then moves *away* from zero, as the number x gets small. It hits zero when $\sin\frac{1}{x} = 0$. It doesn't move far, because y never gets larger than x.

This example does not get *steadily* closer to $L = 0$. It oscillates around its limit. But it converges.

Take $\epsilon = \frac{1}{1000}$ in Questions 1 and 2. Choose δ so that $|f(x) - L| < \frac{1}{1000}$ if $|x| < \delta$.

1. Show that $f(x) = 2x+3$ approaches $L = 3$ as $x \to 0$.

   - Here $|f(x) - L| = |2x|$. We want $|2x| < \epsilon = \frac{1}{1000}$. So we need $|x| < \frac{1}{2000}$. Choose $\delta = \frac{1}{2000}$ or any smaller number like $\delta = \frac{1}{5000}$. Our margin of error on $f(x)$ is $\frac{1}{1000}$ if our margin of error on x is $\frac{1}{2000}$.

2. Show that $\lim_{x\to 0} x^2 = 0$. In other words $x^2 \to 0$ as $x \to 0$. Don't say obvious.

   - $|f(x) - L| = |x^2 - 0| = |x^2|$. We want $|x^2| < \epsilon = \frac{1}{1000}$. This is guaranteed if $|x| < \sqrt{\frac{1}{1000}}$. This square root is a satisfactory δ.

Find the limits in 3–6 if they exist. If direct substitution leads to $\frac{0}{0}$, you need to do more work!

3. $\lim_{x\to 3}\frac{x^2-9}{x-3}$. Substituting $x = 3$ gives $\frac{0}{0}$ which is meaningless.

   - Since $\frac{x^2-9}{x-3} = \frac{(x-3)(x+3)}{(x-3)} = x+3$, substituting $x = 3$ now tells us that the limit is 6.

4. $\lim_{x\to 2}\frac{x^2-4}{\sin(x-2)}$. At $x = 2$ this is $\frac{0}{0}$. But note $x^2 - 4 = (x-2)(x+2)$.

- Write the function as $\frac{(x-2)}{\sin(x-2)}(x+2)$. Since $(x-2) \to 0$ as $x \to 2$, the fraction behaves like $\lim_{\theta\to 0} \frac{\theta}{\sin\theta} = 1$. The other factor $x+2$ goes to 4. The overall limit is $(1)(4) = 4$.

5. $\lim_{x\to 1} \frac{\sqrt{x+8}-3}{x-1}$. Both top and bottom go to zero at $x = 1$.

   - Here is a trick from algebra: Multiply top and bottom by $(\sqrt{x+8}+3)$. This gives $\frac{(x+8)-9}{(x-1)(\sqrt{x+8}+3)}$. The numerator is $x-1$. So cancel that above and below. The fraction approaches $\frac{1}{6}$ when $x \to 1$.

6. $\lim_{x\to 2} \frac{4}{x^2-4} - \frac{1}{x-2}$. Substitution gives "undefined minus undefined." Combine the two fractions into one:

$$\lim_{x\to 2} \frac{4-(x+2)}{(x-2)(x+2)} = \lim_{x\to 2} \frac{-x+2}{(x-2)(x+2)} = \lim_{x\to 2} \frac{-1}{(x+2)} = \frac{-1}{4}.$$

Buried under exercise 34 on page 85 is a handy "important rule" for limits as $x \to \infty$. Use it for 7–8.

7. Find $\lim_{x\to 0} \frac{4x^9 - 2x^7 + 18x}{3x^{12} - 7x^2 + 6}$.

   - When x is large, the expression is very like $\frac{4x^9}{3x^{12}}$. This is $\frac{4}{3x^3}$. As $x \to \infty$, this limit is 0. Whenever the top has lower degree than the bottom, the limit as $x \to \infty$ is 0.

8. Find $\lim_{x\to\infty} \frac{(4x^3-3)^6}{25x^{17}}$. You don't have to multiply out $(4x^3-3)^6$.

   - You just have to know that if you did, *the leading term would be* 4^6x^{18}. The limit of $\frac{4^6x^{18}}{25x^{17}} = \frac{4^6 x}{25}$ is ∞. This is the limit of the original problem.

9. (This is Problem 2.5.35c) Prove that $\lim_{x\to\infty} x \sin\frac{1}{x} = 1$. (The limit at $x = 0$ was zero!)

   - It is useless to say that $x \sin\frac{1}{x} \to \infty \cdot 0$. Infinity times zero is meaningless. The trick is to write $x \sin\frac{1}{x} = \frac{\sin\frac{1}{x}}{\frac{1}{x}}$. In other words, move x into the denominator as $\frac{1}{x}$. Since $\frac{1}{x} \to 0$ as $x \to \infty$, the limit is $\frac{\sin(\frac{1}{x})}{(\frac{1}{x})} \to 1$.

Read-throughs and selected even-numbered solutions :

The limit of $a_n = (\sin n)/n$ is **zero.** The limit of $a_n = n^4/2^n$ is **zero.** The limit of $a_n = (-1)^n$ is **not defined.** The meaning of $a_n \to 0$ is: Only **finitely many** of the numbers $|a_n|$ can be **greater than** ϵ **(an arbitrary positive number).** The meaning of $a_n \to L$ is: For every ϵ there is an **N** such that $|\mathbf{a_n} - \mathbf{L}| < \epsilon$ if $n > \mathbf{N}$. The sequence $1, 1+\frac{1}{2}, 1+\frac{1}{2}+\frac{1}{3}, \cdots$ is not **convergent** because eventually those sums go past **any number L.**

The limit of $f(x) = \sin x$ as $x \to a$ is **sin a.** The limit of $f(x) = x/|x|$ as $x \to -2$ is $\mathbf{-1}$, but the limit as $x \to 0$ does not **exist.** This function only has **one-sided** limits. The meaning of $\lim_{x\to a} f(x) = L$ is: For every ϵ there is a δ such that $|f(x) - L| < \epsilon$ whenever $\mathbf{0} < |\mathbf{x} - \mathbf{a}| < \delta$.

Two rules for limits, when $a_n \to L$ and $b_n \to M$, are $a_n + b_n \to \mathbf{L} + \mathbf{M}$ and $a_n b_n \to \mathbf{LM}$. The corresponding rules for functions, when $f(x) \to L$ and $g(x) \to M$ as $x \to a$, are $\mathbf{f(x)} + \mathbf{g(x)} \to \mathbf{L} + \mathbf{M}$ and $\mathbf{f(x)g(x)} \to \mathbf{LM}$. In all limits, $|a_n - L|$ or $|f(x) - L|$ must eventually go below and **stay below** any positive **number** ϵ.

$A \Rightarrow B$ means that A is a **sufficient** condition for B. Then B is true **if** A is true. $A \Leftrightarrow B$ means that A is a **necessary and sufficient** condition for B. Then B is true **if and only if** A is true.

2 (a) is false when $L = 0$: $\mathbf{a_n} = \frac{\mathbf{1}}{\mathbf{n}} \to 0$ and $\mathbf{b_n} = \frac{\mathbf{1}}{\mathbf{n^2}} \to 0$ but $\frac{a_n}{b_n} = n \to \infty$ (b) It is true that: If $a_n \to L$ then $a_n^2 \to L^2$. It is false that: If $a_n^2 \to L^2$ then $a_n \to L$: $\mathbf{a_n}$ **could approach** $-L$ or $a_n = L, -L, L, -L, \cdots$ has no limit. (c) $a_n = -\frac{1}{n}$ is negative but the limit $L = 0$ is **not** negative (d) $1, \frac{1}{2}, 1, \frac{1}{3}, 1, \frac{1}{4}, \cdots$ has infinitely many a_n in every strip around zero but a_n does not approach zero.

8 No limit **10** Limits equals $\mathbf{f'(1)}$ **if the derivative exists.** **12** $\frac{2x \tan x}{\sin x} = \frac{2x}{\cos x} \to \frac{0}{1} = \mathbf{0}$

14 $|x| = -x$ when x is negative; the limit of $\frac{-x}{x}$ is $\mathbf{-1}$. **16** $\frac{f(c)-f(a)}{c-a} \to \mathbf{f'(a)}$ if the derivative exists.

18 $\frac{x^2-25}{x-5} = x + 5$ approaches **10** as $x \to 5$ **20** $\frac{\sqrt{4-x}}{\sqrt{6+x}}$ approaches $\frac{\sqrt{2}}{\sqrt{8}} = \frac{\sqrt{2}}{2\sqrt{2}} = \frac{\mathbf{1}}{\mathbf{2}}$ as $x \to 2$

22 $\sec x - \tan x = \frac{1-\sin x}{\cos x} = \frac{1-\sin x}{\cos x}\left(\frac{1+\sin x}{1+\sin x}\right) = \frac{1-\sin^2 x}{\cos x(1+\sin x)} = \frac{\cos x}{1+\sin x}$ which approaches $\frac{0}{2} = \mathbf{0}$ at $x = \frac{\pi}{2}$.

24 $\frac{\sin(x-1)}{x-1}\left(\frac{1}{x+1}\right)$ approaches $1 \cdot \frac{1}{2} = \frac{\mathbf{1}}{\mathbf{2}}$ as $x \to 1$

28 Given any $\epsilon > 0$ there is an X such that $|f(x)| < \epsilon$ if $x < X$.

32 The limit is $e = 2.718 \cdots$

2.7 Continuous Functions (page 89)

Notes on the text: "blows up" means "approaches infinity." Even mathematicians use slang. To understand the Extreme Value Property, place two dots on your paper, and connect them with any function you like. Do not lift your pencil from the paper. The left dot is $(a, f(a))$, the right dot is $(b, f(b))$. Since you did not lift your pencil, your function is continuous on $[a, b]$. The function reaches a maximum (high point) and minimum (low point) somewhere on this closed interval. These extreme points are called $(x_{\max}, M)$ and $(x_{\min}, m)$. It is quite possible that the min or max is reached more than once.

The Extreme Value Property states that m and M are reached *at least once*. Now take a ruler and draw a horizontal line anywhere you like between m and M. The Intermediate Value Property says that, because $f(x)$ is continuous, your line and graph cross at least once.

In 1–3, decide if $f(x)$ is continuous for all x. If not, which requirement is not met? Can $f(x)$ be "fixed" to be continuous?

1. $f(x) = \frac{x^2-9}{x+3}$. This is a standard type of example. At $x = -3$ it gives $\frac{0}{0}$.

   - Note that $f(x) = \frac{(x-3)(x+3)}{(x+3)} = x - 3$. We can remove the difficulty (undefined value) by $f(-3) = -6$.

2. The "sign function" is 1 for positive x and -1 for negative x (and $f(0) = 0$ at the jump).

   - The sign function is continuous except at $x = 0$, where it jumps from -1 to 1. There is no way to redefine $f(0)$ to make this continuous. This $f(x)$ is not "continuable."

3. Suppose $f(x) = 3 + |x|$ except $f(0) = 0$. Is this function "continuable"?

   - Yes. At $x = 0$ the limit $L = 3$ does not equal $f(0) = 0$. Change to $f(0) = 3$.

Exercises 1–18 are excellent for understanding continuity and differentiability. A few solutions are worked out here. Find a number c (*if possible*) to make the function continuous and differentiable.

4. Problem 2.7.5 has $f(x) = \begin{cases} c+x & \text{for } x < 0 \\ c^2+x^2 & \text{for } x \geq 0. \end{cases}$ The graph is a straight line then a parabola.

   - For continuity, the line $y = c + x$ must be made to meet the parabola $y = c^2 + x^2$ at $x = 0$. This means $c = c^2$, so $c = 0$ or 1. The slope is 1 from the left and 0 from the right. This function cannot be made differentiable at $x = 0$.

5. Problem 2.7.7 has $f(x) = \begin{cases} 2x, & x < c \\ x+1, & x \geq c. \end{cases}$ A line then another line.

   - $y = 2x$ meets $y = x + 1$ at $x = 1$. If $c = 1$ then $f(x)$ is continuous. It is not differentiable because the lines have different slopes 2 and 1. If $c \neq 1$ then $f(x)$ is not even continuous. The lines don't meet.

6. Problem 2.7.12 has $f(x) = \begin{cases} c & x \leq 0 \\ \sec x & x > 0. \end{cases}$ A constant (horizontal line) and a curve.

   - At $x = 0$ the limit of $\sec x = \frac{1}{\cos x}$ is $\frac{1}{1} = 1$. So if $c = 1$, the function is continuous at $x = 0$. The slope of $f(x) = \sec x$ is $\sec x \tan x$, which is 0 at $x = 0$. Therefore the function is differentiable at 0 if $c = 1$. However, $\sec x$ is undefined at $x = \frac{\pi}{2}, \frac{3\pi}{2}, \cdots$.

7. Problem 2.7.13 has $f(x) = \begin{cases} \frac{x^2+c}{x-1}, & x \neq 1 \\ 2 & x = 1. \end{cases}$ The expression $\frac{x^2+c}{x-1}$ is undefined at $x = 1$.

   - If we choose $c = -1$, the fraction $\frac{x^2+x}{x-1}$ reduces to $x + 1$. This is good. The value at $x = 1$ agrees with $f(1) = 2$. Then $f(x)$ is both continuous and differentiable if $c = -1$. Remember that $f(x)$ must be continuous if it is differentiable. *Not vice versa!*

Read-throughs and selected even-numbered solutions :

Continuity requires the **limit** of $f(x)$ to exist as $x \to a$ and to agree with **f(a)**. The reason that $x/|x|$ is not continuous at $x = 0$ is : **it jumps from −1 to 1.** This function does have **one-sided** limits. The reason that $1/\cos x$ is discontinuous at $\mathbf{x} = \pi/2$ is **that it approaches infinity.** The reason that $\cos(1/x)$ is discontinuous at $x = 0$ is **infinite oscillation.** The function $f(x) = \mathbf{\frac{1}{x-3}}$ has a simple pole at $x = 3$, where f^2 has a **double** pole.

The power x^n is continuous at all x provided n is **positive.** It has no derivative at $x = 0$ when n is **between 0 and 1.** $f(x) = \sin(-x)/x$ approaches **−1** as $x \to 0$, so this is a **continuous** function provided we define $f(0) = \mathbf{-1}$. A "continuous function" must be continuous at all **points in its domain.** A "continuable function" can be extended to every point x so that **it is continuous.**

If f has a derivative at $x = a$ then f is necessarily **continuous** at $x = a$. The derivative controls the speed at which $f(x)$ approaches **f(a).** On a closed interval $[a, b]$, a continuous f has the **extreme** value property and the **intermediate** value property. It reaches its **maximum** M and its **minimum** m, and it takes on every value **in between.**

8 $c > 0$ gives $f(x) = x^c$: For $0 < c < 1$ this is not differentiable at $x = 0$ but is continuous for $(x \geq 0)$. For $c \geq 1$ this is continuous and differentiable where it is defined ($x \geq 0$ for noninteger c).

10 Need $x+c=1$ at $x=c$ which gives $2c=1$ or $\mathbf{c}=\mathbf{\frac{1}{2}}$. Then $x+\frac{1}{2}$ matches 1 at $x=\frac{1}{2}$ (continuous but not differentiable).

16 At $x=c$ continuity requires $c^2=2c$. Then $\mathbf{c=0}$ **or** $\mathbf{2}$. At $x=c$ the derivative jumps from $2x$ to 2.

36 $\cos x$ is greater than $2x$ at $x=0$; $\cos x$ is less than $2x$ at $x=1$. The continuous function $\cos x-2x$ changes from positive to negative. By the **intermediate value theorem** there is a point where $\cos x-2x=0$.

38 $x\sin\frac{1}{x}$ approaches zero as $x\to 0$ (so it is continuous) because $|\sin\frac{1}{x}|<1$. There is no derivative because $\frac{f(h)-f(0)}{h}=\frac{h}{h}\sin\frac{1}{h}=\sin\frac{1}{h}$ has no limit (infinite oscillation).

40 A continuous function is continuous at each point x in its domain (where $f(x)$ is defined). A **continuable** function can be defined at all other points x in such a way that it is continuous there too. $f(x)=\frac{1}{x}$ is continuous away from $x=0$ but not continuable.

42 $f(x)=x$ if x is a fraction, $f(x)=0$ otherwise

44 Suppose L is the limit of $f(x)$ as $x\to a$. To prove continuity we have to show that $f(a)=L$. For any ϵ we can obtain $|f(x)-L|<\epsilon$, and this applies **at** $\mathbf{x=a}$ (since that point is not excluded any more). Since ϵ is arbitrarily small we reach $f(a)=L$: the function has the right value at $x=a$.

2 Chapter Review Problems

Review Problems

R1 The average slope of the graph of $y(x)$ between two points x and $x+\Delta x$ is ____ . The slope at the point x is ____ .

R2 For a distance function $f(t)$, the average velocity between times t and $t+\Delta t$ is ____ . The instantaneous velocity at time t is ____.

R3 Identify these limits as derivatives at specific points and compute them:

(a) $\lim\limits_{x\to 1}\dfrac{x^6-1}{x-1}$ (b) $\lim\limits_{x\to -1}\dfrac{x^8-1}{x+1}$ (c) $\lim\limits_{x\to \pi}\dfrac{\sin x}{x-\pi}$ (d) $\lim\limits_{x\to t}\dfrac{\cos x-\cos t}{x-t}$

R4 Write down the six terms of $(x+h)^5$. Subtract x^5. Divide by h. Set $h=0$ to find ____ .

R5 When u increases by Δu and v increases by Δv, how much does uv increase by? Divide that increase by Δx and let $\Delta x\to 0$ to find $\frac{d}{dx}(uv)=$ ____ .

R6 What is the power rule for the derivative of $1/f(x)$ and specifically of $1/(x^2+1)$?

R7 The tangent line to the graph of $y=\tan x-x$ at $x=\frac{\pi}{4}$ is $y=$ ____ .

R8 Find the slope and the equation of the normal line perpendicular to the graph of $y=\sqrt{1+x^2}$ at $x=3$.

R9 $f(x)=0$ for $x\le 1$ and $f(x)=(x-1)^2$ for $x>1$. Find the derivatives $f'(1)$ and $f''(1)$ if they exist.

R10 The limit as $x\to 2$ of $f(x)=x^2-4x+10$ is 6. Find a number δ so that $|f(x)-6|<.01$ if $|x-2|<\delta$.

Drill Problems Find the derivative $\frac{dy}{dx}$ in **D1 – D7**.

D1 $y = 3x^5 + 8x^2 - \frac{10}{x} + 7$

D2 $y = (\cos 2x)(\tan \frac{x}{8})$

D3 $y = \sin^3(x-3)$

D4 $y = (x^3+2)/(3-x^2)$

D5 $y = \sqrt{1+\sqrt{x}}$

D6 $y = x \tan x \sin x$

D7 $y = \frac{x}{\sin x} + \frac{\sin x}{x}$

D8 Compute $\lim_{x \to 0} \frac{\sin 2x}{x}$ and $\lim_{x\to 0} \frac{\sin x^2}{1-\cos x}$ and $\lim_{x\to 2} \frac{\sin x}{x}$.

D9 Evaluate the limits as $x \to \infty$ of $\frac{5-x^3}{x^2}$ and $\frac{5-x^2}{x^2}$ and $\frac{5-x}{x^2}$.

D10 For what values of x is $f(x)$ continuous? $f(x) = \frac{5x^3+12}{x^2-8}$ and $f(x) = \sqrt{\frac{x-3}{x^2-9}}$ and $f(x) = \frac{x}{|x|}$.

D11 Draw any curve $y = f(x)$ that goes up and down and up again between $x = 0$ and $x = 4$. Then aligned below it sketch the derivative of $f(x)$. Then aligned below that sketch the second derivative.

CHAPTER 3 APPLICATIONS OF DERIVATIVES

3.1 Linear Approximation (page 95)

This section is built on one idea and one formula. The idea is to use the tangent line as an approximation to the curve. The formula is written in several ways, depending which letters are convenient.

$$f(x) \approx f(a) + f'(a)(x-a) \qquad \text{or} \qquad f(x+\Delta x) \approx f(x) + f'(x)\Delta x.$$

In the first formula, a is the "basepoint." We use the function value $f(a)$ and the slope $f'(a)$ at that point. The step is $x - a$. The tangent line $Y = f(a) + f'(a)(x-a)$ approximates the curve $y = f(x)$.

In the other form the basepoint is x and the step is Δx. Remember that the average $\frac{\Delta f}{\Delta x} = \frac{f(x+\Delta x)-f(x)}{\Delta x}$ is close to $f'(x)$. That is all the second form says: Δf *is close to* $f'(x)\Delta x$. Here are the steps to approximate $\frac{1}{4.1}$:

Step 1: What is the function being used? • It is $f(x) = \frac{1}{x}$.

Step 2: At what point "a" near $x = 4.1$ is $f(x)$ exactly known? We know that $\frac{1}{4} = 0.25$. Let $a = 4$.

Step 3: What is the slope of $f(x)$ at $x = a$? The derivative is $f'(x) = -\frac{1}{x^2}$ so $f'(4) = -\frac{1}{4^2} = -\frac{1}{16}$.

Step 4: What is the formula for the approximation? • $f(x) \approx \frac{1}{4} - \frac{1}{16}(x-4)$.

Step 5: Substitute $x = 4.1$ to find $f(4.1) \approx \frac{1}{4} - \frac{1}{16}(4.1-4) = 0.24375$. The calculator gives $\frac{1}{4.1} = .24390\cdots$.

1. Find the linear approximation to $f(x) = \sec x$ near $x = \frac{\pi}{3}$

   • You will need the derivative of $\sec x$. It's good to memorize all six trigonometric derivatives at the bottom of page 76. Since $\sec x = \frac{1}{\cos x}$, you can also use the reciprocal rule. Either way, $f(x) = \sec x$ leads to $f'(x) = \sec x \tan x$ and $f'(\frac{\pi}{3}) = \sec\frac{\pi}{3}\tan\frac{\pi}{3} = (2)(\frac{1}{\sqrt{3}})$. This is the slope of the tangent line.

   The approximation is $f(x) \approx \sec\frac{\pi}{3} + \frac{2}{\sqrt{3}}(x - \frac{\pi}{3})$ or $f(x) \approx 2 + \frac{2}{\sqrt{3}}(x - \frac{\pi}{3})$. The right side is the tangent line $Y(x)$. The other form is $f(\frac{\pi}{3} + \Delta x) \approx 2 + \frac{2}{\sqrt{3}}\Delta x$. The step $x - \frac{\pi}{3}$ is Δx.

2. (This is Problem 3.1.15) Calculate the numerical error in the linear approximation to $(\sin 0.01)^2$ from the base point $a = 0$. Compare the error with the quadratic correction $\frac{1}{2}(\Delta x)^2 f''(x)$.

   • The function being approximated is $f(x) = (\sin x)^2$. We chose $a = 0$ since 0.01 is near that basepoint and $f(0) = 0$ is known. Then $f'(x) = 2\sin x\cos x$ and $f'(0) = 2\sin 0\cos 0 = 0$. The linear approximation is $f(0.01) \approx 0 + 0(.01)$. The tangent line is the x axis! It is like the tangent to $y = x^2$ at the bottom of the parabola, where $x = 0$ and $y = 0$ and slope $= 0$.

   The linear approximation is $(\sin 0.01)^2 \approx 0$. A calculator gives the value $(\sin 0.01)^2 = 9.9997 \times 10^{-5}$ which is close to $10 \times 10^{-5} = 10^{-4}$. To compare with the quadratic correction we need $f''(x)$. From $f'(x) = 2\sin x\cos x$ comes $f''(x) = 2\cos^2 x - 2\sin^2 x$. Then $\frac{1}{2}(\Delta x)^2 f''(0) = \frac{1}{2}(0.01)^2(2) = 10^{-4}$ exactly. (Note: The calculator is approximating too, but it is using much more accurate methods.)

   To summarize: Linear approximation 0, quadratic approximation .0001, calculator approximation .000099997.

3. A melting snowball of diameter six inches loses a half inch in diameter. Estimate its loss in surface area and volume.

   • The area and volume formulas on the inside back cover are $A = 4\pi r^2$ and $V = \frac{4}{3}\pi r^3$. Since $\frac{dA}{dr} = 8\pi r$ and $\frac{dV}{dr} = 4\pi r^2$, the linear corrections (the "differentials") are $dA = 8\pi r dr$ and $dV = 4\pi r^2 dr$. Note that $dr = -\frac{1}{4}$ inches since the *radius* decreased from 3 inches to $2\frac{3}{4}$ inches. Then $dA = 8\pi(3)(-\frac{1}{4}) = -6\pi$ square inches and $dV = 4\pi(3)^2(\frac{-1}{4}) = -9\pi$ cubic inches.

Note on differentials: df is exactly the linear correction $f'(x)dx$. When $f(x)$ is the area $A(r)$, this problem had $dA = 8\pi r dr$. The point about differentials is that they allow us to use an equal sign ($=$) instead of an approximation sign ($\approx$). It is a good notation for the *linear correction term.*

4. Imagine a steel band that fits snugly around the Earth's equator. More steel will be added to the band to make it lie one foot above the equator all the way around. How much more steel will have to be added? Take a guess before you turn to differentials.

   - Did you guess about a couple of yards? Here's the mathematics: $C = 2\pi r$. We want to estimate the change in circumference C when the radius increases by one foot. The differential is $dC = 2\pi dr$. Substitute $dr = 1$ foot to find $dC = 2\pi$ feet ≈ 6.3 feet. Note that the actual radius of the earth doesn't enter into the calculations, so this 2-yard answer is the same on any planet or any sphere.

Read-throughs and selected even-numbered solutions :

On the graph, a linear approximation is given by the **tangent** line. At $x = a$, the equation for that line is $Y = f(a) + \mathbf{f'(a)(x - a)}$. Near $x = a = 10$, the linear approximation to $y = x^3$ is $Y = 1000 + \mathbf{300(x - 10)}$. At $x = 11$ the exact value is $(11)^3 = \mathbf{1331}$. The approximation is $Y = \mathbf{1300}$. In this case $\Delta y = \mathbf{331}$ and $dy = \mathbf{300}$. If we know $\sin x$, then to estimate $\sin(x + \Delta x)$ we add $\mathbf{(\cos x)\Delta x}$.

In terms of x and Δx, linear approximation is $f(x + \Delta x) \approx f(x) + \mathbf{f'(x)\Delta x}$. The error is of order $(\Delta x)^p$ or $(x - a)^p$ with $p = \mathbf{2}$. The differential dy equals **dy/dx** times the differential **dx**. Those movements are along the **tangent** line, where Δy is along the **curve**.

2 $f(x) = \frac{1}{x}$ and $a = 2 : Y = f(a) + f'(a)(x - a) = \frac{1}{2} - \frac{1}{4}(x - 2)$. Tangent line is $Y = 1 - \frac{1}{4}x$.
6 $f(x) = \sin^2 x$ and $a = 0 : Y = \sin^2 0 + 2\sin 0 \cos 0(x - 0) = 0$. Tangent line is x axis.
10 $f(x) = x^{1/4}, Y = 16^{1/4} + \frac{1}{4}16^{-3/4}(15.99 - 16) = 2 + \frac{1}{4}\frac{1}{8}(-.01) = 1.9996875$. Compare $15.99^{1/4} = 1.9996874$.
18 Actual error: $\sqrt{8.99} - (3 + \frac{1}{6}(-.01)) = -4.6322\ \ 10^{-7}$; predicted error for $f = \sqrt{x}$, $f'' = \frac{-1}{4x^{3/2}}$ near $x = 9 : \frac{1}{2}(.01)^2(\frac{-1}{4(9)^{3/2}}) = -4.6296\ \ 10^{-7}$.
26 $V = \pi r^2 h$ so $dV = \pi r^2 dh = \pi(2)^2(0.5) = \pi(.2)$.

3.2 Maximum and Minimum Problems (page 103)

Here is the outstanding application of differential calculus. There are three steps: *Find the function, find its derivative, and solve* $f'(x) = 0$. The first step might come from a word problem – you have to choose a good variable x and find a formula for $f(x)$. The second step is calculus – to produce the formula for $f'(x)$. This may be the easiest step. The third step is fast or slow, according as $f'(x) = 0$ can be solved by a little algebra or a lot of computation. Every textbook gives problems where algebra gets the answer – so you see how the whole method works. We start with those.

A maximum can also occur at a rough point (where f' is not defined) or at an endpoint. Usually those are easier to locate than the stationary points that solve $f'(x) = 0$. They are not only easier to find, they are easier to forget. In Problems 1 – 4 find stationary points, rough points, and endpoints. Decide whether each of these is a local or absolute minimum or maximum.

1. (This is Problem 3.2.5) The function is $f(x) = (x - x^2)^2$ for $-1 \le x \le 1$. The end values are $f(-1) = 4$ and $f(1) = 0$. There are no rough points.

- The derivative by the power rule is $f' = 2(x - x^2)(1 - 2x) = 2x(1 - x)(1 - 2x)$. Then $f'(x) = 0$ at $x = 0$, $x = 1$, and $x = \frac{1}{2}$. Substitute into $f(x)$ to find $f(0) = 0$, $f(1) = 0$ and $f(\frac{1}{2}) = \frac{1}{16}$. Plot all these *critical points* to see that $(-1, 4)$ is the absolute maximum and $(\frac{1}{2}, \frac{1}{16})$ is a relative maximum. There is a tie (0,0) and (1,0) for absolute minimum.

The next function has two parts with two separate formulas. In such a case watch for a rough point at the "breakpoint" between the parts – where the formula changes.

2. (This is Problem 3.2.8) The two-part function is $f(x) = x^2 - 4x$ for $0 \le x \le 1$ and $f(x) = x^2 - 4$ for $1 \le x \le 2$. The breakpoint is at $x = 1$, where $f(x) = -3$. The endpoints are (0,0) and (2,0).

- The slope for $0 < x < 1$ is $f'(x) = 2x - 4$. Although $f'(2) = 0$, this point $x = 2$ is not in the interval $0 \le x \le 1$. For $1 < x < 2$ the slope $f'(x) = 2x$ is never zero.

This function has no stationary points. Its derivative is never zero. Still it has a maximum and a minimum! There is a rough point at $x = 1$ because the slope $2x - 4$ from the left does not equal $2x$ from the right. This is the minimum point. The endpoints where $f = 0$ are absolute maxima.

3. $f(x) = 2 + x^{2/3}$ for $-1 \le x \le 8$.

- The endpoints have $f(-1) = 3$ and $f(8) = 6$. The slope is $f'(x) = \frac{2}{3}x^{1/3}$. Note that $f'(0)$ is not defined, so $x = 0$, $f(x) = 2$ is a rough point. Since $f'(x)$ never equals zero, there are no stationary points. The endpoints are maxima, the rough point is a minimum. All powers x^p with $0 < p < 1$ are zero at $x = 0$ but with infinite slope; x^{p-1} blows up; rough point.

4. $f(x) = 3x^4 - 40x^3 + 150x^2 - 600$ all for x. "For all x" is often written "$-\infty < x < \infty$". No endpoints.

- The slope is $f'(x) = 12x^3 - 120x^2 + 300x = 12x(x^2 - 10x + 25) = 12x(x - 5)^2$. Then $f'(x) = 0$ at $x = 0$ and $x = 5$. Substituting 0 and 5 into $f(x)$ gives -600 and $+25$. That means $(0, -600)$ and (5,25) are stationary points. Since $f'(x)$ is defined for all x, there are no rough points.

Discussion: Look more closely at $f'(x)$. The double factor $(x - 5)^2$ is suspicious. The slope $f'(x) = 12x(x - 5)^2$ is positive on both sides of $x = 5$. The stationary point at $x = 5$ is *not* a maximum or minimum, just a "pause point" before continuing upward.

The graph is rising for $x > 0$ and falling for $x < 0$. The bottom is $(0, -600)$ – an absolute minimum. There is no maximum. Or you could say that the maximum value is *infinite*, as $x \to \infty$ and $x \to -\infty$. The maximum is at the "end" even if there are no endpoints.

5. (This is Problem 3.2.26) A limousine gets $(120–2v)/5$ miles per gallon. The chauffeur costs \$10/hour. Gas is \$1/gallon. Find the cheapest driving speed. What is to be minimized?

- Minimize the *cost per mile*. This is (cost of driver per mile) + (cost of gas per mile). The driver costs $\frac{10 \text{ dollars/hour}}{v \text{ miles/hour}} = \frac{10}{v}$ dollars/mile. The gas costs $\frac{1 \text{ dollar/gallon}}{\frac{120-2v}{5} \text{ miles/gallon}} = \frac{5}{120-2v}$ dollars/mile. Total cost per mile is $C(v) = \frac{10}{v} + \frac{5}{120-2v}$. Note that $0 < v < 60$. The cost blows up at the speed limit $v = 60$. Now that the hard part is over, we do the calculus:

$$\frac{dC}{dv} = \frac{-10}{v^2} + \frac{-5(-2)}{(120 - 2v)^2} = 0 \quad \text{if} \quad \frac{10}{v^2} = \frac{10}{(120 - 2v)^2}. \quad \text{This gives } v^2 = (120 - 2v)^2.$$

Then $v = 40$ or $v = 120$. We reject $v = 120$. The speed with lowest cost is $v = 40$ mph.

6. Form a box with no top by cutting four squares of sides x from the corners of a $12'' \times 18''$ rectangle. What x gives a box with maximum volume?

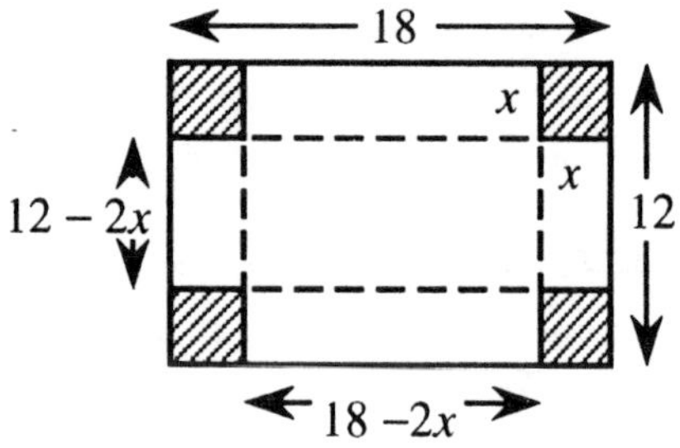

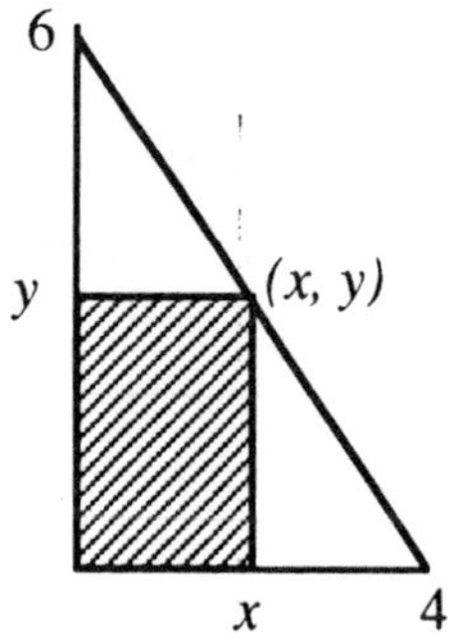

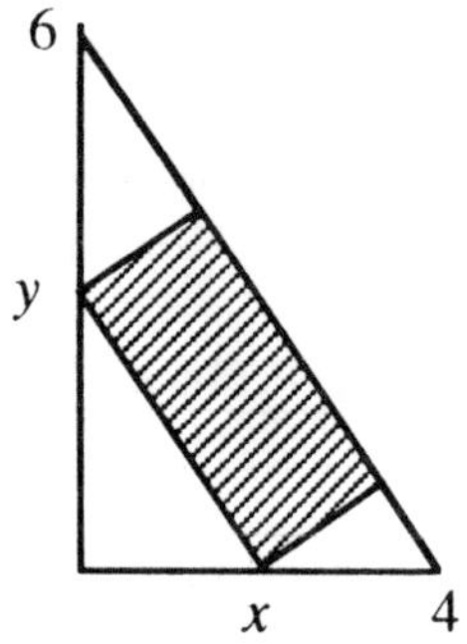

- Cutting out the squares and folding on the dotted lines gives a box x inches high and $12 - 2x$ inches wide and $18 - 2x$ inches long. We want to maximize volume = (height)(width)(length):

$$V(x) = x(12 - 2x)(18 - 2x) = 4x^3 - 60x^2 + 216x.$$

The endpoints are $x = 0$ (no height) and $x = 6$ (no width). Set $\frac{dV}{dx}$ to zero:

$$\frac{dV}{dx} = 12x^2 - 120x + 216 = 12(x^2 - 10x + 18) = 0$$

The quadratic formula gives $x = \frac{10 \pm \sqrt{100 - 4(18)}}{2} = 5 \pm \sqrt{7}$, so that $x \approx 7.6$ or $x \approx 2.35$. Since $x < 6$, the maximum volume is obtained when $x = 5 - \sqrt{7} \approx 2.35$ inches.

7. (This is Problem 3.3.43) A rectangle fits into a triangle with sides $x = 0$, $y = 0$ and $\frac{x}{4} + \frac{y}{6} = 1$. Find the point on the third side which maximizes the area xy. We want to maximize $A = xy$ with $0 < x < 4$ and $0 < y < 6$. Before we can take a derivative, we need to write y in terms of x (or vice versa). The equation $\frac{x}{4} + \frac{y}{6} = 1$ links x and y, and gives $y = 6 - \frac{3}{2}x$. This makes $a = x(6 - \frac{3}{2}x) = 6x - \frac{3}{2}x^2$. Now take the derivative: $\frac{dA}{dx} = 6 - 3x = 0$ when $x = 2$. Then $y = 6 - \frac{3}{2}x = 3$. The point (2,3) gives maximum area 6.

   I believe that a tilted rectangle of the same area also fits in the triangle. Correct?

8. Find the minimum distance from the point (2,4) to the parabola $y = \frac{x^2}{8}$.

   - Of all segments from (2,4) to the parabola, we would like to find the shortest. The distance is $D = \sqrt{(x-2)^2 + (\frac{x^2}{8} - 4)^2}$. Because of the square root, $D'(x)$ is algebraically complicated. We can make the problem easier by working with $f(x) = (x-2)^2 + (\frac{x^2}{8} - 4)^2 =$ *distance squared*. If we can minimize $f = D^2$, we have also minimized D. Using the square rule,

$$\frac{df}{dx} = 2(x-2) + 2(\frac{x^2}{8} - 4)(\frac{x}{4}) = 2x - 4 + \frac{x^3}{16} - 2x = \frac{x^3}{16} - 4.$$

Then $\frac{df}{dx} = 0$ if $x^3 = 64$ or $x = 4$. This gives $y = \frac{x^2}{8} = 2$. The minimum distance is from (2,4) to the point (4,2) on the parabola. The distance is $D = \sqrt{(4-2)^2 + (2-4)^2} = \sqrt{4+4} = 2\sqrt{2}$.

Read-throughs and selected even-numbered solutions :

If $df/dx > 0$ in an interval then $f(x)$ is **increasing**. If a maximum or minimum occurs at x then $f'(x) = \mathbf{0}$. Points where $f'(x) = 0$ are called **stationary** points. The function $f(x) = 3x^2 - x$ has a **(minimum)** at $x = \mathbf{\frac{1}{6}}$. A stationary point that is not a maximum or minimum occurs for $f(x) = \mathbf{x^3}$.

Extreme values can also occur when $\mathbf{f'(x)}$ is not defined or at the **endpoints** of the domain. The minima of $|x|$ and $5x$ for $-2 \le x \le 2$ are at $x = \mathbf{0}$ and $x = \mathbf{-2}$, even though df/dx is not zero. x^* is an absolute **maximum** when $f(x^*) \ge f(x)$ for all x. A **relative** minimum occurs when $f(x^*) \le f(x)$ for all x near x^*.

The minimum of $\frac{1}{2}ax^2 - bx$ is $\mathbf{-b^2/2a}$ at $x = \mathbf{b/a}$.

18 $f'(x) = \cos x - \sin x = 0$ at $x = \frac{\pi}{4}$ and $x = \frac{5\pi}{4}$. At those points $f(\frac{\pi}{4}) = \sqrt{2}$, the maximum, and $f(\frac{5\pi}{4}) = -\sqrt{2}$, the minimum. The endpoints give $f(0) = f(2\pi) = 1$.

28 When the length of day has its maximum and minimum, its derivative is zero (no change in the length of day). In reality the time unit of days is discrete not continuous; then Δf is small instead of $df = 0$.

30 $f'(t) = \frac{(1+3t^2)3-(1+3t)(6t)}{(1+3t^2)^2} = \frac{-9t^2-6t+3}{(1+3t^2)^2}$. Factoring out -3, the equation $3t^2+2t-1 = 0$ gives $t = \frac{-2+\sqrt{16}}{6} = \frac{1}{3}$. At that point $f_{\max} = \frac{2}{4/3} = \frac{3}{2}$. The endpoints $f(0) = 1$ and $f(\infty) = 0$ are minima.

36 Volume of popcorn box $= x(6-x)(12-x) = 72x - 18x^2 + x^3$. Then $\frac{dV}{dx} = 72 - 36x + 3x^2$. Dividing by 3 gives $x^2 - 12x + 24 = 0$ or $x = 6 \pm \sqrt{36-24} = 6 \pm \sqrt{12}$ at stationary points. **Maximum volume is at** $\mathbf{x = 6 - \sqrt{12}}$. ($V$ has a minimum at $x = 6 + \sqrt{12}$, when the box has negative width.)

46 The cylinder has radius r and height h. Going out r and up $\frac{1}{2}h$ brings us to the sphere: $r^2 + (\frac{1}{2}h)^2 = 1$. The volume of the cylinder is $V = \pi r^2 h = \pi[1 - (\frac{1}{2}h)^2]h$. Then $\frac{dV}{dh} = \pi[1 - (\frac{1}{2}h)^2] + \pi(-\frac{1}{2}h)h = 0$ gives $1 = \frac{3}{4}h^2$. The best h is $\mathbf{\frac{2}{\sqrt{3}}}$, so $V = \pi[1 - \frac{1}{3}]\frac{2}{\sqrt{3}} = \mathbf{\frac{4\pi}{3\sqrt{3}}}$. Note: $r^2 + \frac{1}{3} = 1$ gives $r = \sqrt{\frac{2}{3}}$.

56 *First method:* Use the identity $\sin x \sin(10-x) = \frac{1}{2}\cos(2x-10) - \frac{1}{2}\cos 10$. The maximum when $2x = 10$ is $\frac{1}{2} - \frac{1}{2}\cos 10 = .92$. The minimum when $2x - 10 = \pi$ is $-\frac{1}{2} - \frac{1}{2}\cos 10 = -.08$. *Second method:* $\sin x \sin(10-x)$ has derivative $\cos x \sin(10-x) - \sin x \cos(10-x)$ which is $\sin(10-x-x)$. This is zero when $10 - 2x$ equals 0 or π. Then $\sin x \sin(10-x)$ is $(\sin 5)(\sin 5) = .92$ or $\sin(5+\frac{\pi}{2})\sin(5-\frac{\pi}{2}) = -.08$.

62 The squared distance $x^2 + (y-\frac{1}{3})^2 = x^2 + (x^2-\frac{1}{3})^2$ has derivative $2x + 4x(x^2-\frac{1}{3}) = 0$ at $\mathbf{x = 0}$. Don't just cancel the factor x! The nearest point is (0,0). Writing the squared distance as $x^2+(y-\frac{1}{3})^2 = y+(y-\frac{1}{3})^2$ we forget that $y = x^2 \ge 0$. Zero is an **endpoint** and it gives the minimum.

3.3 Second Derivatives: Bending and Acceleration (page 110)

The first derivative gives the slope. The second derivative gives the *change of slope*. When the slope changes, the graph *bends*. When the slope $f'(x)$ does not change, then $f''(x) = 0$. This happens only for straight lines $f(x) = mx + b$.

At a minimum point, the slope is going from $-$ to $+$. Since f' is increasing, f'' must be positive.

At a maximum point, f' goes from $+$ to $-$ so f'' is negative.

At an inflection point, $f'' = 0$ and the graph is momentarily straight.

It makes sense that the approximation to $f(x + \Delta x)$ is better if we include bending. The linear part $f(x) + f'(x)\Delta x$ follows the tangent line. *The term to add is* $\frac{1}{2}f''(x)(\Delta x)^2$. With "$a$" as basepoint this is $\frac{1}{2}f''(a)(x-a)^2$.

The $\frac{1}{2}$ makes this exactly correct when $f = (x-a)^2$, because then $f'' = 2$ cancels the $\frac{1}{2}$.

1. Use $f''(x)$ to decide between maxima and minima of $f(x) = \frac{1}{3}x^3 + x^2 - 3x$.

   - $f'(x) = x^2 + 2x - 3 = (x+3)(x-1)$, so the stationary points are $x = 1$ and $x = -3$. Now compute $f'' = 2x + 2$. At $x = 1$, $f''(x)$ is positive. So this point is a local minimum. At $x = -3$, $f''(x)$ is negative. This point is a local maximum.

2. Find the minimum of $f(x) = x^2 + \frac{54}{x}, x \neq 0$.

   - $f'(x) = 2x - \frac{54}{x^2} = 0$ if $2x^3 = 54$ or $x^3 = 27$. There is a stationary point at $x = 3$. Taking the second derivative we get $f''(x) = 2 + \frac{108}{x^3}$. Since $f''(3) > 0$ the point is a minimum. This test on f'' does not say whether the minimum is relative or absolute.

3. Locate inflection points, if any, of $f(x) = \frac{1}{20}x^5 - \frac{1}{6}x^4 + \frac{1}{6}x^3 - 10x$. Here $f' = \frac{1}{4}x^4 - \frac{2}{3}x^3 + \frac{1}{2}x^2 - 10$.

   - This example has $f''(x) = x^3 - 2x^2 + x = x(x^2 - 2x + 1) = x(x-1)^2$. Inflection points require $f''(x) = 0$, which means that $x = 0$ or $x = 1$. But $x = 1$ is not a true inflection point. The double root from $(x-1)^2$ is again suspicious. The sign of f'' does not change as x passes 1. The bending stays positive and the tangent line stays under the curve. At a true inflection point the curve and line cross.

   The only inflection point is at $x = 0$, where f'' goes from negative to positive: bend down then bend up.

4. Write down the quadratic approximation for the function $y = x^4$ near $x = 1$. Use the formula boxed on page 109 with $a =$ ____, $f(a) =$ ____, $f'(x) =$ ____, $f'(a) =$ ____, $f''(x) =$ ____, $f''(a) = 12$.

   - The quadratic approximation is $x^4 \approx 1 + 4(x-1) + 6(x-1)^2$. Note $f(a) = 1, f'(a) = 4, \frac{1}{2}f''(a) = 6$.

Read-throughs and selected even-numbered solutions :

The direction of bending is given by the sign of $\mathbf{f''(x)}$. If the second derivative is **positive** in an interval, the function is concave up (or convex). The graph bends **upward**. The tangent lines are **below** the graph. If $f''(x) < 0$ then the graph is concave **down,** and the slope is **decreasing**.

At a point where $f'(x) = 0$ and $f''(x) > 0$, the function has a **minimum.** At a point where $\mathbf{f'(x) = 0}$ **and** $\mathbf{f''(x) < 0}$, the function has a maximum. A point where $f''(x) = 0$ is an **inflection** point, provided f'' changes sign. The tangent line **crosses** the graph.

The centered approximation to $f'(x)$ is $[\mathbf{f(x+\Delta x) - f(x-\Delta x)}]/2\Delta x$. The 3-point approximation to $f''(x)$ is $[\mathbf{f(x+\Delta x) - 2f(x) + f(x-\Delta x)}]/(\Delta x)^2$. The second-order approximation to $f(x+\Delta x)$ is $f(x) + f'(x)\Delta x + \mathbf{\frac{1}{2}f''(x)(\Delta x)^2}$. Without that extra term this is just the **linear (or tangent)** approximation. With that term the error is $\mathbf{O((\Delta x)^3)}$.

2 We want inflection points $\frac{d^2y}{dx^2} = 0$ at $x = 0$ and $x = 1$. Take $\frac{d^2y}{dx^2} = x - x^2$. This is positive ($y$ is concave up) between 0 and 1. Then $\mathbf{y = \frac{1}{6}x^3 - \frac{1}{12}x^4}$. (Intermediate step: the first derivative is $\frac{1}{2}x^2 - \frac{1}{3}x^3$). Alternative: $y = -x^2$ for $x < 0$, then $y = +x^2$ up to $x = 1$, then $y = 2 - x^2$ for $x > 1$.

20 $f'(x) = \cos x + 3(\sin x)^2 \cos x$ gives $f''(x) = -\sin x - 3(\sin x)^2 \sin x + 6\ \sin x(\cos x)^2 = \mathbf{5\sin x - 9\sin^3 x}$. Inflection points where $\sin x = 0$ (at $0, \pi, \cdots$) and also where $(\sin x)^2 = \frac{5}{9}$ (an angle x in each quadrant). Concavity is up-down-up from 0 to π. Then down-up-down from π to 2π.

34

$f(x) = \frac{1}{1-x}$	$\frac{f(\Delta x)-f(0)}{\Delta x} - f'(0)$	$\frac{f(\Delta x)-f(-\Delta x)}{2\Delta x} - f'(0)$	$\frac{f(\Delta x)-2f(0)+f(-\Delta x)}{(\Delta x)^2} - f''(0)$	$1 + x + x^2 - \frac{1}{1-x}$
$\Delta x = 1/4$	$1/3 = .333$	$1/15 = .067$	$2/15 = .133$	$-1/48 = -.021$
$\Delta x = 1/8$	$1/7 = .142$	$1/63 = .016$	$2/63 = .032$	$-1/448 = -.002$

36 At $x = 0.1$ the difference $\frac{1}{.9} - (1.11) = .00111\cdots$ comes from the omitted terms $x^3 + x^4 + x^5 + \cdots$.
At $x = 2$ the difference is $\frac{1}{1-2} - (1 + 2 + 4) = -8$. This is large because $x = 2$ is far from the basepoint.

42 $f(1) = 3, f(2) = 2 + 4 + 8 = 14, f(3) = 3 + 9 + 27 = 39$. The second difference is $\frac{39-28+3}{1^2} = 14$. The true $f'' = 2 + 6x$ is also 14. The error involves f'''' which in this example is zero.

3.4 Graphs (page 119)

1. The intercepts or axis crossings of this graph are (____ , ____) and (____ , ____). • (0,0) and (2,0).

2. $f'(x)$ is positive for which x? • $x < -1$ and $x > \frac{1}{2}$.

3. The stationary point with horizontal tangent is (____, ____). • $(\frac{1}{2}, -\frac{4}{3})$.

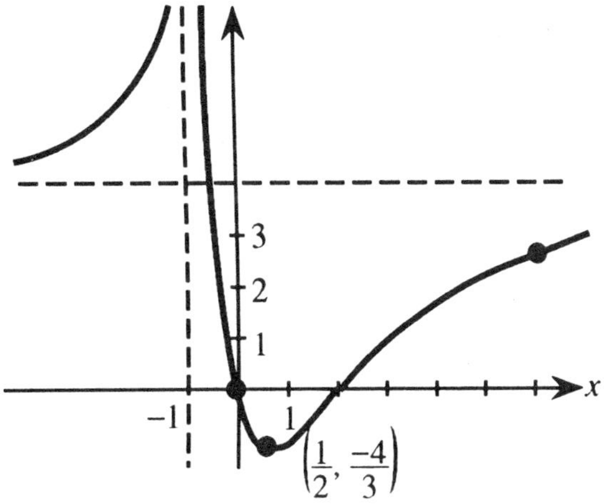

4. $f''(x) > 0$ and the curve bends up when ______. • $x < 2$ omitting $x = -1$.

5. The inflection point where $f''(x) = 0$ is (____ , ____). • (2,0).

6. As $x \to \infty$, the function $f(x)$ approaches ______ . • The limit is 4 (from below).

7. As $x \to -\infty$, the function $f(x)$ approaches ______ . • The limit is 4 (from above).

8. $f(x) \to \infty$ as x approaches ______ . • $x \to -1$ (from the left and from the right).

9. If this $f(x)$ is a ratio of polynomials, the denominator must have what factor? • $(x + 1)^2$ or $(x + 1)^4 \ldots$
Note $\frac{1}{x+1}$ causes blowup at $x = -1$. When it is squared the function goes to $+\infty$ on both sides of $x = -1$.

10. The degree of the numerator is $(<, =, >)$ the degree of the denominator? • *Equal* degrees since $y \to 4$.

Questions 11 - 13 are based on the graph below, which is drawn for $0 < x < 6$. In each question you complete the graph for $-6 < x < 0$.

11. Make $f(x)$ periodic with period 6. **12.** *Make $f(x)$ even.* **13.** *Make $f(x)$ odd.*

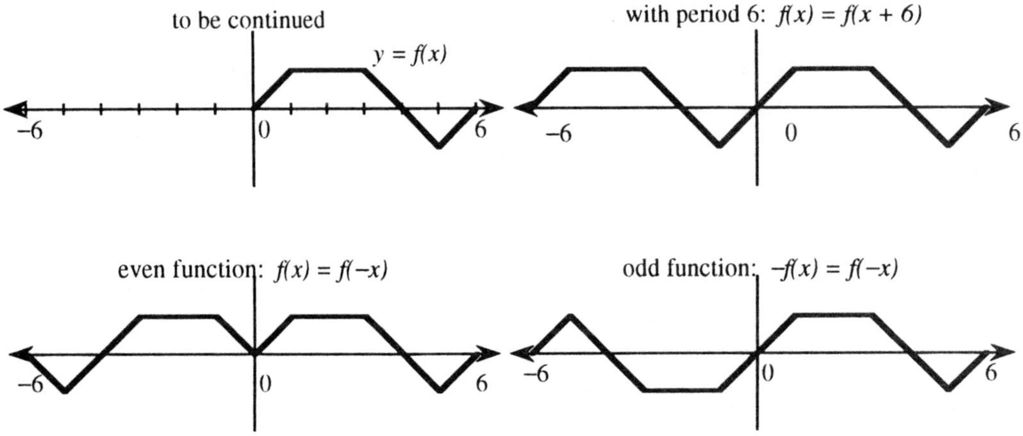

Problems 14 – 17 are designed to be done by hand. To graph a function from its equation, you need to think about x and y intercepts, asymptotes, and behavior as $x \to \infty$ and $x \to -\infty$. Consider also whether the function is even or odd. Use the information given by derivatives. (And it helps to plot a few points!)

14. Sketch $y = \frac{x}{x+1}$. Especially find the asymptotes.

- At $x = 0$ we find the only intercept $y = 0$. Since $x = -1$ makes the denominator zero, the line $x = -1$ is a *vertical asymptote*. To find horizontal asymptotes, let x get very large. Then $\frac{x}{x+1}$ gets very near to $y = 1$. Thus $y = 1$ is a *horizontal asymptote* (also as $x \to -\infty$). The function is neither even nor odd since $f(-x) = \frac{-x}{-x+1}$ is neither $f(x)$ nor $-f(x)$. The slope is $f'(x) = \frac{1}{(x+1)^2}$, so the graph is always rising. Since $f''(x) = \frac{-2}{(x+1)^3}$ the graph is concave up when $x < -1$ and concave down when $x > -1$.

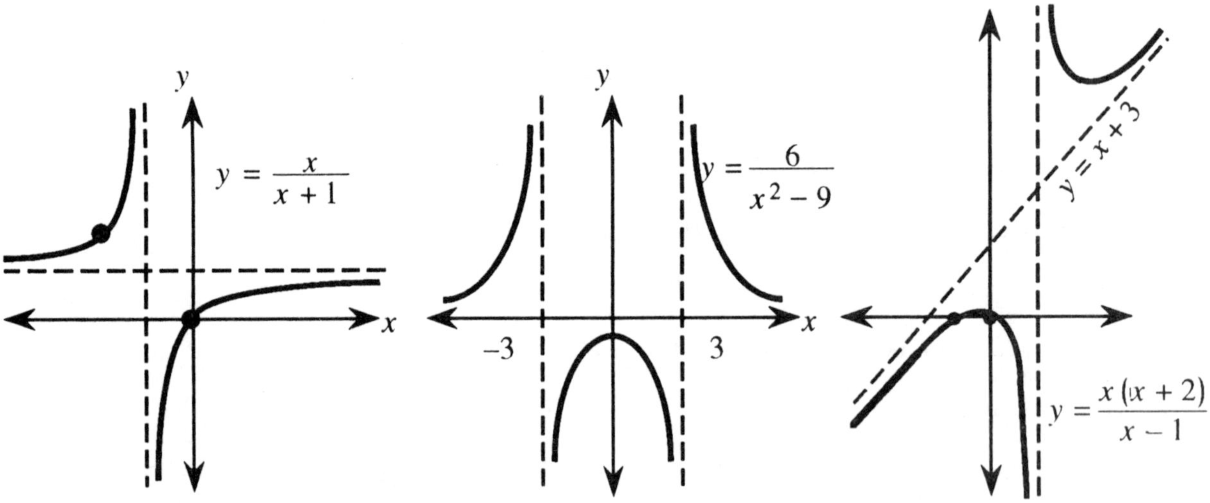

15. Sketch $y = \frac{6}{x^2-9}$. Explain why this function is even.

- Vertical asymptotes are $x = 3$ and $x = -3$, when the denominator $x^2 - 9$ is zero. The graph passes through $x = 0$, $y = -\frac{6}{9}$. There is no solution $y = 0$. In fact $y = 0$ is a horizontal asymptote (because $x^2 - 9 \to +\infty$). Note $f(-x) = \frac{6}{(-x)^2-9} = \frac{6}{x^2-9} = f(x)$ so this function is *even*. Its graph is symmetric about the y axis. The derivative is $f'(x) = -12x(x^2-9)^{-2}$, so there is a stationary point at $x = 0$. That point is a local maximum. *Even functions have odd derivatives. Always max or min at* $x = 0$.

16. Sketch $y = \frac{x^2-4}{x+2}$. It has a straight graph!

- $x = -2$ is *not* an asymptote even though the denominator is zero! We can factor x^2-4 into $(x-2)(x+2)$ and cancel $x+2$ from the top and bottom. (This is illegal at the point where $x+2=0$ and we have $\frac{0}{0}$.) The graph is the *straight line* $y = x - 2$ with a hole at $x = -2, y = -4$.

17. Sketch $y = \frac{x(x+2)}{(x-1)}$. Why will it have a sloping asymptote?

- There are intercepts at (0,0) and $(-2, 0)$. The vertical asymptote is $x = 1$ where the denominator is zero. The numerator has greater degree (2 versus 1) so we divide:

$$y = \frac{x^2+2x}{x-1} = x + 3 + \frac{3}{x-1}. \text{ (You should redo that division.)}$$

As x gets large, the last fraction is small. The line $y = x + 3$ is a *sloping asymptote.* (Sloping asymptotes come when "top degree = 1 + bottom degree.") The derivative is $\frac{dy}{dx} = 1 - \frac{3}{(x-1)^2}$, so there are stationary points at $(x-1)^2 = 3$ or $x = 1 \pm \sqrt{3}$. Since $\frac{d^2y}{dx^2} = 6(x-1)^{-3}$, the graph is concave up if $x > 1$ and concave down if $x < -1$.

Read-throughs and selected even-numbered solutions :

The position, slope, and bending of $y = f(x)$ are decided by $\mathbf{f(x), f'(x)}$, and $\mathbf{f''(x)}$. If $|f(x)| \to \infty$ as $x \to a$, the line $x = a$ is a vertical **asymptote.** If $f(x) \to b$ for large x, then $x = b$ is a **horizontal asymptote.** If $f(x) - mx \to b$ for large x, then $y = mx + b$ is a **sloping asymptote.** The asymptotes of $y = x^2/(x^2-4)$ are $\mathbf{x = 2, x = -2, y = 1}$. This function is even because $y(-x) = \mathbf{y(x)}$. The function $\sin kx$ has period $\mathbf{2\pi/k}$.

Near a point where $dy/dx = 0$, the graph is extremely **flat**. For the model $y = Cx^2, x = .1$ gives $y = \mathbf{.01C}$. A box around the graph looks long and **thin**. We **zoom** in to that box for another digit of x^*. But solving $dy/dx = 0$ is more accurate, because its graph **crosses** the x axis. The slope of dy/dx is $\mathbf{d^2y/dx^2}$. Each derivative is like an **infinite** zoom.

To move (a, b) to (0,0), shift the variables to $X = \mathbf{x - a}$ and $Y = \mathbf{y - b}$. This **centering** transform changes $y = f(x)$ to $Y = \mathbf{-b + f(X + a)}$. The original slope at (a, b) equals the new slope at $\mathbf{(0,0)}$. To stretch the axes by c and d, set $\mathbf{x} = cX$ and $\mathbf{y} = dY$. The **zoom** transform changes $Y = F(X)$ to $\mathbf{y = dF(x/c)}$. Slopes are multiplied by $\mathbf{d/c}$. Second derivatives are multiplied by $\mathbf{d/c^2}$.

12 $\frac{x}{\sin x}$ **is even.** Vertical asymptotes at all multiples $\mathbf{x = n\pi}$, except at $x = 0$ where $f(0) = 1$.
16 $\frac{\sin x + \cos x}{\sin x - \cos x}$ is periodic, not odd or even, vertical asymptotes when $\sin x = \cos x$ at $x = \frac{\pi}{4} + n\pi$.
22 $f(x) = \frac{1}{x} + 2x + 3$
30 (a) False: $\frac{x^4}{x^2+1}$ has no asymptotes (b) True: the second difference on page 108 is even (c) False: $f(x) = 1 + x$ is not even but $f'' = 0$ is even (d) False: $\tan x$ has vertical asymptotes but $\sec^2 x$ is never zero.

38 This is the second difference $\frac{\Delta^2 f}{\Delta x^2} \approx (\sin x)'' = -\sin x$.

40 (a) The asymptotes are $y = 0$ and $x = -3.48$. (b) The asymptotes are $y = 1$ and $x = 1$ (double root).

42 The exact solution is $x^* = \sqrt{3} = 1.73205$. The zoom should find those digits.

48 The exact $x = \sqrt{5}$ solves $\frac{x}{30\sqrt{15+x^2}} = \frac{1}{60}$ or $(2x)^2 = 15 + x^2$ or $x^2 = 5$.

52 $\sqrt{3x+1}$ is defined only for $x \geq -\frac{1}{3}$ (local maximum); minimum near $x = .95$; no inflection point.

58 Inflection points and second derivatives are **harder** to compute than maximum points and first derivatives. (In examples, derivatives seem easier than integrals. For numerical computation **it is the other way:** derivatives are very sensitive, integrals are smooth.)

3.5 Parabolas, Ellipses, and Hyperbolas (page 128)

This section is about "second degree equations," with x^2 and xy and y^2. In geometry, a plane cuts through a cone – therefore *conic* section. In algebra, we try to write the equation so the special points of the curve can be recognized. Also the curve itself has to be recognized – as circle, ellipse, parabola, or hyperbola.

The special point of a circle is its *center*. The special points for an ellipse are the two *foci*. A parabola has only one focus (the other one is at infinity) and we emphasize the *vertex*. A hyperbola has two foci and two vertices and two branches and frequently a *minus sign* in a critical place. But $B^2 - 4AC$ is positive.

1. Write the equation of a parabola with vertex at (1,3) and focus at (1,1).

- The focus is below the vertex, so the parabola opens downward. The distance from the vertex to the focus is $a = -2$, so $\frac{1}{4a} = -\frac{1}{8}$. The equation is $(y-3) = -\frac{1}{8}(x-1)^2$. Notice how the vertex is at (1,3) when the equation has $x - 1$ and $y - 3$. The vertex is at (0,0) when the equation is $y = \frac{1}{4a}x^2$.

2. $4(x+3)^2 + 9(y-2)^2 = 36$ is the equation for an ellipse. Find the center and the two foci.

- Divide by 36 to make the right side equal 1: $\frac{(x+3)^2}{9} + \frac{(y-2)^2}{4} = 1$. We see an ellipse centered at $(-3, 2)$. It has $a = \sqrt{9}$ and $b = \sqrt{4}$. From $c^2 = a^2 - b^2$ we calculate $c = \sqrt{9-4} = \sqrt{5}$. Then the foci are left and right of the center at $(-3-\sqrt{5}, 2)$ and $(-3+\sqrt{5}, 2)$.

3. $9x^2 - 16y^2 + 36x + 32y - 124 = 0$ is the equation of a hyperbola. Find the center, vertices, and foci.

- The x^2 and y^2 terms have opposite signs, which indicates a hyperbola. Regroup the terms:

$$9x^2 + 36x - 16y^2 + 32y = 124 \quad \text{or} \quad 9(x^2 + 4x + \_\_\_) - 16(y^2 - 2y + \_\_\_) = 124.$$

The blanks are for completing the squares. They should be filled by 4 and 1 to give $(x+2)^2$ and $(y-1)^2$. We must also add 9×4 and -16×1 to the right side: $9(x+2)^2 - 16(y-1)^2 = 124 + 36 - 16 = 144$. Dividing by 144 gives the standard form $\frac{(x+2)^2}{16} - \frac{(y-1)^2}{9} = 1$.

Now read off the main points. Center at $(-2, 1)$. Lengths $a^2 = 16$, $b^2 = 9$, $c = \sqrt{16+9} = 5$. Vertices are $a = 4$ from the center, at $(-6, 1)$ and (2,1). Foci are $c = 5$ from the center, at (–7,1) and (3,1).

4. (This is Problem 3.5.33) Rotate the axes of $x^2 + xy + y^2 = 1$. Use equation (7) with $\sin\alpha = \cos\alpha = \frac{1}{\sqrt{2}}$.

- The rotation is $x = \frac{1}{\sqrt{2}}x' - \frac{1}{\sqrt{2}}y'$ and $y = \frac{1}{\sqrt{2}}x' + \frac{1}{\sqrt{2}}y'$. The original equation involves x^2 and y^2 and xy so we compute

$$x^2 = \frac{1}{2}(x')^2 - x'y' + \frac{1}{2}(y')^2 \text{ and } y^2 = \frac{1}{2}(x')^2 + x'y' + \frac{1}{2}(y')^2 \text{ and } xy = \frac{1}{2}(x')^2 - \frac{1}{2}(y')^2.$$

Substitution changes $x^2 + xy + y^2 = 1$ into $\frac{3}{2}(x')^2 + \frac{1}{2}(y')^2 = 1$. This is $\frac{(x')^2}{\frac{2}{3}} + \frac{(y')^2}{2} = 1$, an ellipse.

5. $x - 2y^2 + 12y - 14 = 0$ is the equation for a parabola. Find the vertex and focus.

   - There is no xy term and no x^2 term, so we have a sideways parabola with a horizontal axis. Group the y terms: $2y^2 - 12y = x - 14$. Divide by 2 to get $y^2 - 6y = \frac{1}{2}x - 7$. Add 9 to both sides so that $y^2 - 6y + 9 = (y-3)^2$ is a perfect square. Then $(y-3)^2 = \frac{1}{2}x + 2 = \frac{1}{2}(x+4)$. The vertex of this parabola is at $x = -4$, $y = 3$. At the vertex the equation becomes $0 = 0$.

   The form $X = \frac{1}{4a}Y^2$ is $x + 4 = 2(y-3)^2$. Thus $\frac{1}{4a} = 2$. The focus is $a = \frac{1}{8}$ from the vertex.

6. What is the equation of a hyperbola having a vertex at (7,1) and foci at $(-2,1)$ and $(8,1)$?

   - The center must be at (3, 1), halfway between the foci. The distance from the center to each focus is $c = 5$. The distance to (7, 1) is $a = 4$, and so the other vertex must be at $(-1,1)$. Finally $b = \sqrt{c^2 - a^2} = 3$. Therefore the equation is $\frac{(x-3)^2}{4^2} - \frac{(y-1)^2}{3^2} = 1$.
     The negative sign is with y^2 because the hyperbola opens horizontally. The foci are on the line $y = 1$.

7. Find the ellipse with vertices at (0, 0) and (10, 0), if the foci are at (1,0) and (9,0).

   - The center must be at (5, 0). The distances from it are $a = 5$ and $c = 4$. Since an ellipse has $a^2 = b^2 + c^2$ we find $b = 3$. The equation is $\frac{(x-5)^2}{5^2} + \frac{y^2}{3^2} = 1$. Notice $x - 5$ and $y - 0$.

Read-throughs and selected even-numbered solutions :

The graph of $y = x^2 + 2x + 5$ is a **parabola.** Its lowest point (the vertex) is $(x, y) = (\mathbf{-1, 4})$. Centering by $X = x + 1$ and $Y = \mathbf{y - 4}$ moves the vertex to (0,0). The equation becomes $Y = \mathbf{X^2}$. The focus of this centered parabola is $(\mathbf{0, \frac{1}{4}})$. All rays coming straight down are **reflected** to the focus.

The graph of $x^2 + 4y^2 = 16$ is an **ellipse.** Dividing by **16** leaves $x^2/a^2 + y^2/b^2 = 1$ with $a = \mathbf{4}$ and $b = \mathbf{2}$. The graph lies in the rectangle whose sides are $\mathbf{x = \pm 4, y = \pm 2}$. The area is $\pi ab = \mathbf{8\pi}$. The foci are at $x = \pm c = \pm\sqrt{\mathbf{12}}$. The sum of distances from the foci to a point on the ellipse is always **8**. If we rescale to $X = x/4$ and $Y = y/2$ the equation becomes $\mathbf{X^2 + Y^2 = 1}$ and the graph becomes a **circle**.

The graph of $y^2 - x^2 = 9$ is a **hyperbola**. Dividing by 9 leaves $y^2/a^2 - x^2/b^2 = 1$ with $a = \mathbf{3}$ and $b = \mathbf{3}$. On the upper branch $y \geq \mathbf{0}$. The asymptotes are the lines $\mathbf{y = \pm x}$. The foci are at $y = \pm c = \pm\sqrt{\mathbf{18}}$. The **difference** of distances from the foci to a point on this hyperbola is **6**.

All these curves are conic sections – the intersection of a **plane** and a **cone.** A steep cutting angle yields a **hyperbola.** At the borderline angle we get a **parabola.** The general equation is $Ax^2 + \mathbf{Bxy + Cy^2 + Dx + Ey + F = 0}$. If $D = E = 0$ the center of the graph is at **(0,0)**. The equation $Ax^2 + Bxy + Cy^2 = 1$ gives an ellipse when $\mathbf{4AC > B^2}$. The graph of $4x^2 + 5xy + 6y^2 = 1$ is an **ellipse.**

14 $xy = 0$ gives the two lines $x = 0$ and $y = 0$, a degenerate hyperbola with vertices and foci all at (0,0).
16 $y = x^2 - x$ has vertex at $(\frac{1}{2}, -\frac{1}{4})$. To move the vertex to (0,0) set $X = x - \frac{1}{2}$ and $Y = y + \frac{1}{4}$. Then $Y = X^2$.

20 The path $x = t,\ y = t - t^2$ starts with $\frac{dx}{dt} = \frac{dy}{dt} = 1$ at $t = 0$ ($45°$ angle). Then $y_{\max} = \frac{1}{4}$ at $t = \frac{1}{2}$. The path is the parabola $y = x - x^2$.

32 The square has side s if the point $(\frac{s}{2}, \frac{s}{2})$ is on the ellipse. This requires $\frac{1}{a^2}(\frac{s}{2})^2 + \frac{1}{b^2}(\frac{s}{2})^2 = 1$ or $s^2 = 4(\frac{1}{a^2} + \frac{1}{b^2})^{-1}$ = area of square.

34 The Earth has $a = 149,597,870$ kilometers (Problem 19 on page 469 says $1.5 \cdot 10^8$ km). The eccentricity $e = \frac{c}{a}$ is 0.167 (or .02 on page 356). Then $c = 2.5 \cdot 10^6$ and $b = \sqrt{a^2 - c^2}$. This b is very near a; our orbit is nearly a circle. Use $\sqrt{a^2 - c^2} \approx a - \frac{c^2}{2a} \approx a - 2 \cdot 10^4$ km.

40 Complete squares: $y^2 + 2y = (y+1)^2 - 1$ and $x^2 + 10x = (x+5)^2 - 25$. Then $Y = y + 1$ and $X = x + 5$ satisfy $Y^2 - 1 = X^2 - 25$: the hyperbola is $\mathbf{X^2 - Y^2 = 24}$.

46 The quadratic $ax^2 + bx + c$ has two real roots if $b^2 - 4ac$ is positive and no real roots if $b^2 - 4ac$ is negative. Equal roots if $b^2 = 4ac$.

3.6 Iterations $x_{n+1} = F(x_n)$ (page 136)

This is not yet a standard topic in calculus. Or rather, the title "Iterations" is not widely used. Certainly Newton's method with $x_1 = F(x_0)$ and $x_2 = F(x_1)$ is always taught (as it should be). Other iterations should be presented too!

Main ideas: The *fixed points* satisfy $x^* = F(x^*)$. The iteration is fixed there if it starts there. This fixed point is *attracting* if $|F'(x^*)| < 1$. It is *repelling* if $|F'(x^*)| > 1$. The derivative F' at the fixed point decides whether $x_{n+1} - x^*$ is smaller than $x_n - x^*$, which means attraction. The "cobweb" spirals inward.

We gave iteration a full section because it is a very important application of calculus.

1. Problem 3.6.5 studies $x_{n+1} = 3x_n(1 - x_n)$. Start from $x_0 = .6$ and $x_0 = 2$. Compute $x_1, x_2, x_3, \cdots$ to test convergence. Then check dF/dx at the fixed points. Are they attracting or repelling?

   - This is one of the many times when a calculator comes in handy. From $x_0 = 0.6$ it gives $x_1 = 3(0.6)(1 - 0.6) = 0.72$ and $x_2 = 3(0.72)(1 - 0.72) = 0.6048$. On the TI-81 I use 3 ANS (1–ANS) and just push ENTER. Then $x_3 = 0.71705088$ and eventually $x_6 = 0.6118732$. Many iterations lead to $x = 0.6666\cdots$ (this looks like $x^* = \frac{2}{3}$). On the other hand $x_0 = 2$ leads to $x_1 = 3(2)(1-2) = -6$ and $x_2 = 3(6)(1 - (-6)) = 126$ and $x_3 = 3(126)(1 - 126) = -47250$. The sequence is diverging.

     To find the fixed points algebraically, solve $x^* = 3x^*(1 - x^*)$. This gives $x^* = 3x^* - 3(x^*)^2$, or $3(x^*)^2 = 2x^*$. The solutions $x^* = 0$ and $x^* = \frac{2}{3}$ are fixed points. If we start at $x_0 = 0$, then $x_1 = 0$ and we stay at 0. If we start at $x_0 = \frac{2}{3}$, then $x_1 = \frac{2}{3}$ and we stay at that fixed point.

     To decide whether the fixed points attract or repel, take the derivative of $F(x) = 3x(1-x)$. This is $3 - 6x$. At $x^* = 0$ the derivative is 3, so $x^* = 0$ is repelling. At $x^* = \frac{2}{3}$ the derivative is -1 so we are on the borderline. That value $|F'| = 1$ is why the calculator showed very slow attraction to $x^* = \frac{2}{3}$.

2. (This is Problem 3.6.15) Solve $x = \cos\sqrt{x}$ by iteration. Get into radian mode!

   - By sketching $y = x$ and $y = \cos\sqrt{x}$ together you can see that there is just one solution, somewhere in the interval $\frac{1}{2} < x^* < \frac{3}{4}$. Pick a starting point like $x_0 = 0.6$. Press $\sqrt{\ }$, cos, = and repeat. On the TI-81 press cos, $\sqrt{\ }$, ANS. Then ENTER many times to reach three-place accuracy $x^* = 0.679$.

3. Write out the first few steps of Newton's method for solving $x^3 - 7 = 0$.

   - Note $x^* = \sqrt[3]{7}$ so $x_0 = 2$ is a good place to start. Follow the steps in this table:

	x_n	$f(x_n) = x_n^3 - 7$	$c = \frac{1}{f'(x_n)} = \frac{1}{3(x_n)^2}$	$x_{n+1} = x_n - cf(x_n)$
$n = 0$	2	1	$\frac{1}{12}$	$2 - 1(\frac{1}{12}) = \frac{23}{12}$
$n = 1$	$\frac{23}{12}$	0.041088	0.0907372	1.9129385
$n = 2$	1.9129385	0.00007987	0.0910912	1.9129312

Since x_2 and x_3 begin with 1.9129, those digits are almost certainly correct in $x^* = \sqrt[3]{7} = 1.912931\cdots$.

4. (This is Problem 3.6.26) Show that both fixed points of $x_{n+1} = x_n^2 + x_n - 3$ are repelling.

   - The fixed points are solutions of $x^* = (x^*)^2 + x^* - 3$. This gives $(x^*)^2 - 3 = 0$ and $x^* = \sqrt{3}$ or $-\sqrt{3}$. Let $F(x) = x^2 + x - 3$. Then $F'(x) = 2x + 1$. Since $F'(\sqrt{3}) = 2\sqrt{3} + 1 > 1$ and $F'(-\sqrt{3}) = -2\sqrt{3} + 1 < -1$, both fixed points are repelling.

The iteration has nowhere to settle down. It probably blows up but it might cycle and it might be chaotic.

Read-throughs and selected even-numbered solutions :

$x_{n+1} = x_n^3$ describes an **iteration**. After one step $x_1 = \mathbf{x_0^3}$. After two steps $x_2 = F(x_1) = \mathbf{x_1^3} = \mathbf{x_0^9}$. If it happens that input = output, or $x^* = \mathbf{F(x^*)}$, then x^* is a **fixed** point. $F = x^3$ has **three** fixed points, at $x^* = \mathbf{0, 1 \text{ and } -1}$. Starting near a fixed point, the x_n will converge to it if $|\mathbf{F'(x^*)}| < 1$. That is because $x_{n+1} - x^* = F(x_n) - F(x^*) \approx \mathbf{F'(x^*)(x_n - x^*)}$. The point is called **attracting**. The x_n are repelled if $|\mathbf{F'(x^*)}| > \mathbf{1}$. For $F = x^3$ the fixed points have $F' = \mathbf{0 \text{ or } 3}$. The cobweb goes from (x_0, x_0) to (x_0, x_1) to $\mathbf{(x_1, x_1)}$ and converges to $(x^*, x^*) = \mathbf{(0, 0)}$. This is an intersection of $y = x^3$ and $y = \mathbf{x}$, and it is super-attracting because $\mathbf{F' = 0}$.

$f(x) = 0$ can be solved iteratively by $x_{n+1} = x_n - cf(x_n)$, in which case $F'(x^*) = \mathbf{1 - cf'(x^*)}$. Subtracting $x^* = x^* - cf(x^*)$, the error equation is $x_{n+1} - x^* \approx m(\mathbf{x_n - x^*})$. The multiplier is $m = \mathbf{1 - cf'(x^*)}$. The errors approach zero if $\mathbf{-1 < m < 1}$. The choice $c_n = \mathbf{1/f'(x^*)}$ produces Newton's method. The choice $c = 1$ is "successive **substitution**" and $c = \mathbf{1/f'(x_0)}$ is modified Newton. Convergence to x^* is **not** certain.

We have three ways to study iterations $x_{n+1} = F(x_n)$: (1) compute $x_1, x_2, \cdots$ from different x_0 (2) find the fixed points x^* and test $|dF/dx| < 1$ (3) draw cobwebs.

10 $x_0 = -1, x_1 = 1, x_2 = -1, x_3 = 1, \cdots$ The double step $x_{n+2} = x_n^9$ has fixed points $x^* = (x^*)^9$, which allows $x^* = 1$ and $x^* = -1$.

18 At $x^* = (a-1)/a$ the derivative $f' = a - 2ax$ equals $f'(x^*) = a - 2(a-1) = 2 - a$. Convergence if $|F'(x^*)| < 1$ or $\mathbf{1 < a < 3}$. (For completeness check $a = 1$: convergence to zero. Also check $a = 3$: with $x_0 = .66666$ my calculator gives back $x_2 = .66666$. Apparently period 2.)

20 $x^* = (x^*)^2 - \frac{1}{2}$ gives $x_+^* = \frac{1+\sqrt{3}}{2}$ and $x_-^* = \frac{1-\sqrt{3}}{2}$. At these fixed points $F' = 2x^*$ equals $1 + \sqrt{3}$ (greater than 1 so x_+^* repels) and $F'(x_-^*) = 1 - \sqrt{3}(x_-^*$ attracts). Cobwebs show convergence to x_-^* if $|x_0| < |x_+^*|$, convergence to x_+^* if $|x_0| = |x_+^*|$, divergence to ∞ if $|x_0| > |x_+^*|$.

26 The fixed points satisfy $x^* = (x^*)^2 + x^* - 3$ or $(x^*)^2 = 3$; thus $x^* = \sqrt{3}$ or $x^* = -\sqrt{3}$. The derivative $2x^* + 1$ equals $2\sqrt{3} + 1$ or $-2\sqrt{3} + 1$; both have $|F'| > 1$. The iterations blow up.

28 (a) Start with $x_0 > 0$. Then $x_1 = \sin x_0$ is less than x_0. The sequence $x_0, \sin x_0, \sin(\sin x_0) \cdots$ decreases to zero (**convergence:** also if $x_0 < 0$.) On the other hand $x_1 = \tan x_0$ is larger than x_0. The sequence $x_0, \tan x_0, \tan(\tan x_0), \cdots$ is increasing (slowly repelled from 0). Since $(\tan x)' = \sec^2 x \geq 1$ there

is no attractor (**divergence**). (b) F'' is $(\sin x)'' = -\sin x$ and $(\tan x)'' = 2\sec^2 x \tan x$.
Theory: When F'' changes from + to − as x passes x_0, the curve stays closer to the axis than the 45° line (convergence). Otherwise divergence. See Problem 22 for $F'' = 0$.

42 The graphs of $\cos x, \cos(\cos x), \cos(\cos(\cos x))$ are approaching the **horizontal line y = .7391**$\cdots$ (where $x^* = \cos x^*$). For every x this number is the limit.

3.7 Newton's Method and Chaos (page 145)

1. Use Newton's method to approximate $\sqrt[4]{20} = (20)^{1/4}$. Choose $f(x) = x^4 - 20$ with $f'(x) = 4x^3$.

$$\textbf{\textit{Newton's method is}} \quad x_{n+1} = x_n - \frac{f(x_n)}{f'(x_n)} = x_n - \frac{x_n^4 - 20}{4x_n^3}.$$

It helps to simplify the expression on the right:

$$x_{n+1} = \frac{4x_n^4 - (x_n^4 - 20)}{4x_n^3} = \frac{3x_n^4 + 20}{4x_n^3} = \frac{3}{4}x_n + \frac{5}{x_n^3}.$$

Choose a starting point, say $x_0 = 2$. Then $x_1 = \frac{3}{4}(2) + \frac{5}{2^3} = \frac{3}{2} + \frac{5}{8} = 2.125$. The next point is $x_2 = \frac{3}{4}(2.125) + \frac{5}{(2.125)^3} = 2.1148$. Then $x_3 = 2.11474$. Comparing x_2 and x_3 indicates that $\sqrt[4]{20} = 2.115$ is correct to three places.

2. (This is Problem 3.7.4) Show that Newton's method for $f(x) = x^{1/3} = 0$ gives $x_{n+1} = 2x_n$. Draw a graph to show the iterations. They don't converge to $x^* = 0$.

 - Newton's method for the equation $x^{1/3} = 0$ gives $x_{n+1} = x_n - \frac{x_n^{1/3}}{\frac{1}{3}x_n^{-2/3}}$. The fraction is made simpler by multiplying top and bottom by $3x_n^{2/3}$ so $x_{n+1} = x_n - \frac{3x_n}{1} = -2x_n$. Every step just multiplies by -2. The tangent line at each x hits the axis at the next point $-2x$.

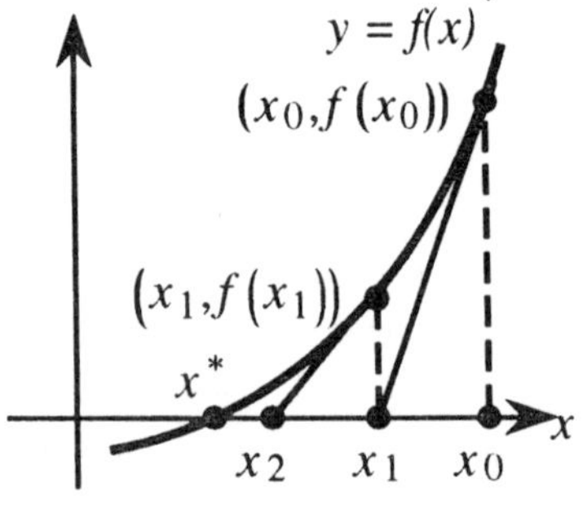

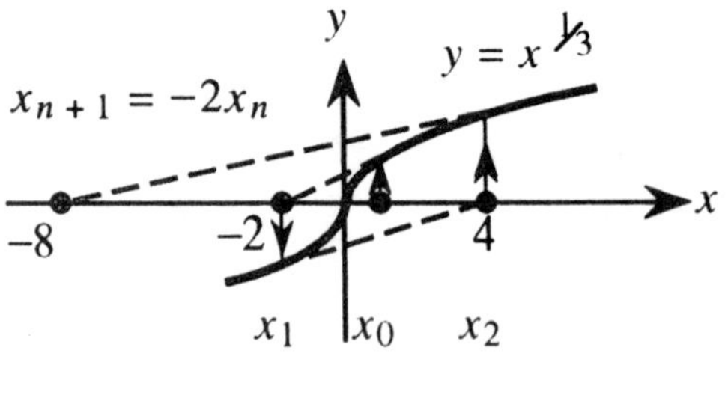

3. The *secant method* is good for those occasions when you don't know $f'(x)$. Use this method to solve $\sin x = 0.5$. (Of course we do know $(\sin x)' = \cos x$!). Start with $x_0 = 0.5$ and $x_1 = 0.6$ (*two* start points).

 - With $f(x) = \sin x - 0.5$, this table gives x_{n+1} by the secant method (formula at end of the line)

n	x_{n-1}	$f(x_{n-1})$	x_n	$f(x_n)$	$(\frac{\Delta f}{\Delta x})_n = \frac{f(x_n)-f(x_{n-1})}{x_n - x_{n-1}}$	$x_{n+1} = x_n - \frac{f(x_n)}{(\frac{\Delta f}{\Delta x})n}$
1	0.5	−0.0206	0.6	0.0646	0.8522	0.5241
2	0.6	0.0646	0.5241	0.00047	0.8449	0.5236
3	0.5241	0.00047	0.5236	−.000001	0.8659	0.5236

Since $x_3 = x_4 = 0.5236$, this should solve $\sin x = 0.5$ to four places. Actually $\sin^{-1} .5 = \frac{\pi}{6} \approx 0.523599$.

Read-throughs and selected even-numbered solutions :

When $f(x) = 0$ is linearized to $f(x_n) + f'(x_n)(x - x_n) = 0$, the solution $x = \mathbf{x_n - f(x_n)/f'(x_n)}$ is Newton's x_{n+1}. The **tangent line** to the curve crosses the axis at x_{n+1}, while the **curve** crosses at x^*. The errors at x_n and x_{n+1} are normally related by $(\text{error})_{n+1} \approx M(\mathbf{error})^{\mathbf{2}}_{\mathbf{n}}$. This is **quadratic** convergence. The number of correct decimals **doubles** at every step.

For $f(x) = x^2 - b$, Newton's iteration is $x_{n+1} = \mathbf{\frac{1}{2}(b + \frac{x_n}{b})}$. The x_n converge to $\mathbf{\sqrt{b}}$ if $x_0 > 0$ and to $\mathbf{-\sqrt{b}}$ if $x_0 < 0$. For $f(x) = x^2 + 1$, the iteration becomes $x_{n+1} = \mathbf{\frac{1}{2}(x_n - x_n^{-1})}$. This cannot converge to $\mathbf{i = \sqrt{-1}}$. Instead it leads to chaos. Changing to $z = 1/(x^2 + 1)$ yields the parabolic iteration $z_{n+1} = \mathbf{4z_n - 4z_n^2}$.

For $a \le 3, z_{n+1} = az_n - az_n^2$ converges to a single **fixed point.** After $a = 3$ the limit is a 2-cycle, which means **that the z's alternate between two values.** Later the limit is a Cantor set, which is a one-dimensional example of a **fractal.** The Cantor set is self-**similar.**

2 $f(x) = \frac{x-1}{x+1}$ has $f'(x) = \frac{(x+1)-(x-1)}{(x+1)^2} = \frac{2}{(x+1)^2}$ so Newton's formula is $x_{n+1} = x_n - \frac{(x_n+1)^2}{2}\frac{x_n-1}{x_n+1} = x_n - \frac{x_n^2-1}{2}$. The fixed points of this F satisfy $x^* = x^* - \frac{(x^*)^2-1}{2}$ which gives $x^* = 1$ and $x^* = -1$. The derivatives $F' = 1 - x^*$ are 0 and -2. So the sequence approaches $x^* = 1$, the correct zero of $f(x)$.

6 $f(x) = x^3 - 3x - 1 = 0$: roots near 1.9, −.5, −1.6

10 Newton's method for $f(x) = x^4 - 100$ approaches $x^* = \sqrt{10}$ if $x_0 > 0$ and $x^* = -\sqrt{10}$ if $x_0 < 0$. In this case the error at step $n+1$ equals $\frac{3}{2x^*}$ times (error at step n)^2. In Problem 9 the multiplier is $\frac{1}{2x^*}$ and convergence is quicker. Note to instructors: The multiplier is $\frac{f''(x^*)}{2f'(x^*)}$ (this is $\frac{1}{2}F''(x^*)$: see Problem 31 of Section 3.8)

24 $\theta = \frac{\pi}{9}, \frac{2\pi}{9}, \frac{4\pi}{9}, \frac{8\pi}{9}, \frac{16\pi}{9}, \frac{32\pi}{9}, \frac{64\pi}{9} = 7\pi + \frac{\pi}{9}$; this happened at step 6 so $x_6 = x_0$.

26 If $z_0 = \sin^2\theta$ then $z_1 = 4z_0 - 4z_0^2 = 4\sin^2\theta - 4\sin^4\theta = 4\sin^2\theta(1 - \sin^2\theta) = 4\sin^2\theta\cos^2\theta = (2\sin\theta\cos\theta)^2 = \sin^2 2\theta$.

44 A Newton step goes from $x_0 = .308$ to $x_1 = x_0 + \frac{\cos x_0}{\sin x_0} = 3.45143$. Then $\frac{\Delta f}{\Delta x} = \frac{\cos x_1 - \cos x_0}{x_1 - x_0} = -.606129$ and a secant step leads to $x_2 = x_1 + \frac{\cos x_1}{.606129} = 1.88$.

3.8 The Mean Value Theorem and l'Hôpital's Rule (page 152)

Be careful to apply l'Hôpital's Rule only when the limit of $\frac{f(x)}{g(x)}$ leads to $\frac{0}{0}$ or $\frac{\infty}{\infty}$. (The limit can be as $x \to a$ or $x \to \infty$ or $x \to -\infty$. If you get a form "$\infty - \infty$" or "$0 \cdot \infty$" use algebra to transform it into $\frac{0}{0}$.

1. Find $\lim_{x\to 0} \frac{\sin^2 x}{x}$.

• This approaches $\frac{0}{0}$. The rule takes the derivative of $\sin^2 x$ and x to find $2\sin x\cos x$ and 1. Then $\lim_{x\to 0} \frac{2\sin x\cos x}{1} = \frac{0}{1} = 0$. The limiting answer is 0.

Direct answer without l'Hôpital: $\sin x \to 0$ times $\frac{\sin x}{x} \to 1$ gives $\frac{\sin^2 x}{x} \to (0)(1) = 0$.

2. Find the limit of $\cot x - \csc x$ as $x \to \pi$.

• This has the form "$\infty - \infty$". Rewrite $\cot x - \csc x$ as $\frac{\cos x}{\sin x} - \frac{1}{\sin x} = \frac{\cos x - 1}{\sin x}$. Now $\lim_{x\to\pi} \frac{\cos x-1}{\sin x}$ is of the form $\frac{0}{0}$. We can apply l'Hôpital's rule to get $\lim_{x\to\pi} \frac{-\sin x}{\cos x} = 0$. The answer is 0.

3. Find the limit of $x - \sqrt{x^2+4}$ as $x \to \infty$.

• This is also of the form $\infty - \infty$. An algebraic trick works (no l'Hôpital):

$$x - \sqrt{x^2-4} = \frac{(x-\sqrt{x^2+4})(x+\sqrt{x^2+4})}{(x+\sqrt{x^2+4})} = \frac{x^2-(x^2+4)}{x+\sqrt{x^2+4}} = \frac{-4}{x+\sqrt{x^2+4}}.$$

As $x \to \infty$, the numerator stays at -4 while the denominator goes to ∞. The limit is 0.

4. Find $\lim_{x\to\frac{\pi}{4}} \sec 2x(1-\cot x)$. Since $\sec 2(\frac{\pi}{4})$ is infinite and $1 - \cot\frac{\pi}{4} = 0$, we have $\infty \cdot 0$.

• Rewrite the problem to make it $\frac{0}{0}$, by changing $\sec 2x$ to $\frac{1}{\cos 2x}$. Then use l'Hôpital's rule:

$$\lim_{x\to\frac{\pi}{4}} \frac{1-\cot x}{\cos 2x} = \lim_{x\to\frac{\pi}{4}} \frac{\csc^2 x}{-2\sin 2x} = \frac{2}{-2} = -1$$

5. $\lim_{x\to 1} \frac{x^3-x^2-x+1}{x^3-1}$ has the form $\frac{0}{0}$. One application of l'Hôpital's rule gives $\lim_{x\to 1} \frac{3x^2-2x-1}{3x^2}$. Don't be tempted to apply the rule again! The limit is $\frac{0}{3} = 0$.

Questions 6 – 8 are about the Mean Value Theorem.

6. What does the Mean Value Theorem say about the function $f(x) = x^3$ on the interval $[-1, 2]$?

• The graph tells the story. The secant line connecting $(-1, 1)$ to $(2, 8)$ has slope $\frac{8-1}{2-(-1)} = \frac{7}{3}$. The MVT says that there is some c between -1 and 2 where the slope of the tangent line at $x = c$ is $\frac{7}{3}$.

7. What is the number c in question 6?

• The slope of $y = x^3$ is $\frac{dy}{dx} = 3x^2$. The slope of the tangent at $x = c$ is $3c^2 = \frac{7}{3}$. So $c^2 = \frac{7}{9}$ and $c \approx \pm 0.882$. There are actually *two* values of c where the tangent line is parallel to the secant. (The Mean Value Theorem states there is at least one c, not "exactly" one.)

8. Use the Mean Value Theorem to show that $\tan a > a$ when $0 < a < \frac{1}{2}\pi$. Choose $f(x) = \tan x$.

• On the interval $[0, a]$, the Theorem says that there exists a point c where $f'(c) = \sec^2 c = \frac{\tan a - \tan 0}{a - 0} = \frac{\tan a}{a}$. Since $\sec^2 c$ always exceeds 1 inside the interval we have $\frac{\tan a}{a} > 1$ and $\tan a > a$.

Read-throughs and selected even-numbered solutions :

The Mean Value Theorem equates the average slope $\Delta f/\Delta x$ over an **interval** $[a, b]$ to the slope df/dx at an unknown **point**. The statement is $\mathbf{\Delta f/\Delta x = f'(x)}$ **for some point** $\mathbf{a < c < b}$. It requires $f(x)$ to be **continuous** on the **closed** interval $[a, b]$, with a **derivative** on the open interval (a, b). Rolle's theorem is the special case when $f(a) = f(b) = 0$, and the point c satisfies $\mathbf{f'(c) = 0}$. The proof chooses c as the point where f reaches its **maximum or minimum.**

Consequences of the Mean Value Theorem include: If $f'(x) = 0$ everywhere in an interval then $f(x) =$ **constant**. The prediction $f(x) = f(a) + \mathbf{f'(c)}(x - a)$ is exact for some c between a and x. The quadratic prediction $f(x) = f(a) + f'(x)(x - a) + \mathbf{\frac{1}{2}f''(c)}(x - a)^2$ is exact for another c. The error in $f(a) + f'(a)(x - a)$ is less than $\frac{1}{2}M(x - a)^2$ where M is the maximum of $|\mathbf{f''}|$.

A chief consequence is l'Hôpital's Rule, which applies when $f(x)$ and $g(x) \to \mathbf{0}$ as $x \to a$. In that case the limit of $f(x)/g(x)$ equals the limit of $\mathbf{f'(x)/g'(x)}$, provided this limit exists. Normally this limit is $f'(a)/g'(a)$. If this is also $0/0$, go on to the limit of $\mathbf{f''(x)/g''(x)}$.

2 $\sin 2\pi - \sin 0 = (\pi \cos \pi c)(2 - 0)$ when $\cos \pi c = 0$: then $\mathbf{c = \frac{1}{2}}$ or $\mathbf{c = \frac{3}{2}}$.

10 $f(x) = \frac{1}{x^2}$ has $f(1) = 1$ and $f(-1) = 1$, but no point c has $f'(c) = 0$. MVT does not apply because $f(x)$ is **not continuous** in this interval.

12 $\frac{d}{dx}\csc^2 x = 2\csc x(-\csc x \cot x)$ is equal to $\frac{d}{dx}\cot^2 x = 2\cot x(-\csc^2 x)$. Then $f(x) = \csc^2 x - \cot^2 x$ has $f' = 0$ at every point c. By the MVT $f(x)$ must have the same value at every pair of points a and b. By trigonometry $\csc^2 x - \cos^2 x = \frac{1}{\sin^2 x} - \frac{\cos^2 x}{\sin^2 x} = \frac{\sin^2 x}{\sin^2 x} = 1$ at all points.

16 l'Hôpital's Rule does not apply because $\sqrt{1 - \cos x}$ has no derivative at $x = 0$. (There is a corner.) The knowledge that $1 - \cos x \approx \frac{x^2}{2}$ gives $\sqrt{1 - \cos x} \approx \frac{|x|}{\sqrt{2}}$. Then $\frac{|x|}{\sqrt{2}}x$ has one-sided limits. The limit from the right (where $x > 0$) is $\mathbf{L = \frac{1}{\sqrt{2}}}$. The limit from the left (where $x < 0$) is $\mathbf{L = -\frac{1}{\sqrt{2}}}$. These limits also come from "one-sided l'Hôpital rules."

18 $\lim_{x\to 1} \frac{x-1}{\sin x} = \frac{0}{\sin 1} = \mathbf{0}$ (not an application of l'Hôpital's Rule).

20 $\lim_{x\to 0} \frac{(1+x)^n - 1 - nx}{x^2} = \lim_{x\to 0} \frac{n(1+x)^{n-1} - n}{2x} =$ (l'Hôpital again) $\lim_{x\to 0} \frac{n(n-1)(1+x)^{n-2}}{2} = \mathbf{\frac{n(n-1)}{2}}$.

30 Mean Value Theorem: $f(x) - f(y) = f'(c)(x - y)$. Therefore $|f(x) - f(y)| = |f'(c)||x - y| \le |x - y|$ since we are given that $|f'| \le 1$ at all points. Geometric interpretation: If the tangent slope stays between -1 and 1, so does the slope of any secant line.

32 No: The converse of Rolle's theorem is false. The function $f(x) = x^3$ has $f' = 0$ at $x = 0$ (horizontal tangent). But there are no two points where $f(a) = f(b)$ (no horizontal secant line).

3 Chapter Review Problems

Computing Problems

C1 Find the fixed points of $f(x) = 1.2 \sin x$. Which are attracting and which are repelling?

C2 Find the real roots of $f(x) = x^3 + 2x + 1$ using Newton's method, correct to four places.

C3 Solve $f(x) = 2\sin x - 1 = 0$ correct to four places.

C4 Minimize $y = \frac{2x^{1/3}}{2-x}$ (you will find a *relative* minimum: why?).

Review Problems

R1 Linear Approximation: Sketch a function $f(x)$ and a tangent line through $(a, f(a))$. Label $dy, dx, \Delta y, \Delta x$. Based on Figure 3.2 indicate the error between the curve and its tangent line.

R2 To find the maximum (or minimum) of a function, what are the three types of critical point to consider? Sketch a function having maxima or minima at each type of critical point.

R3 Explain the second derivative test for local maxima and minima of $f(x)$.

R4 Are stationary points always local maxima or minima? Give an example or a counterexample.

R5 Show by a picture why a fixed point of $y = F(x)$ is attracting if $|\frac{dF}{dx}| < 1$ and repelling if $|\frac{dF}{dx}| > 1$.

R6 For $f(x) = x^3 - 3x + 1 = 0$, Newton's method is $x_{n+1} =$ ______.

R7 Draw a picture to explain the Mean Value Theorem. With another sketch, show why this theorem may not hold if $f(x)$ is not differentiable on (a, b). With a third sketch, show why the theorem may not hold if $f(x)$ is defined on (a, b) but not at the endpoints.

R8 State l'Hôpital's Rule. How do you know when to use it ? When do you use it twice? Illustrate with $\lim_{x\to 0} \frac{\tan x - x}{x^3}$.

Drill Problems

D1 Find a linear approximation for $y = 1 + \frac{1}{x}$ near $a = 2$. Then find a quadratic approximation.

D2 Approximate $\sqrt[3]{500}$ using a linear approximation.

D3 The volume of a cylinder is $V = \pi r^2 h$. What is the percent change in volume if the radius decreases by 3% and the height remains the same? Is this exact or approximate?

In **4** – **8** find stationary points, rough points, and endpoints. Also relative and absolute maxima and minima.

D4 $y = 3x^2 - 2x^3$ **D5** $y = x^4 - 8x^2 + 16$ **D6** $y = \frac{1}{x+1}, 0 < x < 99$

D7 $y = \sin x + |\sin x|, \ -2\pi \le x \le 2\pi$ **D8** $f(x) = \frac{1}{x} - \frac{x^2}{16}$ on $[-4, -1]$

D9 Where are these functions concave up and concave down? Sketch their graphs.

(a) $(x^2 - 1)^3$ (b) $12x^{2/3} - 4x$ (c) $x^2 + \frac{2}{x}$ (d) $\frac{2x^2}{x^2+3}$

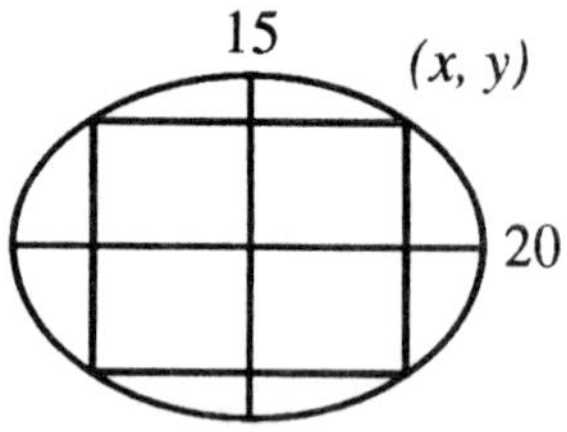

D10 A 20 meter wire is formed into a rectangle. Maximize the area.

D11 A level rectangle is to be inscribed in the ellipse $\frac{x^2}{400} + \frac{y^2}{225} = 1$. Find the rectangle of maximum area. The ellipse is drawn above.

D12 Write these equations in standard form and identify parabola, ellipse, or hyperbola. Draw rough graphs.

(a) $4x^2 + 9y^2 = 36$ (b) $16x^2 - 7y^2 + 112 = 0$ (c) $x^2 - 2x + y^2 + 6y + 2 = 0$ (d) $x^2 = 2y + 2x$

D13 Compute $\lim_{x\to 0} \frac{x+\sin 2x}{x-\sin 2x}$ and $\lim_{x\to\pi} \frac{\sin 2x}{x-\pi}$.

D14 Find $\lim_{x\to\infty} \frac{\sqrt{2+x^2}}{x}$ and $\lim_{x\to 0^+} x^2 \cot x$ and $\lim_{x\to 0^+}(\frac{1}{x} - \frac{1}{\sqrt{x}})$.

D15 Find the "c" guaranteed by the Mean Value Theorem for $y = x^3 + 2x$ on the interval $[-1, 2]$.

CHAPTER 4 DERIVATIVES BY THE CHAIN RULE

4.1 The Chain Rule (page 158)

The function $\sin(3x+2)$ is "composed" out of two functions. The *inner* function is $u(x) = 3x+2$. The *outer* function is $\sin u$. I don't write $\sin x$ because that would throw me off. The derivative of $\sin(3x+2)$ is not $\cos x$ or even $\cos(3x+2)$. The chain rule produces the extra factor $\frac{du}{dx}$, which in this case is the number 3. *The derivative of* $\sin(3x+2)$ *is* $\cos(3x+2)$ *times* 3.

Notice again: Because the sine was evaluated at u (not at x), its derivative is also evaluated at u. We have $\cos(3x+2)$ not $\cos x$. The extra factor 3 comes because u changes as x changes:

$$(\textit{algebra}) \quad \frac{\Delta y}{\Delta x} = \frac{\Delta y}{\Delta u}\frac{\Delta u}{\Delta x} \quad \text{approaches} \quad \frac{dy}{dx} = \frac{dy}{du}\frac{du}{dx} \quad (\textit{calculus}).$$

These letters can and will change. Many many functions are chains of simpler functions.

1. Rewrite each function below as a composite function $y = f(u(x))$. Then find $\frac{dy}{dx} = f'(u)\frac{du}{dx}$ or $\frac{dy}{du}\frac{du}{dx}$.

   (a) $y = \tan(\sin x)$ (b) $y = \cos(3x^4)$ (c) $y = \frac{1}{(2x-5)^2}$

   - $y = \tan(\sin x)$ is the chain $y = \tan u$ with $u = \sin x$. The chain rule gives $\frac{dy}{du}\frac{du}{dx} = (\sec^2 u)(\cos x)$. Substituting back for u gives $\frac{dy}{dx} = \sec^2(\sin x)\cos x$.
   - $\cos(3x^4)$ separates into $\cos u$ with $u = 3x^4$. Then $\frac{dy}{du}\frac{du}{dx} = (-\sin u)(12x^3) = -12x^3\sin(3x^4)$.
   - $y = \frac{1}{(2x-5)^2}$ is $y = \frac{1}{u^2}$ with $u = 2x-5$. The chain rule gives $\frac{dy}{dx} = (-2u^{-3})(2) = -4(2x-5)^{-3}$. Another perfectly good "decomposition" is $y = \frac{1}{u}$, with $u = (2x-5)^2$. Then $\frac{dy}{du} = -\frac{1}{u^2}$ and $\frac{du}{dx} = 2(2x-5)(2)$ (really another chain rule). The answer is the same: $\frac{dy}{dx} = \frac{-1}{[(2x-5)^2]^2}\cdot 4(2x-5) = \frac{-4}{(2x-5)^3}$.

2. Write $y = \sin\sqrt{3x^2-5}$ and $y = \frac{1}{1-\frac{1}{x}}$ as triple chains $y = f(g(u(x)))$. Then find $\frac{dy}{dx} = f'(g(u))\cdot g'(u)\cdot\frac{du}{dx}$. You could write the chain as $y = f(w)$, $w = g(u)$, $u = u(x)$. Then you see the slope as a product of *three factors*, $\frac{dy}{dx} = (\frac{dy}{dw})(\frac{dw}{du})(\frac{du}{dx})$.

   - For $y(x) = \sin\sqrt{3x^2-5}$ the triple chain is $y = \sin w$, where $w = \sqrt{u}$ and $u = 3x^2-5$. The chain rule is $\frac{dy}{dx} = (\frac{dy}{dw})(\frac{dw}{du})(\frac{du}{dx}) = (\cos w)(\frac{1}{2\sqrt{u}})(6x)$. Substitute to get back to x:

   $$\frac{dy}{dx} = \cos\sqrt{3x^2-5}\cdot\frac{1}{2\sqrt{(3x^2-5)}}\cdot 6x = \frac{6x\cos\sqrt{3x^2-5}}{2\sqrt{3x^2-5}}.$$

   - For $y(x) = \frac{1}{1-\frac{1}{x}}$ let $u = \frac{1}{x}$. Let $w = 1-u$. Then $y = \frac{1}{w}$. The derivative is

   $$\frac{dy}{dx} = (\frac{dy}{dw})(\frac{dw}{du})(\frac{du}{dx}) = (-\frac{1}{w^2})(-1)(\frac{-1}{x^2}) = \frac{-1}{(1-u)^2x^2} = \frac{-1}{(1-\frac{1}{x})^2x^2} = \frac{-1}{(x-1)^2}.$$

With practice, you should get to the point where it is not necessary to write down u and w in full detail. Try this with exercises 1 – 22, doing as many as you need to get good at it. Problems 45 – 54 are excellent practice, too.

Questions 3 – 6 are based on the following table, which gives the values of functions f and f' and g and g' at a few points. You do not know what these functions are!

x	$f(x)$	$f'(x)$	$g(x)$	$g'(x)$
0	1	-1	0	undefined
$\frac{1}{3}$	$\frac{3}{4}$	$-\frac{9}{4}$	$\frac{\sqrt{3}}{3}$	$\frac{\sqrt{3}}{2}$
$\frac{1}{2}$	$\frac{2}{3}$	$-\frac{4}{9}$	$\frac{\sqrt{2}}{2}$	$\frac{\sqrt{2}}{2}$
1	$\frac{1}{2}$	$-\frac{1}{4}$	1	$\frac{1}{2}$

x	$f(x)$	$f'(x)$	$g(x)$	$g'(x)$
2	$\frac{1}{3}$	$-\frac{1}{9}$	$\sqrt{2}$	$\frac{\sqrt{2}}{4}$
3	$\frac{1}{4}$	$-\frac{1}{16}$	$\sqrt{3}$	$\frac{\sqrt{3}}{6}$
4	$\frac{1}{5}$	$-\frac{1}{25}$	2	$\frac{1}{4}$
9	$\frac{1}{10}$	$-\frac{1}{100}$	3	$\frac{1}{6}$

3. Find: $f(g(4))$ and $f(g(1))$ and $f(g(0))$.
   - $g(4) = 2$ and $f(2) = \frac{1}{3}$ so $f(g(4)) = \frac{1}{3}$. Also $g(1) = 1$ so $f(g(1)) = f(1) = \frac{1}{2}$. Then $f(g(0)) = f(0) = 0$.
4. Find: $g(f(1))$ and $g(f(2))$ and $g(f(0))$.
   - Since $f(1) = \frac{1}{2}$, the chain $g(f(1))$ is $g(\frac{1}{2}) = \frac{\sqrt{2}}{2}$. Also $g(f(2)) = g(\frac{1}{3}) = \frac{\sqrt{3}}{3}$. Then $g(f(0)) = g(1) = 1$.

     Note that $g(f(1))$ does not equal $f(g(1))$. Also $g(f(0)) \neq f(g(0))$. This is normal. Chains in a different order are different chains.
5. If $y = f(g(x))$ find $\frac{dy}{dx}$ at $x = 9$.
   - The chain rule says that $\frac{dy}{dx} = f'(g(x)) \cdot g'(x)$. At $x = 9$ we have $g(9) = 3$ and $g'(9) = \frac{1}{6}$. At $g = 3$ we have $f'(3) = -\frac{1}{16}$. Therefore at $x = 9$, $\frac{dy}{dx} = f'(g(9)) \cdot g'(9) = -\frac{1}{16} \cdot \frac{1}{6} = -\frac{1}{96}$.
6. If $y = g(f(x))$ find $\frac{dy}{dx}(1)$. Note that $f(1) = \frac{1}{2}$.
   - $g'(f(1)) \cdot f'(1) = g'(\frac{1}{2}) \cdot f'(1) = \frac{\sqrt{2}}{2}(-\frac{1}{4}) = \frac{-\sqrt{2}}{8}$.
7. If $y = f(f(x))$ find $\frac{dy}{dx}$ at $x = 2$. This chain repeats the same function ($f = g$). It is "iteration."
   - If you let $u = f(x)$, then $\frac{dy}{dx} = \frac{dy}{du} \cdot \frac{du}{dx}$ becomes $\frac{dy}{dx} = f'(u) \cdot f'(x)$. At $x = 2$ the table gives $u = \frac{1}{3}$. Then $\frac{dy}{dx} = f'(\frac{1}{3}) \cdot f'(2) = (-\frac{9}{4})(-\frac{1}{9}) = \frac{1}{4}$. Note that $(f'(2))^2 = (-\frac{1}{9})^2$. The derivative of $f(f(x))$ is *not* $(f'(x))^2$. And it is *not* the derivative of $(f(x))^2$.

Read-throughs and selected even-numbered solutions :

$z = f(g(x))$ comes from $z = f(y)$ and $y = \mathbf{g(x)}$. At $x = 2$ the chain $(x^2 - 1)^3$ equals $\mathbf{3^3} = \mathbf{27}$. Its inside function is $y = \mathbf{x^2 - 1}$, its outside function is $z = \mathbf{y^3}$. Then dz/dx equals $\mathbf{3y^2dy/dx}$. The first factor is evaluated at $y = \mathbf{x^2 - 1}$ (not at $y = x$). For $z = \sin(x^4 - 1)$ the derivative is $\mathbf{4x^3 \cos(x^4 - 1)}$. The triple chain $z = \mathbf{\cos(x+1)^2}$ has a shift and a **square** and a cosine. Then $dz/dx = \mathbf{2}\cos(\mathbf{x+1})(-\mathbf{\sin(x+1)})$.

The proof of the chain rule begins with $\Delta z/\Delta x = (\mathbf{\Delta z/\Delta y})\,(\mathbf{\Delta y/\Delta x})$ and ends with $\mathbf{dz/dx = (dz/dy)(dy/dx)}$. Changing letters, $y = \cos u(x)$ has $dy/dx = -\mathbf{\sin u(x)\frac{du}{dx}}$. The power rule for $y = [u(x)]^n$ is the chain rule $dy/dx = \mathbf{nu^{n-1}\frac{du}{dx}}$. The slope of $5g(x)$ is $\mathbf{5g'(x)}$ and the slope of $g(5x)$ is $\mathbf{5g'(5x)}$. When f = cosine and g = sine and $x = 0$, the numbers $f(g(x))$ and $g(f(x))$ and $f(x)g(x)$ are **1 and sin 1 and 0**.

18 $\frac{dz}{dx} = \frac{\cos(x+1)}{2\sqrt{\sin(x+1)}}$ **20** $\frac{dz}{dx} = \frac{\cos(\sqrt{x}+1)}{2\sqrt{x}}$ **22** $\frac{dz}{dx} = 4x(\sin x^2)(\cos x^2)$

28 $f(y) = y + 1; h(y) = \sqrt[3]{y}; k(y) \equiv 1$

38 For $g(g(x)) = x$ the graph of g should be **symmetric across the 45° line:** If the point (x, y) is on the graph so is (y, x). Examples: $g(x) = -\frac{1}{x}$ or $-x$ or $\sqrt[3]{1 - x^3}$.

40 **False** (The chain rule produces -1 : so derivatives of even functions are odd functions)
False (The derivative of $f(x) = x$ is $f'(x) = 1$) **False** (The derivative of $f(1/x)$ is $f'(1/x)$ times $-1/x^2$)

True (The factor from the chain rule is 1) **False** (see equation (8)).

42 From $x = \frac{\pi}{4}$ go up to $y = \sin\frac{\pi}{4}$. Then go **across** to the parabola $z = y^2$. Read off $z = (\sin\frac{\pi}{4})^2$ on the horizontal z axis.

4.2 Implicit Differentiation and Related Rates (page 163)

Questions 1 – 5 are examples using *implicit differentiation* (**ID**).

1. Find $\frac{dy}{dx}$ from the equation $x^2 + xy = 2$. Take the x derivative of all terms.

   - The derivative of x^2 is $2x$. The derivative of xy (a product) is $x\frac{dy}{dx} + y$. The derivative of 2 is 0. Thus $2x + x\frac{dy}{dx} + y = 0$, and $\frac{dy}{dx} = -\frac{y+2x}{x}$.

     In this example the original equation can be solved for $y = \frac{1}{x}(2 - x^2)$. Ordinary *explicit* differentiation yields $\frac{dy}{dx} = \frac{-2}{x^2} - 1$. This must agree with our answer from **ID**.

2. Find $\frac{dy}{dx}$ from $(x+y)^3 = x^4 + y^4$. This time we cannot solve for y.

   - The chain rule tells us that the x-derivative of $(x+y)^3$ is $3(x+y)^2(1 + \frac{dy}{dx})$. Therefore **ID** gives $3(x+y)^2(1 + \frac{dy}{dx}) = 4x^3 + 4y^3\frac{dy}{dx}$. Now algebra separates out $\frac{dy}{dx} = \frac{3(x+y)^2 - 4y^3}{4x^3 - 3(x+y)^2}$.

3. Use **ID** to find $\frac{dy}{dx}$ for $y = x\sqrt{1-x}$.

   - Implicit differentiation (**ID** for short) is not necessary, but you might appreciate how it makes the problem easier. Square both sides to eliminate the square root: $y^2 = x^2(1-x) = x^2 - x^3$, so that

$$2y\frac{dy}{dx} = 2x - 3x^2 \quad \text{and} \quad \frac{dy}{dx} = \frac{2x - 3x^2}{2y} = \frac{2x - 3x^2}{2x\sqrt{1-x}} = \frac{2 - 3x}{2\sqrt{1-x}}.$$

4. Find $\frac{d^2y}{dx^2}$ when $xy + y^2 = 1$. Apply **ID** twice to this equation.

   - First derivative: $x\frac{dy}{dx} + y + 2y\frac{dy}{dx} = 0$. Rewrite this as $\frac{dy}{dx} = \frac{-y}{x+2y}$. Now take the derivative again. The second form needs the quotient rule , so I prefer to use **ID** on the first derivative equation:

$$x\frac{d^2y}{dx^2} + \frac{dy}{dx} + \frac{dy}{dx} + 2y\frac{d^2y}{dx^2} + 2(\frac{dy}{dx})^2 = 0 \quad \text{or} \quad \frac{d^2y}{dx^2} = -2\frac{\frac{dy}{dx} + (\frac{dy}{dx})^2}{x + 2y}.$$

   Now substitute $\frac{-y}{x+2y}$ for $\frac{dy}{dx}$ and simplify the answer to $\frac{d^2y}{dx^2} = \frac{2}{(x+2y)^3}$.

5. Find the equation of the tangent line to the ellipse $x^2 + xy + y^2 = 1$ through the point (1,0).

   - The line has equation $y = m(x-1)$ where m is the slope at (1,0). To find that slope, apply **ID** to the equation of the ellipse: $2x + x\frac{dy}{dx} + y + 2y\frac{dy}{dx} = 0$. Do not bother to solve this for $\frac{dy}{dx}$. Just plug in $x = 1$ and $y = 0$ to obtain $2 + \frac{dy}{dx} = 0$. Then $m = \frac{dy}{dx} = -2$ and the tangent equation is $y = -2(x-1)$.

Questions 6–8 are problems about ***related rates***. The slope of one function is known, we want the slope of a *related* function. Of course slope = *rate* = derivative. You must find the relation between functions.

6. Two cars leave point A at the same time $t = 0$. One travels north at 65 miles/hour, the other travels east at 55 miles/hour. How fast is the distance D between the cars changing at $t = 2$?

   - The distance satisfies $D^2 = x^2 + y^2$. This is the relation between our functions! Find the rate of change (take the derivative): $2D\frac{dD}{dt} = 2x\frac{dx}{dt} + 2y\frac{dy}{dt}$. We need to know $\frac{dD}{dt}$ at $t = 2$. We already know $\frac{dx}{dt} = 55$ and $\frac{dy}{dt} = 65$. At $t = 2$ the cars have traveled for two hours: $x = 2(55) = 110$, $y = 2(65) = 130$ and $D = \sqrt{110^2 + 130^2} \approx 170.3$.

     Substituting these values gives $2(170.3)\frac{dD}{dt} = 2(110)(55) + 2(130)(65)$, so $\frac{dD}{dt} \approx 85$ miles/hour.

7. Sand pours out from a conical funnel at the rate of 5 cubic inches per second. The funnel is 6″ wide at the top and 6″ high. At what rate is the sand height falling when the remaining sand is 1″ high?

   - Ask yourself what rate(s) you know and what rate you want to know. In this case you know $\frac{dV}{dt} = -5$ (V is the volume of the sand). You want to know $\frac{dh}{dt}$ when $h = 1$ (h is the height of the sand). Can you get an equation *relating* V and h? This is usually the crux of the problem.

     The volume of a cone is $V = \frac{1}{3}\pi r^2 h$. If we could eliminate r, then V would be related to h. Look at the figure. By similar triangles $\frac{r}{h} = \frac{3}{6}$, so $r = \frac{1}{2}h$. This means that $V = \frac{1}{3}\pi(\frac{h}{2})^2 h = \frac{1}{12}\pi h^3$.

     Now take the t derivative: $\frac{dV}{dt} = \frac{1}{12}\pi(3h^2)\frac{dh}{dt}$. *After* the derivative has been taken, substitute what is known at $h = 1$: $-5 = \frac{1}{12}\pi(3)\frac{dh}{dt}$, so $\frac{dh}{dt} = \frac{-20}{\pi}$ in/sec ≈ -6.4 in/sec.

8. (This is Problem 4.2.21) The bottom of a 10-foot ladder moves away from the wall at 2 ft/sec. How fast is the top going down the wall when the top is (a) 6 feet high? (b) 5 feet high? (c) zero feet high?

   - We are given $\frac{dx}{dt} = 2$. We want to know dy/dt. *The equation relating x and y is* $x^2 + y^2 = 100$. This gives $2x\frac{dx}{dt} + 2y\frac{dy}{dt} = 0$. Substitute $\frac{dx}{dt} = 2$ to find $\frac{dy}{dt} = -\frac{2x}{y}$.

   (a) If $y = 6$, then $x = 8$ (use $x^2 + y^2 = 100$) and $\frac{dy}{dt} = -\frac{8}{3}$ ft/sec.

   (b) If $y = 5$, then $x = 5\sqrt{3}$ (use $x^2 + y^2 = 100$) and $\frac{dy}{dt} = -2\sqrt{3}$ ft/sec.

   (c) If $y = 0$, then we are dividing by zero: $\frac{dy}{dx} = -\frac{2x}{0}$. Is the speed infinite? How is this possible?

Read-throughs and selected even-numbered solutions :

For $x^3 + y^3 = 2$ the derivative dy/dx comes from **implicit** differentiation. We don't have to solve for **y**. Term by term the derivative is $3x^2 + \mathbf{3y^2\frac{dy}{dx}} = 0$. Solving for dy/dx gives $\mathbf{-x^2/y^2}$. At $x = y = 1$ this slope is $\mathbf{-1}$. The equation of the tangent line is $y - 1 = \mathbf{-1(x-1)}$.

A second example is $y^2 = x$. The x derivative of this equation is $\mathbf{2y\frac{dy}{dx}} = 1$. Therefore $dy/dx = \mathbf{1/2y}$. Replacing y by $\sqrt{x}$ this is $dy/dx = \mathbf{1/2\sqrt{x}}$.

In related rates, we are given dg/dt and we want df/dt. We need a relation between f and **g**. If $f = g^2$, then $(df/dt) = \mathbf{2g}(dg/dt)$. If $f^2 + g^2 = 1$, then $df/dt = -\mathbf{\frac{g}{f}\frac{dg}{dt}}$. If the sides of a cube grow by $ds/dt = 2$, then its volume grows by $dV/dt = \mathbf{3s^2(2)} = \mathbf{6s^2}$. To find a number (8 is wrong), you also need to know **s**.

6 $f'(x) + F'(y)\frac{dy}{dx} = y + x\frac{dy}{dx}$ so $\frac{dy}{dx} = \mathbf{\frac{y - f'(x)}{F'(y) - x}}$

12 $2(x-2) + 2y\frac{dy}{dx} = 0$ gives $\frac{dy}{dx} = \mathbf{1}$ at $(1,1)$; $2x + 2(y-2)\frac{dy}{dx} = 0$ also gives $\frac{dy}{dx} = \mathbf{1}$.

20 x is a constant (fixed at 7) and therefore **a change Δ x is not allowed**

24 Distance to you is $\sqrt{x^2+8^2}$, rate of change is $\frac{x}{\sqrt{x^2+8^2}}\frac{dx}{dt}$ with $\frac{dx}{dt} = 560$. (a) Distance $= 16$ and $x = 8\sqrt{3}$ and rate is $\frac{8\sqrt{3}}{16}(560) = \mathbf{280\sqrt{3}}$; (b) $x = 8$ and rate is $\frac{8}{\sqrt{8^2+8^2}}(560) = \mathbf{280\sqrt{2}}$; (c) $x = 0$ and rate $= 0$.

28 Volume $= \frac{4}{3}\pi r^3$ has $\frac{dV}{dt} = 4\pi r^2\frac{dr}{dt}$. If this equals twice the surface area $4\pi r^2$ (with minus for evaporation) than $\frac{dr}{dt} = \mathbf{-2}$.

4.3 Inverse Functions and Their Derivatives (page 170)

The vertical line test and the horizontal line test are good for visualizing the meaning of "function" and "invertible." If a vertical line hits the graph twice, we have two y's for the same x. *Not a function.* If a horizontal line hits the graph twice, we have two x's for the same y. *Not invertible.* This means that the inverse is not a function.

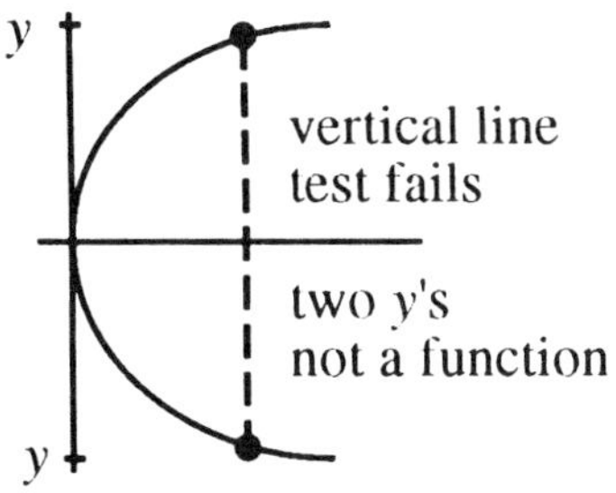

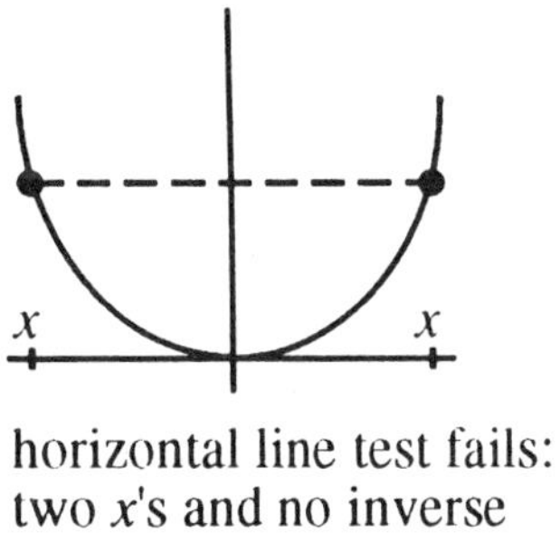

These tests tell you that the sideways parabola $x = y^2$ does not give y as a function of x. (Vertical lines intersect the graph twice. There are two square roots $y = \sqrt{x}$ and $y = -\sqrt{x}$.) Similarly the function $y = x^2$ has no inverse. This is an ordinary parabola – horizontal lines cross it twice. If $y = 4$ then $x = f^{-1}(4)$ has two answers $x = 2$ and $x = -2$. In questions 1 – 2 find the inverse function $x = f^{-1}(y)$.

1. $y = x^2 + 2$. This function fails the horizontal line test. It has no inverse. Its graph is a parabola opening upward, which is crossed twice by some horizontal lines (and not crossed at all by other lines).

   Here's another way to see why there is no inverse: $x^2 = y - 2$ leads to $x = \pm\sqrt{y-2}$. Then $x+ = \sqrt{y-2}$ represents the right half of the parabola, and $x = -\sqrt{y-2}$ is the left half. We can get an inverse by reducing the domain of $y = x^2 + 2$ to $x \geq 0$. With this restriction, $x = f^{-1}(y) = \sqrt{y-2}$. The positive square root is the inverse. The domain of $f(x)$ matches the range of $f^{-1}(y)$.

2. $y = f(x) = \frac{x}{x-1}$. (This is Problem 4.3.4) Find x as a function of y.

   - Write $y = \frac{x}{x-1}$ as $y(x-1) = x$ or $yx - y = x$. ***We always have to solve for*** x. We have $yx - x = y$ or $x(y-1) = y$ or $x = \frac{y}{y-1}$. Therefore $f^{-1}(y) = \frac{y}{y-1}$.

   Note that f and f^{-1} are the same! If you graph $y = f(x)$ and the line $y = x$ you will see that $f(x)$ is symmetric about the 45° line. In this unusual case, $x = f(y)$ when $y = f(x)$.

You might wonder at the statement that $f(x) = \frac{x}{x-1}$ is the same as $g(y) = \frac{y}{y-1}$. The definition of a function does not depend on the particular choice of letters. The functions $h(r) = \frac{r}{r-1}$ and $F(t) = \frac{t}{t-1}$ and $G(z) = \frac{z}{z-1}$ are also the same. To graph them, you would put r, t, or z on the horizontal axis–they are the input (domain) variables. Then $h(r)$, $F(t)$, $G(z)$ would be on the vertical axis as output variables.

The function $y = f(x) = 3x$ and its inverse $x = f^{-1}(y) = \frac{1}{3}y$ (***absolutely not*** $\frac{1}{3y}$) are graphed on page 167. For $f(x) = 3x$, the domain variable x is on the horizontal axis. For $f^{-1}(y) = \frac{1}{3}y$, the *domain variable for* f^{-1} is y.

This can be confusing since we are so accustomed to seeing x along the horizontal axis. The advantage of $f^{-1}(x) = \frac{1}{3}x$ is that it allows you to keep x on the horizontal and to stick with x for domain (input). The advantage of $f^{-1}(y) = \frac{1}{3}y$ is that it emphasizes: f takes x to y and f^{-1} takes y back to x.

3. (This is 4.3.34) Graph $y = |x| - 2x$ and its inverse on separate graphs.

   - $y = |x| - 2x$ should be analyzed in two parts: positive x and negative x. When $x \geq 0$ we have $|x| = x$. The function is $y = x - 2x = -x$. When x is negative we have $|x| = -x$. Then $y = -x - 2x = -3x$. Then $y = -x$ on the right of the y axis and $y = -3x$ on the left. Inverses $x = -y$ and $x = -\frac{y}{3}$. The second graph shows the inverse function.

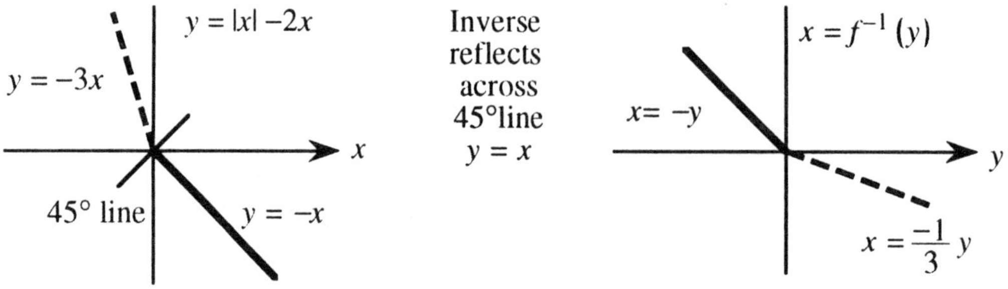

4. Find $\frac{dx}{dy}$ when $y = x^2 + x$. Compare implicit differentiation with $\frac{1}{dy/dx}$.

   - The x derivative of $y = x^2 + x$ is $\frac{dy}{dx} = 2x + 1$. Therefore $\frac{dx}{dy} = \frac{1}{2x+1}$.
   - The y derivative of $y = x^2 + x$ is $1 = 2x\,\frac{dx}{dy} + \frac{dx}{dy} = (2x+1)\frac{dx}{dy}$. This also gives $\frac{dx}{dy} = \frac{1}{2x+1}$.
   - It might be desirable to know $\frac{dx}{dy}$ as a function of y, not x. In that case solve the quadratic equation $x^2 + x - y = 0$ to get $x = \frac{-1\pm\sqrt{1+4y}}{2}$. Substitute this into $\frac{dx}{dy} = \frac{1}{2x+1} = \frac{\pm 1}{\sqrt{1+4y}}$.
   - Now we know $x = \frac{-1\pm\sqrt{1+4y}}{2}$ (this is the inverse function). So we can directly compute $\frac{dx}{dy} = \pm\frac{1}{2}\cdot\frac{1}{2}(1+4y)^{-1/2}\cdot 4 = \frac{\pm 1}{\sqrt{1+4y}}$. Same answer four ways!

5. Find $\frac{dx}{dy}$ at $x = \pi$ for $y = \cos x + x^2$.

   $\frac{dy}{dx} = -\sin x + 2x$. Substitute $x = \pi$ to find $\frac{dy}{dx} = -\sin\pi + 2\pi = 2\pi$. Therefore $\frac{dx}{dy} = \frac{1}{2\pi}$.

Read-throughs and selected even-numbered solutions :

The functions $g(x) = x - 4$ and $f(y) = y + 4$ are **inverse** functions, because $f(g(x)) = \mathbf{x}$. Also $g(f(y)) = \mathbf{y}$. The notation is $f = g^{-1}$ and $g = \mathbf{f^{-1}}$. The composition **of f and** $\mathbf{f^{-1}}$ is the identity function. By definition

$x = g^{-1}(y)$ if and only if $y = \mathbf{g(x)}$. When y is in the range of g, it is in the **domain** of g^{-1}. Similarly x is in the **domain** of g when it is in the **range** of g^{-1}. If g has an inverse then $g(x_1) \neq g(x_2)$ at any two points. The function g must be steadily **increasing** or steadily **decreasing**.

The chain rule applied to $f(g(x)) = x$ gives $(df/dy)(\mathbf{dg/dx}) = \mathbf{1}$. The slope of g^{-1} times the slope of g equals **1**. More directly $dx/dy = 1/(\mathbf{dy/dx})$. For $y = 2x + 1$ and $x = \frac{1}{2}(y-1)$, the slopes are $dy/dx = \mathbf{2}$ and $dx/dy = \mathbf{\frac{1}{2}}$. For $y = x^2$ and $x = \sqrt{y}$, the slopes are $dy/dx = \mathbf{2x}$ and $dx/dy = \mathbf{1/2\sqrt{y}}$. Substituting x^2 for y gives $dx/dy = \mathbf{1/2x}$. Then $(dx/dy)(dy/dx) = \mathbf{1}$.

The graph of $y = g(x)$ is also the graph of $x = \mathbf{g^{-1}(y)}$, but with x across and y up. For an ordinary graph of g^{-1}, take the reflection in the line $y = x$. If (3,8) is on the graph of g, then its mirror image **(8,3)** is on the graph of g^{-1}. Those particular points satisfy $8 = 2^3$ and $3 = \log_2 \mathbf{8}$.

The inverse of the chain $z = h(g(x))$ is the chain $x = \mathbf{g^{-1}(h^{-1}(z))}$. If $g(x) = 3x$ and $h(y) = y^3$ then $z = (\mathbf{3x})^{\mathbf{3}} = \mathbf{27x^3}$. Its inverse is $x = \mathbf{\frac{1}{3}z^{1/3}}$, which is the composition of $\mathbf{g^{-1}(y)} = \mathbf{\frac{1}{3}y}$ and $\mathbf{h^{-1}(z)} = \mathbf{z^{1/3}}$.

4 $x = \frac{y}{y-1}$ (f^{-1} matches f)

14 f^{-1} does not exist because $f(3)$ is the same as $f(5)$.

16 No two x's give the same y. **22** $\frac{dy}{dx} = -\frac{1}{(x-1)^2}$; $\frac{dx}{dy} = -\mathbf{\frac{1}{y^2}} = -(x-1)^2$.

44 **First proof** Suppose $y = f(x)$. We are given that $y > x$. This is the same as $y > f^{-1}(y)$.

Second proof The graph of $f(x)$ is above the 45° line, because $f(x) > x$. The mirror image is below the 45° line so $f^{-1}(y) < y$.

48 $g(x) = x + 6, f(y) = y^3, g^{-1}(y) = y - 6, f^{-1}(z) = \sqrt[3]{z}; \mathbf{x} = \mathbf{\sqrt[3]{z} - 6}$

4.4 Inverses of Trigonometric Functions (page 175)

The table on page 175 summarizes what you need to know – the six inverse trig functions, their domains, and their derivatives. The table gives you $\frac{dx}{dy}$ since the inverse functions have input y and output x. The input y is a *number* and the output x is an *angle*. Watch the restrictions on y and x (to permit an inverse).

1. Compute (a) $\sin^{-1}(\sin\frac{\pi}{4})$ (b) $\cos^{-1}(\sin\frac{\pi}{3})$ (c) $\sin^{-1}(\sin\pi)$ (d) $\tan^{-1}(\cos 0)$ (e) $\cos^{-1}(\cos(-\frac{\pi}{2}))$

- (a) $\sin\frac{\pi}{4}$ is $\frac{\sqrt{2}}{2}$ and $\sin^{-1}\frac{\sqrt{2}}{2}$ brings us back to $\frac{\pi}{4}$.
- (b) $\sin\frac{\pi}{3} = \frac{1}{2}$ and then $\cos^{-1}(\frac{1}{2}) = +\frac{2\pi}{3}$. Note that $\frac{\pi}{3} + \frac{2\pi}{3} = \frac{\pi}{2}$. The angles $\frac{\pi}{3}$ and $\frac{2\pi}{3}$ are complementary (they add to 90° or $\frac{\pi}{2}$). Always $\sin^{-1}y + \cos^{-1}y = \frac{\pi}{2}$.
- (c) $\sin^{-1}(\sin\pi)$ *is not* π! Certainly $\sin\pi = 0$. But $\sin^{-1}(0) = 0$. The $\sin^{-1}$ function or arcsin function only yields angles between $-\frac{\pi}{2}$ and $\frac{\pi}{2}$.
- (d) $\tan^{-1}(\cos 0) = \tan^{-1} 1 = \frac{\pi}{4}$
- (e) $\cos^{-1}(\cos(-\frac{\pi}{2}))$ looks like $-\frac{\pi}{2}$. But $\cos(-\frac{\pi}{2}) = 0$ and then $\cos^{-1}(0) = \frac{\pi}{2}$.

2. Find $\frac{dx}{dy}$ if $x = \sin^{-1} 3y$. What are the restrictions on y?

   We know that $x = \sin^{-1} u$ yields $\frac{dx}{du} = \frac{1}{\sqrt{1-u^2}}$. Set $u = 3y$ and use the chain rule: $\frac{dx}{du}\frac{du}{dy} = \frac{3}{\sqrt{1-u^2}} = \frac{3}{\sqrt{1-9y^2}}$. The restriction $|u| \le 1$ on sines means that $|3y| \le 1$ and $|y| \le \frac{1}{3}$.

3. Find $\frac{dz}{dx}$ when $z = \cos^{-1}(\frac{1}{x})$. What are the restrictions on x?

   $\cos^{-1}$ accepts inputs between -1 and 1, inclusive. For this reason $|\frac{1}{x}| \le 1$ and $|x| \ge 1$. To find the derivative, use the chain rule with $z = \cos^{-1} u$ and $u = \frac{1}{x}$:

$$\frac{dz}{dx} = \frac{dz}{du}\frac{du}{dx} = \frac{-1}{\sqrt{1-u^2}} \cdot \frac{-1}{x^2} = \frac{1}{x\sqrt{x^2 - x^2u^2}} = \frac{1}{x\sqrt{x^2-1}}.$$

4. Find $\frac{dy}{dx}$ when $y = \sec^{-1}\sqrt{x^2+1}$. (This is Problem 4.4.23)

   - The derivative of $y = \sec^{-1} u$ is $\frac{1}{|u|\sqrt{u^2-1}}$. In this problem $u = \sqrt{x^2+1}$. Then

$$\frac{dy}{dx} = \frac{dy}{du}\frac{du}{dx} = \frac{1}{|u|\sqrt{u^2-1}} \frac{x}{\sqrt{x^2+1}} = \text{(substitute for } u) = \frac{x}{(x^2+1)|x|} = \pm\frac{1}{x^2+1}.$$

   Here is another way to do this problem. Since $y = \sec^{-1}\sqrt{x^2+1}$, we have $\sec y = \sqrt{x^2+1}$ and $\sec^2 y = x^2+1$. *This is a trig identity provided* $x = \pm\tan y$. Then $y = \pm\tan^{-1} x$ and $\frac{dy}{dx} = \pm\frac{1}{x^2+1}$.

5. Find $\frac{dy}{dx}$ if $y = \tan^{-1}\frac{2}{x} - \cot^{-1}\frac{x}{2}$. Explain zero.

   - The derivative of $\tan^{-1}\frac{2}{x}$ is $\frac{1}{1+(\frac{2}{x})^2} \cdot \frac{-2}{x^2} = \frac{-2}{x^2+4}$. The derivative of $\cot^{-1}\frac{x}{2}$ is $-\frac{1}{1+(\frac{x}{2})^2} \cdot \frac{1}{2} = -\frac{2}{x^2+4}$. By subtraction $\frac{dy}{dx} = 0$. *Why do* $\tan^{-1}\frac{2}{x}$ *and* $\cot^{-1}\frac{x}{2}$ *have the same derivative? Are they equal?* Think about domain and range before you answer that one.

The relation $x = \sin^{-1} y$ means that **y** is the sine of **x**. Thus x is the angle whose sine is **y**. The number y lies between $\mathbf{-1}$ and $\mathbf{1}$. The angle x lies between $-\pi/\mathbf{2}$ and $\pi/\mathbf{2}$. (If we want the inverse to exist, there cannot be two angles with the same sine.) The cosine of the angle $\sin^{-1} y$ is $\sqrt{\mathbf{1}-\mathbf{y^2}}$. The derivative of $x = \sin^{-1} y$ is $dx/dy = \mathbf{1}/\sqrt{\mathbf{1}-\mathbf{y^2}}$.

The relation $x = \cos^{-1} y$ means that y equals $\mathbf{\cos x}$. Again the number y lies between $\mathbf{-1}$ and $\mathbf{1}$. This time the angle x lies between $\mathbf{0}$ and π (so that each y comes from only one angle x). The sum $\sin^{-1} y + \cos^{-1} y = \pi/\mathbf{2}$. (The angles are called **complementary**, and they add to a **right** angle.) Therefore the derivative of $x = \cos^{-1} y$ is $dx/dy = \mathbf{-1}/\sqrt{\mathbf{1}-\mathbf{y^2}}$, the same as for $\sin^{-1} y$ except for a **minus** sign.

The relation $x = \tan^{-1} y$ means that $y = \mathbf{\tan x}$. The number y lies between $-\infty$ and ∞. The angle x lies between $-\pi/\mathbf{2}$ and $\pi/\mathbf{2}$. The derivative is $dx/dy = \mathbf{1}/(\mathbf{1}+\mathbf{y^2})$. Since $\tan^{-1} y + \cot^{-1} y = \pi/\mathbf{2}$, the derivative of $\cot^{-1} y$ is the same except for a **minus** sign.

The relation $x = \sec^{-1} y$ means that $\mathbf{y} = \mathbf{\sec x}$. The number y *never* lies between $\mathbf{-1}$ and $\mathbf{1}$. The angle x lies between $\mathbf{0}$ and π, but never at $x = \pi/\mathbf{2}$. The derivative of $x = \sec^{-1} y$ is $dx/dy = \mathbf{1}/\mid \mathbf{y} \mid \sqrt{\mathbf{y^2}-\mathbf{1}}$.

10 The sides of the triangle are $y, \sqrt{1-y^2}$, and 1. The tangent is $\frac{\mathbf{y}}{\sqrt{\mathbf{1-y^2}}}$.

14 $\frac{d(\sin^{-1} y)}{dy}|_{x=0} = \mathbf{1}$; $\frac{d(\cos^{-1} y)}{dy}|_{x=0} = -\infty$; $\frac{d(\tan^{-1} y)}{dy}|_{x=0} = \mathbf{1}$; $\frac{d(\sin^{-1} y)}{dy}|_{x=1} = \frac{\mathbf{1}}{\mathbf{\cos 1}}$; $\frac{d(\cos^{-1} y)}{dy}|_{x=1}$ $= -\frac{\mathbf{1}}{\mathbf{\sin 1}}$; $\frac{d(\tan^{-1} y)}{dy}|_{x=1} = \frac{\mathbf{1}}{\mathbf{\sec^2 1}}$.

16 $\cos^{-1}(\sin x)$ is the complementary angle $\frac{\pi}{2} - x$. The tangent of that angle is $\frac{\mathbf{\cos x}}{\mathbf{\sin x}} = \mathbf{\cot x}$.

34 The requirement is $u' = \frac{1}{1+t^2}$. To satisfy this requirement take $u = \mathbf{\tan^{-1} t}$.

36 $u = \tan^{-1} y$ has $\frac{du}{dy} = \frac{1}{1+y^2}$ and $\frac{d^2u}{dy^2} = \frac{\mathbf{-2y}}{\mathbf{(1+y^2)^2}}$.

42 By the product rule $\frac{dz}{dx} = (\cos x)(\sin^{-1} x) + (\sin x)\frac{1}{\sqrt{1-x^2}}$. Note that $z \neq x$ and $\frac{dz}{dx} \neq 1$.

48 $u(x) = \mathbf{\frac{1}{2}\tan^{-1} 2x}$ (need $\frac{1}{2}$ to cancel 2 from the chain rule).

50 $u(x) = \frac{x-1}{x+1}$ has $\frac{du}{dx} = \frac{(x+1)-(x-1)}{(x+1)^2} = \frac{2}{(x+1)^2}$. Then $\frac{d}{dx}\tan^{-1} u(x) = \frac{1}{1+u^2}\frac{du}{dx} = \frac{1}{1+(\frac{x-1}{x+1})^2}\frac{2}{(x+1)^2} =$ $\frac{2}{(x+1)^2+(x-1)^2} = \frac{\mathbf{1}}{\mathbf{x^2+1}}$. This is also the derivative of $\mathbf{\tan^{-1} x}$! So $\tan^{-1} u(x) - \tan^{-1} x$ is a **constant**.

4 Chapter Review Problems

Review Problems

R1 Give the domain and range of the six inverse trigonometric functions.

R2 Is the derivative of $u(v(x))$ ever equal to the derivative of $u(x)v(x)$?

R3 Find y' and the second derivative y'' by implicit differentiation when $y^2 = x^2 + xy$.

R4 Show that $y = x + 1$ is the tangent line to the graph of $y = x + \cos xy$ through the point (0,1).

R5 If the graph of $y = f(x)$ passes through the point (a, b) with slope m, then the graph of $y = f^{-1}(x)$ passes through the point ____ with slope ____ .

R6 Where does the graph of $y = \cos x$ intersect the graph of $y = \cos^{-1} x$? Give an equation for x and show that $x = .7391$ in Section 3.6 is a solution.

R7 Show that the curves $xy = 4$ and $x^2 - y^2 = 15$ intersect at right angles.

R8 "The curve $y^2 + x^2 + 1 = 0$ has $2y\frac{dy}{dx} + 2x = 0$ so its slope is $-x/y$." What is the problem with that statement?

R9 Gas is escaping from a spherical balloon at 2 cubic feet/minute. How fast is the surface area shrinking when the area is 576π square feet?

R10 A 50 foot rope goes up over a pulley 18 feet high and diagonally down to a truck. The truck drives away at 9 ft/sec. How fast is the other end of the rope rising from the ground?

R11 Two concentric circles are expanding, the outer radius at 2 cm/sec and the inner radius at 5 cm/sec. When the radii are 10 cm and 3 cm, how fast is the area between them increasing (or decreasing)?

R12 A swimming pool is 25 feet wide and 100 feet long. The bottom slopes steadily down from a depth of 3 feet to 10 feet. The pool is being filled at 100 cubic feet/minute. How fast is the water level rising when it is 6 feet deep at the deep end?

R13 A five-foot woman walks at night toward a 12-foot street lamp. Her speed is 4 ft/sec. Show that her shadow is shortening by $\frac{20}{7}$ ft/sec when she is 3 feet from the lamp.

R14 A 40 inch string goes around an 8 by 12 rectangle – but we are changing its shape (same string). If the 8 inch sides are being lengthened by 1 inch/second, how fast are the 12 inch sides being shortened? Show that the area is increasing at 4 square inches per second. (For some reason it will take *two* seconds before the area increases from 96 to 100.)

R15 The volume of a sphere (when we know the radius) is $V(r) = 4\pi r^3/3$. The radius of a sphere (when we know the volume) is $r(V) = (3V/4\pi)^{1/3}$. This is the inverse! The surface area of a sphere is $A(r) = 4\pi r^2$. The radius (when we know the area) is $r(A) =$ ___ . The chain $r(A(r))$ equals ___ .

R16 The surface area of a sphere (when we know the volume) is $A(V) = 4\pi(3V/4\pi)^{2/3}$. The volume (when we know the area) is $V(A) =$ ___ .

Drill Problems (Find dy/dx in Problems **D1** to **D6**).

D1 $y = t^3 - t^2 + 2$ with $t = \sqrt{x}$

D2 $y = \sin^3(2x - \pi)$

D3 $y = \tan^{-1}(4x^2 + 7x)$

D4 $y = \csc\sqrt{x}$

D5 $y = \sin(\sin^{-1} x)$ for $|x| \le 1$

D6 $y = \sin u \cos u$ with $u = \cos^{-1} x$

In **D7** to **D10** find y' by implicit differentiation.

D7 $x^2 - 2xy + y^2 = 4$

D8 $y = \sin(xy) + x$

D9 $9x^2 + 16y^2 = 144$

D10 $9y - 6x + y^4 = 0$

D11 The area of a circle is $A(r) = \pi r^2$. Find the radius r when you know the area A. (This is the inverse function $r(A)$!). The derivative of $A = \pi r^2$ is $dA/dr = 2\pi r$. Find dr/dA.

CHAPTER 5 INTEGRALS

5.1 The Idea of the Integral (page 181)

Problems 1–3 review sums and differences from Section 1.2. This chapter goes forward to integrals and derivatives.

1. If $f_0, f_1, f_2, f_3, f_4 = 0, 2, 6, 12, 20$, find the differences $v_j = f_j - f_{j-1}$ and the sum of the v's.

   - The differences are $v_1, v_2, v_3, v_4 = 2, 4, 6, 8$. The sum is $2 + 4 + 6 + 8 = 20$. This equals $f_4 - f_0$.

2. If $v_1, v_2, v_3, v_4 = 3, 3, 3, 3$ and $f_0 = 5$, find the f's. Show that $f_4 - f_0$ is the sum of the v's.

   - Each new f_j is $f_{j-1} + v_j$. So $f_1 = f_0 + v_1 = 5 + 3 = 8$. Similarly $f_2 = f_1 + v_2 = 8 + 3 = 11$. Then $f_3 = 14$ and $f_4 = 17$. The sum of the v's is 12. The difference between f_{last} and f_{first} is also $17 - 5 = 12$.

3. (This is Problem 5.1.5) Show that $f_j = \frac{r^j}{r-1}$ has differences $v_j = f_j - f_{j-1} = r^{j-1}$.

   - The formula gives $f_0 = \frac{r^0}{r-1} = \frac{1}{r-1}$. Then $f_1 = \frac{r}{r-1}$ and $f_2 = \frac{r^2}{r-1}$. Now find the differences:

   $$v_1 = \frac{r}{r-1} - \frac{1}{r-1} = \frac{r-1}{r-1} = 1 \text{ and } v_2 = \frac{r^2}{r-1} - \frac{r}{r-1} = \frac{r^2-r}{r-1} = r.$$

   In general $f_j - f_{j-1} = r^{j-1}$. This is v_j. Adding the v's gives the ***geometric series*** $1+r+r^2+\cdots+r^{n-1}$. Its sum is $f_n - f_0 = \frac{r^n}{r-1} - \frac{1}{r-1} = \frac{r^n-1}{r-1}$.

4. Suppose $v(x) = 2x$ for $0 < x < 3$ and $v(x) = 6$ for $x > 3$. Sketch and find the area from 0 to x under the graph of $v(x)$.

   - There are really two cases to think about. If $x < 3$, the shaded triangle with base x and height $2x$ has area $= \frac{1}{2}(\text{base})\ (\text{height}) = \frac{1}{2}x(2x) = x^2$. If $x > 3$ the area is that of a triangle plus a rectangle. The triangle has base 3 and height 6 and area 9. The rectangle has base $(x-3)$ and height 6. Total area $= 9 + 6(x-3) = 6x - 9$. The area $f(x)$ has a two-part formula:

   $$f(x) = \begin{cases} x^2, & 0 \le x \le 3 \\ 6x - 9, & x > 3. \end{cases}$$

Areas under v(x) give f(x)

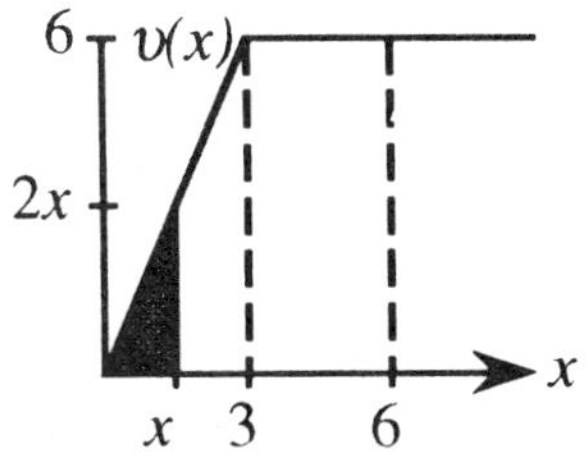

small triangle $\frac{1}{2}x(2x) = x^2$

complete triangle $\frac{1}{2}3(6) = 9$

add rectangle $6(x-3) = 6x - 18$

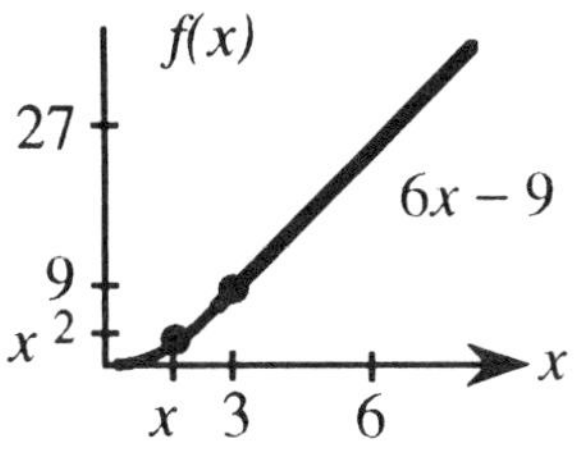

5. Use four rectangles to approximate the area under the curve $y = \frac{1}{2}x^2$ from $x = 0$ to $x = 4$. Then do the same using eight rectangles.

   - The heights of the four rectangles are $f(1) = \frac{1}{2}$, $f(2) = 2$, $f(3) = \frac{9}{2}$, $f(4) = 8$. The width of each rectangle is one. The sum of the four areas is $1 \cdot \frac{1}{2} + 1 \cdot 2 + 1 \cdot \frac{9}{2} + 1 \cdot 8 = 15$. The sketch shows that the actual area under the curve is less than 15.

The second figure shows eight rectangles. Their total area is still greater than the curved area, but less than 15. (Can you see why?) Each rectangle has width $\frac{1}{2}$. Their heights are determined by $y = \frac{1}{2}x^2$, where $x = \frac{1}{2}, 1, \frac{3}{2}, 2, \frac{5}{2}, 3, \frac{7}{2}, 4$. The total area of the rectangles is $\frac{1}{2}(\frac{1}{8} + \frac{1}{2} + \frac{9}{8} + 2 + \frac{25}{8} + \frac{9}{2} + \frac{49}{8} + 8) =$ 12.75. If we took 16 rectangles we would get an even closer estimate. ***The actual area is*** $10\frac{2}{3}$.

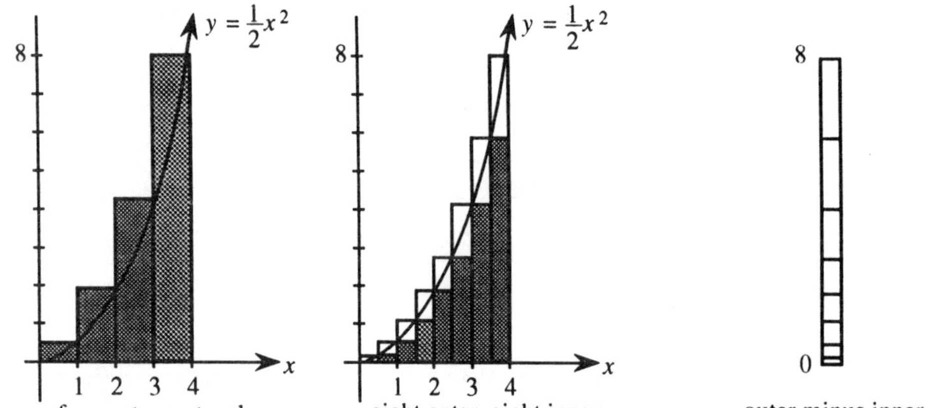

four outer rectangles — eight outer, eight inner — outer minus inner

6. Use the same curve $y = \frac{1}{2}x^2$ as in Problem 5. This time *inscribe* eight rectangles – the rectangles should touch the curve but remain inside it. What is the area approximation using this method? Give bounds on the same true area A under the curve.

   - The eight inscribed rectangles are shown, counting the first one with height zero. The total area is $\frac{1}{2}[f(0) + f(\frac{1}{2}) + \cdots + f(\frac{7}{2})] = \frac{1}{2}[0 + \frac{1}{8} + \frac{1}{2} + \frac{9}{8} + 2 + \frac{25}{8} + \frac{9}{2} + \frac{49}{8}] = 8.75$. The true area A is greater than 8.75 and less than 12.75.

   The differences between the outer and the inner rectangles add to a **single rectangle** with base $\frac{1}{2}$ and height $f(4) = 8$. Difference in areas $= \frac{1}{2}[f(4) - f(0)] = 4$. This is $12.75 - 8.75$.

   Hint for Exercises 5.1.11 – 14: These refer to an optimist and a pessimist. The pessimist would *under*estimate income, using inner rectangles. The optimist *over*estimates. She prefers *outer* rectangles that contain the graph.

Read-throughs and selected even-numbered solutions :

The problem of summation is to add $v_1 + \cdots + v_n$. It is solved if we find f's such that $v_j = \mathbf{f_j - f_{j-1}}$. Then $v_1 + \cdots + v_n$ equals $\mathbf{f_n - f_0}$. The cancellation in $(f_1 - f_0) + (f_2 - f_1) + \cdots + (f_n - f_{n-1})$ leaves only $\mathbf{f_n}$ and $\mathbf{-f_0}$. Taking sums is the **reverse (or inverse)** of taking differences.

The differences between 0, 1, 4, 9 are $v_1, v_2, v_3 = \mathbf{1, 3, 5}$. For $f_j = j^2$ the difference between f_{10} and f_9 is $v_{10} = \mathbf{19}$. From this pattern $1 + 3 + 5 + \cdots + 19$ equals **100**.

For functions, finding the integral is the reverse of **finding the derivative.** If the derivative of $f(x)$ is $v(x)$, then the **integral** of $v(x)$ is $f(x)$. If $v(x) = 10x$ then $f(x) = \mathbf{5x^2}$. This is the **area** of a triangle with base x and height $10x$.

Integrals begin with sums. The triangle under $v = 10x$ out to $x = 4$ has area **80**. It is approximated by four rectangles of heights 10, 20, 30, 40 and area **100**. It is better approximated by eight rectangles of heights $\mathbf{5, 10}, \cdots, \mathbf{40}$ and area **90**. For n rectangles covering the triangle the area is the sum of $\mathbf{\frac{4}{n}(\frac{40}{n} + \frac{80}{n}} + \cdots + \mathbf{40)} =$ $\mathbf{80 + \frac{80}{n}}$. As $n \to \infty$ this sum should approach the number **80**. *That is the integral of* $v = 10x$ *from 0 to 4.*

6 $f_0 = \frac{1-1}{r-1} = 0; 1 + r + \cdots + r^n = f_n = \mathbf{\frac{r^n - 1}{r-1}}$.

8 The f's are $\mathbf{0, 1, -1, 2, -2}, \cdots$ Here $v_j = (-1)^{j+1} j$ or $v_j = \begin{cases} j & j \text{ odd} \\ -j & j \text{ even} \end{cases}$ and $f_j = \begin{cases} \frac{j+1}{2} & j \text{ odd} \\ \frac{-j}{2} & j \text{ even} \end{cases}$

12 The last rectangle for the pessimist has height $\sqrt{\frac{15}{4}}$. Since the optimist's last rectangle of area $\frac{1}{4}\sqrt{\frac{16}{4}} = \frac{1}{2}$ is missed, the total area is reduced by $\frac{1}{2}$.

16 Under the $\sqrt{x}$ curve, the first triangle has base 1, height 1, area $\frac{1}{2}$. To its right is a rectangle of area 3. Above the rectangle is a triangle of base 3, height 1, area $\frac{3}{2}$. The total area $\frac{1}{2}+3+\frac{3}{2}=5$ is below the curve.

5.2 Antiderivatives (page 186)

The text on page 186 explains the difference between antiderivative, indefinite integral, and definite integral. If you're given a function $y = v(x)$ and asked for an *antiderivative*, you give any function whose derivative is $v(x)$. If you need the *indefinite integral*, you take an antiderivative and add $+C$ for any arbitrary constant. If you need a definite integral you will be given two endpoints $x = a$ and $x = b$. Find an antiderivative $y = f(x)$ and then compute $f(b) - f(a)$. In this problem your answer is a number, not a function.

The definite integral gives the exact area under the curve $v(x)$ from $x = a$ to $x = b$. (This has not been proved in the text yet, but this is the direction we are headed.) Finding antiderivatives is like solving puzzles–it takes a bit of insight as well as experience. Questions 1–5 are examples of this process. In each case, find the indefinite integral $f(x)$ of $v(x)$.

1. $v(x) = 9x^3 - 8$

   - An antiderivative of x^n is $\frac{1}{n+1}x^{n+1}$. For x^3 the power is $n = 3$. An antiderivative of x^3 is $\frac{1}{4}x^4$. Multiplying by 9 gives $\frac{9}{4}x^4$ as an antiderivative of $9x^3$. For the term -8, an antiderivative is $-8x$. Put these together to get the indefinite integral $f(x) = \frac{9}{4}x^4 - 8x + C$. ***Check*** $f'(x) = x^3 - 8 + 0 = v(x)$.

2. $v(x) = \sqrt{2x+5}$

   - The square root is the "$\frac{1}{2}$ power." Adding 1 gives $y = (2x+5)^{3/2}$. But now $\frac{dy}{dx}$ is $\frac{3}{2}(2x+5)^{1/2} \cdot 2 = 3\sqrt{2x+5}$. The first guess was off by a factor of 3. Correct this by $f(x) = \frac{1}{3}(2x+5)^{3/2} + C$. Check by taking $f'(x)$.

3. $v(x) = 3\sec 4x \tan 4x$.

   - We know that $y = \sec u$ has $\frac{dy}{du} = \sec u \tan u$. So the first guess at an antiderivative is $y = \sec 4x$. This gives $\frac{dy}{dx} = 4\sec 4x \tan 4x$. (The 4 comes from the chain rule: $u = 4x$ and $\frac{du}{dx} = 4$.) Everything is right but the leading 4, where we want 3. Adjust to get $f(x) = \frac{3}{4}\sec 4x + C$.

4. (This is 5.3.11) $v(x) = \sin x \cos x$.

   - There are two approaches, each illuminating in its own way. The first is to recognize $v(x)$ as $\frac{1}{2}\sin 2x$. Since the derivative of the cosine is minus the sine, $y = -\cos 2x$ is a first guess. That gives $\frac{dy}{dx} = 2\sin 2x$. To get $\frac{1}{2}$ instead of 2, divide by 4. The correct answer is $f(x) = -\frac{1}{4}\cos 2x + C$.

     A second approach is to recognize $v(x) = \sin x \cos x$ as $u\frac{du}{dx}$, where $u = \sin x$. An antiderivative for $u\frac{du}{dx}$ is $\frac{1}{2}u^2$. So $f(x) = \frac{1}{2}(\sin x)^2 + C$. (Find $f'(x)$ to check.)

Wait a minute. How can $f(x) = -\frac{1}{4}\cos 2x + C$ and $f(x) = \frac{1}{2}\sin^2 x + C$ both be correct? The answer to this reasonable question lies in the identity $\cos 2x = 1 - 2\sin^2 x$. The first answer $-\frac{1}{4}\cos 2x$ is $-\frac{1}{4} + \frac{1}{2}\sin^2 x$. The two answers are **the same**, except for the added constant $-\frac{1}{4}$. *For any $v(x)$ all antiderivatives are the same except for constants!*

5. $v(x) = \frac{1}{1+9x^2}$.

   - This seems impossible unless you remember that $\frac{1}{1+u^2}$ is the derivative of $\tan^{-1} u$. Our problem has $u^2 = 9x^2$ and $u = 3x$. If we let $y = \tan^{-1} 3x$, we get $\frac{dy}{dx} = \frac{1}{1+9x^2} \cdot 3$. (The 3 comes from the chain rule.) The coefficient should be adjusted to give $f(x) = \frac{1}{3}\tan^{-1} 3x + C$.

   Questions 6 and 7 use definite integrals to find area under curves.

6. Find the area under the parabola $v = \frac{1}{2}x^2$ from $x = 0$ to $x = 4$. (Section 5.1 of this Guide outlined ways to estimate this area. Now you can find it exactly.)

   - Area $= \int_a^b v(x)dx = \int_0^4 \frac{1}{2}x^2 dx$. We need an antiderivative of $\frac{1}{2}x^2$. It is $\frac{1}{2}$ times $\frac{1}{3}x^3$, or $\frac{x^3}{6}$:

$$\text{Area } = \int_0^4 \frac{1}{2}x^2 dx = \frac{4^3}{6} - \frac{0^3}{6} = \frac{64}{6} = 10\frac{2}{3}.$$

7. (This is 5.2.23) For the curve $y = \frac{1}{\sqrt{x}}$ the area between $x = 0$ and $x = 1$ is 2. Verify that $f(1) - f(0) = 2$.

   - Since $\frac{1}{\sqrt{x}} = x^{-1/2}$, an antiderivative is $f(x) = 2x^{1/2}$. Therefore $\int_0^1 \frac{1}{\sqrt{x}}dx = 2(1)^{1/2} - 2(0)^{1/2} = 2 - 0 = 2$. The unusual part of this problem is that the curve goes up to $y = \infty$ when $x = 0$. But the area underneath is still 2.

8. (This is 5.2.26) Draw $y = v(x)$ so that the area $f(x)$ from 0 to x increases until $x = 1$, stays constant to $x = 2$, and decreases to $f(3) = 1$.

   - Where the area increases, $v(x)$ is positive (from 0 to 1). There is no new area from $x = 1$ to $x = 2$, so $v(x) = 0$ on that interval. The area decreases after $x = 2$, so $v(x)$ must go below the x axis. The total area from 0 to 3 is $f(3) = 1$, so the area from 2 to 3 is one unit less than the area from 0 to 1. One solution has $v(x) = 2$ then 0 then -1 in the three intervals. Can you find a *continuous* $v(x)$?

Read-throughs and selected even-numbered solutions :

Integration yields the **area** under a curve $y = v(x)$. It starts from rectangles with the base $\mathbf{\Delta x}$ and heights $v(x)$ and areas $\mathbf{v(x)\Delta x}$. As $\Delta x \to 0$ the area $v_1\Delta x + \cdots + v_n\Delta x$ becomes the **integral** of $v(x)$. The symbol for the indefinite integral of $v(x)$ is $\int \mathbf{v(x)dx}$.

The problem of integration is solved if we find $f(x)$ such that $\mathbf{\frac{df}{dx} = v(x)}$. Then f is the **antiderivative** of v, and $\int_2^6 v(x)dx$ equals $\mathbf{f(6)}$ minus $\mathbf{f(2)}$. The limits of integration are **2 and 6.** This is a **definite** integral, which is a **number** and not a function $f(x)$.

The example $v(x) = x$ has $f(x) = \mathbf{\frac{1}{2}x^2}$. It also has $f(x) = \mathbf{\frac{1}{2}x^2 + 1}$. The area under $v(x)$ from 2 to 6 is **16.** The constant is canceled in computing the difference $\mathbf{f(6)}$ minus $\mathbf{f(2)}$. If $v(x) = x^8$ then $f(x) = \mathbf{\frac{1}{9}x^9}$.

The sum $v_1 + \cdots + v_n = f_n - f_0$ leads to the Fundamental Theorem $\int_a^b v(x)dx = \mathbf{f(b) - f(a)}$. The **indefinite** integral is $f(x)$ and the **definite** integral is $f(b) - f(a)$. Finding the **area** under the v-graph is the opposite of finding the **slope** of the f-graph.

6 $\frac{x^{1/3}}{x^{2/3}} = x^{-1/3}$ which has antiderivative $f(x) = \frac{3}{2}x^{2/3}$; $f(1) - f(0) = \mathbf{\frac{3}{2}}$.

10 $f(x) = \sin x - x\cos x$; $f(1) - f(0) = \mathbf{\sin 1 - \cos 1}$

18 Areas 0, 1, 2, 3 add to $A_4 = \mathbf{6}$. Each rectangle misses a triangle of base $\frac{4}{N}$ and height $\frac{4}{N}$. There are N triangles of total area $N \cdot \frac{1}{2}(\frac{4}{N})^2 = \frac{8}{N}$. So the N rectangles have area $\mathbf{8 - \frac{8}{N}}$.

22 Two rectangles have base $\frac{1}{2}$ and heights 2 and 1, with area $\mathbf{\frac{3}{2}}$. Four rectangles have base $\frac{1}{4}$ and heights 4, 3, 2, 1 with area $\frac{10}{4} = \mathbf{\frac{5}{2}}$. Eight rectangles have area $\frac{7}{2}$. The limiting area under $y = \frac{1}{x}$ is **infinite.**

28 The area $f(4) - f(3)$ is $-\frac{1}{2}$, and $f(3) - f(2)$ is -1, and $f(2) - f(1)$ is $\frac{1}{2}(\frac{2}{3})(2) - \frac{1}{2}(\frac{1}{3})(1)$. Total $\mathbf{-1}$. The graph of f_4 is x^2 to $x = 1$.

5.3 Summation Versus Integration (page 194)

Problems 1–8 offer practice with sigma notation. The summation limits $\sum_{k=1}^{5}$ are like the integration limits $\int_{x=a}^{b}$.

1. Write out the terms of $\sum_{k=1}^{5}(x+k) = S$.

   - Do not make any substitutions for x. Just let k go from 1 to 5 and add: $S = (x+1) + (x+2) + (x+3) + (x+4) + (x+5) = 5x + 15$.

2. Write out the first three terms and the last term of $S = \sum_{i=0}^{n}(-1)^i 2^{i+1}$.

   - $S = (-1)^0 2^{0+1} + (-1)^1 2^{1+1} + (-1)^2 2^{2+1} + \cdots + (-1)^n 2^{n+1} = 2 - 2^2 + 2^3 - \cdots + (-1)^n 2^{n+1}$.

3. Write this sum using sigma notation: $243 - 81 + 27 - 9 + 3 - 1$.

   - The first thing to notice is that the terms are alternating positive and negative. That means a factor of either $(-1)^i$ or $(-1)^{i-1}$. The first term 243 is positive. If we start the sum with $i = 1$, then $(-1)^{i-1}$ is the right choice. This gives a positive sign for $i = 1, 3, 5, \cdots$.

     Now look at the numbers 243, 81, 27, 9, 3, and 1. Notice the pattern $81 = \frac{243}{3}$ and $27 = \frac{243}{3^2}$. In general the "ith" term is $\frac{243}{3^{i-1}}$. So we can write $\sum_{i=1}^{6}(-1)^{i-1}\frac{243}{3^{i-1}} = \sum_{i=1}^{6} 243(-\frac{1}{3})^{i-1}$.
   - Second solution: Start the sum at $i = 0$. We'll use a new dummy variable j, to avoid confusion between the "old i" and the "new i." We want $j = 0$ when $i = 1$. *This means* $j = i - 1$. Since i goes from 1 to 6, j goes from 0 to 5. This gives $\sum_{j=0}^{5} 243(-\frac{1}{3})^j$.

5. Rewrite $\sum_{i=3}^{2n} a^i b^{i-3}$ so that the indexing starts at $j = 0$ instead of $i = 3$.

   - Changing the dummy variable will require changes in four places: $\sum_{[]}^{[]} a^{[]} b^{[]}$. Replace $i = 3$ with $j = 0$. This means $j = i - 3$ and $i = j + 3$. The indexing ends at $i = 2n$, which means $j = 2n - 3$. Also a^i becomes a^{j+3}. The factor b^{i-3} becomes b^j. The answer is $\sum_{j=0}^{2n-3} a^{j+3} b^j$.

6. Compute the sum $3 + 6 + 9 + \cdots + 3000 = S$.

   - Don't use your calculator! Use your head! Each term is a multiple of 3, so $S = 3(1 + 2 + \cdots + 1000)$. The part in parentheses is the sum of the first 1000 integers. This is $\frac{1}{2}(1000)(1001)$ by equation 2 on page 189. Then $S = 3[\frac{1}{2}(1000)(1001)] = 1,501,500$.

7. Compute the sum $S = \sum_{i=1}^{100}(i+50)^2$. Write out the first and last terms to get a feel for what's happening.

- The sum is $51^2 + 52^2 + \cdots + 149^2 + 150^2$. Equation 7 on page 191 gives the sum of the first n squares. Then

$$\begin{aligned} S &= [\text{sum of the first 150 squares}] - [\text{sum of the first 50 squares}] \\ &= [\tfrac{1}{3}(150)^3 + \tfrac{1}{2}(150)^2 + \tfrac{1}{6}(150)] - [\tfrac{1}{3}(50)^3 + \tfrac{1}{2}(50)^2 + \tfrac{1}{6}(50)] \\ &= 1136275 - 42925 = 1093350. \end{aligned}$$

8. Compute $S = \sum_{i=1}^{n}(\frac{1}{2i} - \frac{1}{2i+2})$. The answer is a function of n. What happens as $n \to \infty$?

- Writing the first few terms and the last term almost always makes the sum clearer. Take $i = 1, 2, 3$ to find $(\frac{1}{2} - \frac{1}{4}) + (\frac{1}{4} - \frac{1}{6}) + (\frac{1}{6} - \frac{1}{8})$. Regroup those terms so that $\frac{1}{4}$ cancels $-\frac{1}{4}$ and $\frac{1}{6}$ cancels $-\frac{1}{6}$:

$$S = \frac{1}{2} + (-\frac{1}{4} + \frac{1}{4}) + (-\frac{1}{6} + \frac{1}{6}) + \cdots + (-\frac{1}{2n} + \frac{1}{2n}) - \frac{1}{2n-2} = \frac{1}{2} - \frac{1}{2n-2}.$$

This kind of series is called ***"telescoping"***. *It is like our sums of differences.* The middle terms collapse to leave $S = \frac{1}{2} - \frac{1}{2n-2}$. As $n \to \infty$ this approaches $\frac{1}{2}$. The sum of the infinite series is $\frac{1}{2}$.

9. (This is 5.4.19) Prove by induction that $f_n = 1 + 3 + \cdots + (2n-1) = n^2$.

In words, the sum of the first n odd integers is n^2. ***A proof by induction has two parts.*** First part: Check the formula for the first value of n. Since the first odd integer is $1 = 1^2$, this part is easy and f_1 is correct. The second part of the proof is to check that $f_n = f_{n-1} + v_n$. Then a formula that is correct for f_{n-1} remains correct at the next step. *When v_n is added, f_n is still correct:*

$$f_{n-1} + v_n = (n-1)^2 + (2n-1) = n^2 - 2n + 1 + 2n - 1 = n^2.$$

The induction proof is complete. The correctness for $n = 1$ leads to $n = 2$, $n = 3$, $\cdots$, and every n.

10. Find q in the formula $1^6 + 2^6 + 3^6 + \cdots + n^6 = qn^7 +$ correction.

- The correction will involve powers below 7. We are concerned here with the coefficient of n^7. The sum $1^6 + 2^6 + 3^6 + \cdots + n^6$ is like adding n rectangles of width 1 under the curve $y = x^6$. The sum is like the area under $y = x^6$ from 0 to n. This is given by the definite integral $\int_0^n x^6 dx$. Since $\frac{1}{7}x^7$ is an antiderivative for $v(x) = x^6$, the definite integral is $\frac{1}{7}(n)^7 - \frac{1}{7}(0)^7 = \frac{1}{7}n^7$. The coefficient q is $\frac{1}{7}$. (This is informal reasoning, not proof.)

11. (This is 5.4.23) Add $n = 400$ to the table for $S_n = 1 + \cdots + n$ and find the relative error E_n. Guess a formula for E_n.

- The question refers to the table on page 193. The sum of the first n integers, $S_n = 1 + 2 + \cdots + n$, is being approximated by the area $I_n = \int_0^n x\,dx$, which is under the line $y = x$ from $x = 0$ to $x = n$. The error D_n is the difference between the true sum and the integral. "Relative error" is the proportional error $E_n = D_n/I_n$.

  Now for specifics: The first 400 integers add to $\frac{1}{2}n(n+1) = \frac{1}{2}(400)(401) = 80,200$. The integral is

$$\begin{aligned} I_{400} &= \int_0^{400} x\,dx = \tfrac{1}{2}(400)^2 - \tfrac{1}{2}0^2 = 80,000 \\ D_{400} &= 80,200 - 80,000 = 200 \text{ and } E_{400} = \tfrac{200}{80,000} = \tfrac{1}{400}. \end{aligned}$$

The table shows $E_{100} = \frac{1}{100}$ and $E_{200} = \frac{1}{200}$. It seems that $E_n = \frac{1}{n}$.

Read-throughs and selected even-numbered solutions :

The Greek letter $\sum$ indicates summation. In $\sum_1^n v_j$ the dummy variable is **j**. The limits are **j = 1 and j = n**, so the first term is $\mathbf{v_1}$ and the last term is $\mathbf{v_n}$. When $v_j = j$ this sum equals $\mathbf{\frac{1}{2}n(n+1)}$. For $n = 100$ the leading term is $\mathbf{\frac{1}{2}100^2 = 5000}$. The correction term is $\mathbf{\frac{1}{2}n = 50}$. The leading term equals the integral of $v = x$ from 0 to 100, which is written $\int_0^{100} \mathbf{x\,dx}$. The sum is the total **area** of 100 rectangles. The correction term is the area between the **sloping line** and the **rectangles.**

The sum $\sum_{i=3}^{6} i^2$ is the same as $\sum_{j=1}^{4} (\mathbf{j+2})^2$ and equals **86**. The sum $\sum_{i=4}^{5} v_i$ is the same as $\sum_{\mathbf{i=0}}^{\mathbf{1}} v_{i+4}$ and equals $\mathbf{v_4 + v_5}$. For $f_n = \sum_{j=1}^{n} v_j$ the difference $f_n - f_{n-1}$ equals $\mathbf{v_n}$.

The formula for $1^2 + 2^2 + \cdots + n^2$ is $f_n = \mathbf{\frac{1}{6}n(n+1)(2n+1) = \frac{1}{3}n^3 + \frac{1}{2}n^2 + \frac{1}{6}n}$. To prove it by mathematical induction, check $f_1 = \mathbf{1}$ and check $f_n - f_{n-1} = \mathbf{n^2}$. The area under the parabola $v = x^2$ from $x = 0$ to $x = 9$ is $\mathbf{\frac{1}{3}9^3}$. This is close to the area of $\mathbf{9/\Delta x}$ rectangles of base Δx. The correction terms approach zero very **slowly.**

16 $\sum_{i=1}^{n} v_i = \sum_{j=0}^{n-1} \mathbf{v_{j+1}}$ and $\sum_{i=0}^{6} i^2 = \sum_{i=2}^{8} (\mathbf{i-2})^2$.
20 $f_1 = \frac{1}{4}(1)^2(2)^2 = 1$; $f_n - f_{n-1} = \frac{1}{4}n^2(n+1)^2 - \frac{1}{4}(n-1)^2n^2 = \frac{1}{4}n^2(4n) = n^3$.
24 $S_{50} = 42925$; $I_{50} = 41666\frac{2}{3}$; $D_{50} = 1258\frac{1}{3}$; $E_{50} = 0.0302$; E_n is approximately $\frac{1.5}{n}$ and exactly $\frac{1.5}{n} + \frac{1}{2n^2}$.
28 $xS = x + x^2 + x^3 + \cdots$ equals $S - 1$. Then $S = \frac{1}{1-x}$. If $x = 2$ the sums are $S = \infty$.
36 The rectangular area is $\Delta x \sum_{j=1}^{1/\Delta x} v((j-1)\Delta x)$ or $\Delta x \sum_{i=0}^{(1/\Delta x)-1} v(i\Delta x)$.

5.4 Indefinite Integrals and Substitutions (page 200)

All the integrals in this section are either direct applications of the known forms on page 196 or have substitutions which lead to those forms. The trick, which takes experience, is choosing the right u. As you gain skill, you will do many of these substitutions in your head. To start, write them out.

1. Find the indefinite integral $\int 2x\sqrt{x^2-5}\,dx$.

   - Take $u = x^2 - 5$. Then $\frac{du}{dx} = 2x$, or $du = 2x\,dx$. This is great because we can take out "$2x\,dx$" in the problem and substitute "du". The problem is now $\int u^{1/2}du$ which equals $\frac{2}{3}u^{3/2} + C$. The final step is to put x back. The answer is $\frac{2}{3}u^{3/2} + C = \frac{2}{3}(x^2-5)^{3/2} + C$.

2. Find $\int \frac{\cos x}{\sin^4 x}dx$.

   - Choose $u = \sin x$ because then $\frac{du}{dx} = \cos x$. We can replace "$\cos x\,dx$" by "du". The problem is now $\int \frac{du}{u^4} = \int u^{-4}du = -\frac{1}{3}u^{-3} + C = -\frac{1}{3}(\sin x)^{-3} + C$.

3. (This is 5.4.14) Find $\int t^3\sqrt{1-t^2}dt$.

   - It's hard to know what substitution to make, but the expression under the radical is a good candidate. Let $u = 1-t^2$. Then $du = -2t\,dt$. Decompose the problem this way: $\int (t^2)(\sqrt{1-t^2})(t\,dt)$. Substitute

$t^2 = 1 - u$, $\sqrt{1-t^2} = u^{1/2}$ and $t\ dt = -\frac{1}{2}du$. The problem is now

$$-\frac{1}{2}\int (1-u)u^{1/2}du = -\frac{1}{2}\int (u^{1/2} - u^{3/2})du = -\frac{1}{2}\int u^{1/2}du + \frac{1}{2}\int u^{3/2}du.$$

- The last step used the linearity property of integrals.
- The solution is $-\frac{1}{3}u^{3/2} + \frac{1}{5}u^{5/2} + C = -\frac{1}{3}(1-t^2)^{3/2} + \frac{1}{5}(1-t^2)^{5/2} + C$.

4. Find $\int \frac{dx}{\sqrt{1-4x^2}}$. Choosing $u = 1 - 4x^2$ fails because $du = -8x\ dx$ is not present in the problem.

- Mentally sifting through the "known" integrals, we remember $\int \frac{du}{\sqrt{1-u^2}} = \sin^{-1} u + C$. The original problem $\int \frac{dx}{\sqrt{1-4x^2}}$ has this form if $u = 2x$. Next step: $du = 2\ dx$. Replace "dx" with "$\frac{1}{2}du$":

$$\int \frac{dx}{\sqrt{1-4x^2}} = \int \frac{\frac{1}{2}du}{\sqrt{1-u^2}} = \frac{1}{2}\sin^{-1} u + C = \frac{1}{2}\sin^{-1} 2x + C.$$

5. Find a function $y(x)$ that solves the differential equation $\frac{dy}{dx} = \sqrt{xy}$. (This is 5.4.26)

- Here are two methods. The first is "guess and check," the second is "separation of variables."

  "*Guess and check*" We guess $y = cx^n$ and try to figure out c and n. Since $\frac{dy}{dx} = ncx^{n-1}$ and we are given $\frac{dy}{dx} = \sqrt{xy}$, we must have $ncx^{n-1} = x^{1/2}(cx^n)^{1/2}$. The power of x on the left is $n-1$ and the power on the right is $\frac{1}{2} + \frac{n}{2}$. They are equal for $n = 3$. The left side is now $3cx^2$ and the right is $c^{1/2}x^2$. This means $3c = c^{1/2}$ and $c = \frac{1}{9}$. The method gives $y = \frac{1}{9}x^3$ and we check $\frac{dy}{dx} = \sqrt{xy}$.

  "*Separation of variables*" Starting with $\frac{dy}{dx} = x^{1/2}y^{1/2}$, first get x and dx on one side of the equal sign and y and dy on the other: $\boldsymbol{y^{-1/2}\ dy = x^{1/2}dx}$. Now integrate both sides: $2y^{1/2} + C_1 = \frac{2}{3}x^{3/2} + C_2$. Divide by 2 and let "$C$" replace the awkward constant $\frac{C_2}{2} - \frac{C_1}{2}$. Then $y^{1/2} = \frac{1}{3}x^{3/2} + C$, and $y = (\frac{1}{3}x^{3/2} + C)^2$. (Again, check $\frac{dy}{dx} = \sqrt{xy}$.)

  Let $C = 0$ and you get the answer $\frac{1}{9}x^3$ obtained by "guess and check." The former is just one solution; the latter is a whole class of solutions, one for every C.

6. Solve the differential equation $\frac{d^2y}{dx^2} = x$.

   The second derivative is x, so the first derivative must be $\frac{1}{2}x^2 + C_1$. The reason for writing C_1 instead of C becomes clear at the next step: Integrate $\frac{dy}{dx} = \frac{1}{2}x^2 + C_1$ to get $y = \frac{1}{6}x^3 + C_1x + C_2$. Each integration gives another "constant of integration."

   The functions $y = \frac{1}{6}x^3$, $y = \frac{1}{6}x^3 - x + 27$, and $y = \frac{1}{6}x^3 - 8$ are all solutions to $y'' = x$.

Read-throughs and selected even-numbered solutions :

Finding integrals by substitution is the reverse of the **chain** rule. The derivative of $(\sin x)^3$ is $\mathbf{3(\sin x)^2\cos x}$. Therefore the antiderivative of $\mathbf{3(\sin x)^2\cos x}$ is $\mathbf{(\sin x)^3}$. To compute $\int (1+\sin x)^2 \cos x\ dx$, substitute $u = \mathbf{1 + \sin x}$. Then $du/dx = \mathbf{\cos x}$ so substitute $du = \mathbf{\cos x\ dx}$. In terms of u the integral is $\int \mathbf{u^2\ du} = \mathbf{\frac{1}{3}u^3}$. Returning to x gives the final answer.

The best substitutions for $\int \tan(x+3)\sec^2(x+3)dx$ and $\int (x^2+1)^{10}x\ dx$ are $u = \mathbf{\tan(x+3)}$ and $u = \mathbf{x^2+1}$. Then $du = \mathbf{\sec^2(x+3)dx}$ and $\mathbf{2x\ dx}$. The answers are $\mathbf{\frac{1}{2}\tan^2(x+3)}$ and $\mathbf{\frac{1}{22}(x^2+1)^{11}}$. The antiderivative of $v\ dv/dx$ is $\mathbf{\frac{1}{2}v^2}$. $\int 2x\ dx/(1+x^2)$ leads to $\int \frac{\mathbf{du}}{\mathbf{u}}$, which we don't yet know. The integral $\int dx/(1+x^2)$ is known immediately as $\mathbf{tan^{-1}x}$.

6 $\frac{-2}{9}(1-3x)^{3/2}+C$ **10** $\cos^3 x \sin 2x$ equals $2\cos^4 x \sin x$ and its integral is $\frac{-2}{5}\cos^5 x + C$

16 The integral of $x^{1/2}+x^2$ is $\frac{2}{3}x^{3/2}+\frac{1}{3}x^3+C$.

20 Write $\sin^3 x$ as $(1-\cos^2 x)\sin x$. The integrals of $-\cos^2 x \sin x$ and $\sin x$ give $\frac{1}{3}\cos^3 x - \cos x + C$.

26 $dy/dx = x/y$ gives $y\ dy = x\ dx$ or $y^2 = x^2 + C$ or $y = \sqrt{x^2+C}$.

28 $y = \frac{1}{120}x^5 + C_1x^4 + C_2x^3 + C_3x^2 + C_4x + C_5$

34 (a) **False:** The derivative of $\frac{1}{2}f^2(x)$ is $f(x)\frac{df}{dx}$ (b) **True:** The chain rule gives $\frac{d}{dx}f(v(x)) = \frac{df}{dx}(v(x))$ times $\frac{dv}{dx}$ (c) **False:** These are inverse *operations* not inverse functions and (d) is **True.**

5.5 The Definite Integral (page 205)

1. Make a sketch to show the meaning of $\int_1^3 \sqrt{x}\ dx$. Then find the value.

   - $\int_1^3 \sqrt{x}\ dx$ is the area under the curve $y = \sqrt{x}$ from $x=1$ to $x=3$. Since $f(x) = \frac{2}{3}x^{3/2}$ is an antiderivative for $v(x) = \sqrt{x}$, the definite integral is

   $$\int_1^3 \sqrt{x}\ dx = \frac{2}{3}x^{3/2}]_1^3 = \frac{2}{3}(3)^{3/2} - \frac{2}{3}(1)^{3/2} = 2\sqrt{3} - \frac{2}{3} \approx 2.797.$$

2. Use integral notation to describe the area between $y = \sqrt{x}$ and $y = 0$ and $x = 2$ and $x = 4$.

   - The area under $y = \sqrt{x}$ from $x = 2$ to $x = 4$ is $\int_2^4 \sqrt{x}\ dx$. Don't leave off the dx!

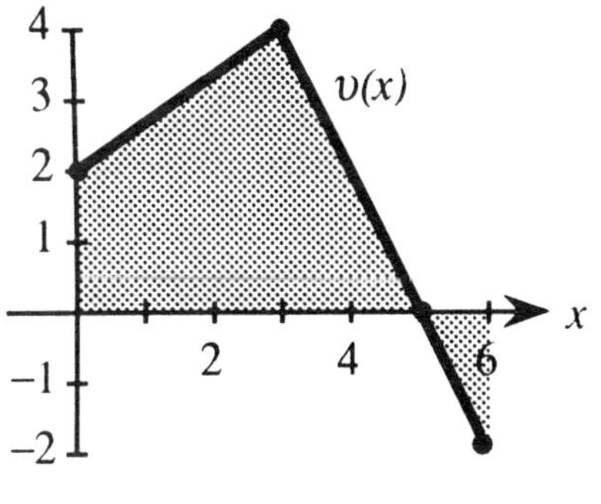

Quick question
Which x gives the maximum area $f(x)$?
$f(x)$ is a maximum when $v(x) = \frac{df}{dx} =$ ____.

3. The function $f(x) = \int_0^x v(t)dt$ gives the area under the curve $y = v(t)$ from $t = 0$ to $t = x$. From the graph of v, find $f(3), f(5), f(6)$, and $f(0)$. Note: This is not a graph of $f(x)$! The function $f(x)$ is the *area* under this graph.

   - You could write a formula for $f(x)$ but it's just as easy to read off the areas. $f(3)$ is the trapezoid area with base 3: $f(3) = \frac{1}{2}(2+4)3 = 9$. To find $f(5)$, add the area of the triangle with base 2: $f(5) = 9+\frac{1}{2}(4)(2) = 13$. The next triangle has area 1 but it is below the x axis. This area is negative so that $f(6) = f(5) - 1 = 12$. Finally $f(0) = 0$ since the area from $t = 0$ to $t = 0$ is zero!

   Problems 4 and 5 will give practice with definite integrals – especially changes of limits.

4. $\int_0^{\pi/4} \sec^3 x \tan x \, dx = \int_?^? u^2 \, du$ (with $u = \sec x, du = \sec x \tan x \, dx$).

- The lower limit is the value of u when $x = 0$ (the original lower limit). Thus $u(0) = \sec 0 = 1$. The upper limit is the value of u when $x = \frac{\pi}{4}$ (the original upper limit). This is $\sec \frac{\pi}{4} = \sqrt{2}$.
- Now $\int_1^{\sqrt{2}} u^2 du = \frac{1}{3} u^3]_1^{\sqrt{2}} = \frac{2\sqrt{2}}{3} - \frac{1}{3} \approx 0.609$.

5. $\int_{-2}^{0} x^5 \sqrt{1 - x^3} \, dx = \int_?^? (1-u)\sqrt{u} \, \frac{du}{-3}$ (with $u = 1 - x^3$).

- We replaced x^3 with $1 - u$ and $x^2 dx$ with $-\frac{1}{3} du$. The lower limit is $u(-2) = 1 - (-2)^3 = 9$ and the upper limit is $u(0) = 1$. The integral is

$$-\frac{1}{3}\int_9^1 (u^{1/2} - u^{3/2}) du = -\frac{1}{3}\left(\frac{2}{3}u^{3/2} - \frac{2}{5}u^{5/2}\right)]_9^1 = -\frac{1}{3}\left(\left(\frac{2}{3} - \frac{2}{5}\right) - \left(\frac{2}{3} \cdot 27 - \frac{2}{5} \cdot 243\right)\right) \approx -26.5.$$

Problems 6-7 help flesh out the idea of the maximum M_k, minimum m_k, and sums S and s.

6. Find the maximum and minimum of $y = \sin \pi x$ in the intervals from 0 to $\frac{1}{4}$, $\frac{1}{4}$ to $\frac{1}{2}$, $\frac{1}{2}$ to $\frac{3}{4}$, $\frac{3}{4}$ to 1. Use these M's and m's in the upper sum S and lower sum s enclosing $\int_0^1 \sin \pi x \, dx$.

- The maximum values M in the four intervals are $\frac{\sqrt{2}}{2}$, 1, 1, $\frac{\sqrt{2}}{2}$. The upper sum with $\Delta x = \frac{1}{4}$ is $S = \frac{1}{4}(\frac{\sqrt{2}}{2} + 1 + 1 + \frac{\sqrt{2}}{2}) \approx .85$. The minimum values m_k are $0, \frac{\sqrt{2}}{2}, \frac{\sqrt{2}}{2}, 0$. The lower sum is $s = \frac{1}{4}(0 + \frac{\sqrt{2}}{2} + \frac{\sqrt{2}}{2} + 0) \approx .35$. The integral is between .35 and .85. (It is exactly $[-\frac{\cos \pi x}{\pi}]_0^1 = \frac{2}{\pi} \approx .64$.)

7. Repeat Problem 6 with $\Delta x = \frac{1}{8}$ and eight intervals. Compare the answers.

- Adding $\frac{1}{8}$ times the maximum values M_k gives $S \approx .75$. Adding $\frac{1}{8}$ times the minimum values m_k gives $s \approx .50$. As expected, s has grown and S is smaller. The correct value .64 is still in between.

Read-throughs and selected even-numbered solutions :

If $\int_a^x v(x) dx = f(x) + C$, the constant C equals $-\mathbf{f(a)}$. Then at $x = a$ the integral is **zero.** At $x = b$ the integral becomes $\mathbf{f(b) - f(a)}$. The notation $f(x)]_a^b$ means $\mathbf{f(b) - f(a)}$. Thus $\cos x]_0^\pi$ equals $\mathbf{-2}$. Also $[\cos x + 3]_0^\pi$ equals $\mathbf{-2}$, which shows why the antiderivative includes an arbitrary **constant.** Substituting $u = 2x - 1$ changes $\int_1^3 \sqrt{2x-1} \, dx$ into $\int_1^5 \frac{1}{2}\sqrt{u} \, du$ (with limits on u).

The integral $\int_a^b v(x) dx$ can be defined for any **continuous** function $v(x)$, even if we can't find a simple **antiderivative.** First the meshpoints $x_1, x_2, \cdots$ divide $[a, b]$ into subintervals of length $\Delta x_k = \mathbf{x_k - x_{k-1}}$. The upper rectangle with base Δx_k has height $M_k =$ **maximum of v(x) in interval k.**
The upper sum S is equal to $\mathbf{\Delta x_1 M_1 + \Delta x_2 M_2} + \cdots$. The lower sum s is $\mathbf{\Delta x_1 m_1 + \Delta x_2 m_2} + \cdots$. The **area** is between s and S. As more meshpoints are added, S **decreases** and s **increases.** If S and s approach the same **limit,** that defines the integral. The intermediate sums S^*, named after **Riemann,** use rectangles of height $v(x_k^*)$. Here x_k^* is any point between $\mathbf{x_{k-1}}$ **and** $\mathbf{x_k}$, and $S^* = \sum \mathbf{\Delta x_k v(x_k^*)}$ approaches the area.

4 $C = -f(\sin \frac{\pi}{2}) = \mathbf{-f(1)}$ so that $\int v(u) du = f(u) + C$
6 $C = \mathbf{0}$. No constant in the derivative!

10 Set $x = 2t$ and $dx = 2dt$. Then $\int_{x=0}^{2} v(x)dx = \int_{t=0}^{1} v(2t)(2dt)$ so $C = \mathbf{2}$.

16 Choose $u = x^2$ with $du = 2x\,dx$ and $u = 0$ at $x = 0$ and $u = 1$ at $x = 1$. Then $\int_0^1 \frac{du}{2\sqrt{1-u}} = -\sqrt{1-u}]_0^1 = +\mathbf{1}$. (Could also choose $u = 1 - x^2$.)

22 Maximum of x in the four intervals is: $M_k = -\frac{1}{2}, 0, \frac{1}{2}, 1$. Minimum is $m_k = -1, -\frac{1}{2}, 0, \frac{1}{2}$. Then $S = \frac{1}{2}(-\frac{1}{2} + 0 + \frac{1}{2} + 1) = \mathbf{\frac{1}{2}}$ and $s = \frac{1}{2}(-1 - \frac{1}{2} + 0 + \frac{1}{2}) = -\mathbf{\frac{1}{2}}$.

5.6 Properties of the Integral and Average Value (page 212)

Properties 1–7 are elegantly explained by the accompanying figures (5.11 and 5.12). Focus on them.

1. Explain how property 4 applies to $\int_{-\pi/3}^{\pi/3} \sec x\, dx$. Note that $\sec(-x) = \sec x$.

2. Explain how property 4 applies to $\int_{-\sqrt{3}}^{\sqrt{3}} \tan^{-1} x\, dx$. Note that $\tan^{-1}(-x) = -\tan^{-1} x$.

- Solution to Problem 1: Since $y = \sec x$ is an ***even*** function, the area on the left of the y axis equals the area on the right. Property 4 states that $\int_{-\pi/3}^{\pi/3} \sec x\, dx = 2\int_0^{\pi/3} \sec x\, dx$.
- Solution to Problem 2: Since $y = \tan^{-1} x$ is an ***odd*** function, the area on the left is equal in absolute value to the area on the right. One area is negative while the other is positive. Property 4 says that $\int_{-\sqrt{3}}^{\sqrt{3}} \tan^{-1} x\, dx = 0$.

3. For the function $y = x^3 - 3x + 3$, compute the average value on $[-2, 2]$. Then draw a sketch like those in Figure 5.13 showing the area divided into two equal portions by a horizontal line at the average value.

- Equation 3 on page 208 gives average value as $\frac{1}{b-a}\int_a^b v(x)dx$. Since the interval is $[-2, 2]$ we have $a = -2$ and $b = 2$. The function we're averaging is $v(x) = x^3 - 3x + 3$. The average value is

$$\frac{1}{2-(-2)}\int_{-2}^{2}(x^3 - 3x + 3)dx = \frac{1}{4}[\frac{1}{4}x^4 - \frac{3}{2}x^2 + 3x]_{-2}^{2} = \frac{1}{4}(4+8) = 3.$$

- ***Quick reason*** : x^3 and $3x$ are *odd functions* (average zero). The average of 3 is 3.

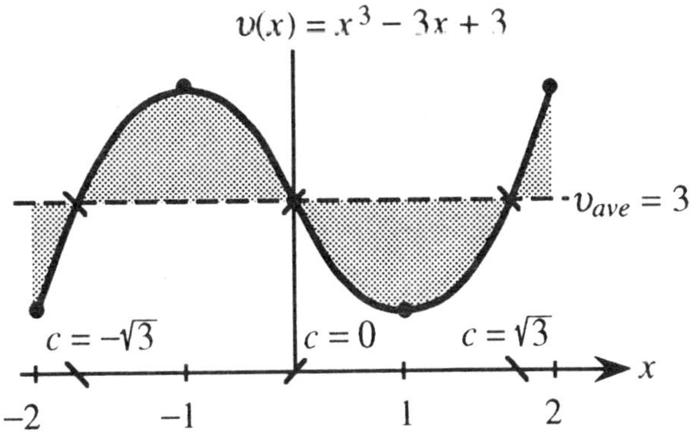

In this sketch it is easy to see that the two shaded areas above the dotted line fit into the shaded areas below the line. The area under the curve equals that of a rectangle whose height is the average value.

4. (Refer to Problem 3) The Mean Value Theorem says that the function $v(x) = x^3 - 3x + 3$ actually takes on its average value somewhere in the interval $(-2, 2)$. Find the value(s) c for which $v(c) = v_{ave}$.

   - Question 3 found $v_{ave} = 3$. Question 4 is asking where the graph crosses the dotted line. Solve $v(x) = v_{ave}$. There are three solutions (three c's):
   - $x^3 - 3x + 3 = 3$ or $x^3 - 3x = 0$ or $x(x^2 - 3) = 0$. Then $c = 0$ or $c = \pm\sqrt{3}$.

5. (This is 5.6.25) Find the average distance from $x = a$ to points in the interval $0 \le x \le 2$.

   The full solution depends on where a is. First assume $a > 2$. The distance to x is $a - x$. The *average* distance is $\frac{1}{2-0}\int_0^2 (a - x)dx = a - 1$. This is the distance to the middle point $x = 1$.

   If $a < 0$, the distance to x is $x - a$ and the average distance is $\frac{1}{2-0}\int_0^2 (x - a)dx = 1 - a$.

   Now suppose a lies in the interval $[0, 2]$. The distance $|x - a|$ is never negative. This is best considered in two sections, $x < a$ and $x > a$. The average of $(x - a)$ is

$$\frac{1}{2-0}\int_0^2 |x-a|dx = \frac{1}{2}\left(\int_0^a (a-x)dx + \int_a^2 (x-a)dx\right) = 1 - a + \frac{a^2}{2}.$$

6. A point P is chosen randomly along a semicircle (equal probability for equal arcs). What is the average distance y from the x axis? The radius is 1. (This is 5.6.27, drawn on page 212)

   - Since the problem states "equal probability for equal arcs" we imagine the semicircle divided by evenly spaced radii (like equal slices of pie). The central angle of each arc is $\Delta\theta$, and the total angle is π. The probability of a particular wedge being chosen is $\frac{\Delta\theta}{\pi}$. A point P on the unit circle has height $y = \sin\theta$.

     For a finite number of arcs, the average distance would be $\sum \sin\theta \frac{\Delta\theta}{\pi}$. As we let $\Delta\theta \to 0$, the sum becomes an integral ($\Delta\theta$ becomes $d\theta$). The average distance is $\int_0^\pi \sin\theta \frac{d\theta}{\pi} = \frac{1}{\pi}(-\cos\theta)]_0^\pi = \frac{2}{\pi}$.

Read-throughs and selected even-numbered solutions :

The integrals $\int_0^b v(x)dx$ and $\int_b^5 v(x)dx$ add to $\mathbf{\int_0^5 v(x)dx}$. The integral $\int_3^1 v(x)\,dx$ equals $-\mathbf{\int_1^3 v(x)dx}$. The reason is **that the steps Δx are negative.** If $v(x) \le x$ then $\int_0^1 v(x)\,dx \le \mathbf{\frac{1}{2}}$. The average value of $v(x)$ on the interval $1 \le x \le 9$ is defined by $\mathbf{\frac{1}{8}\int_1^9 v(x)\,dx}$. It is equal to $v(c)$ at a point $x = c$ which is **between 1 and 9.** The rectangle across the interval with height $v(c)$ has the same area as **the region under v(x)**. The average value of $v(x) = x + 1$ on the interval $1 \le x \le 9$ is **6.**

If x is chosen from 1,3,5,7 with equal probabilities $\frac{1}{4}$, its expected value (average) is **4.** The expected value of x^2 is **21.** If x is chosen from $1, 2, \cdots, 8$ with probabilities $\frac{1}{8}$, its expected value is **4.5.** If x is chosen from $1 \le x \le 9$, the chance of hitting an integer is **zero.** The chance of falling between x and $x + dx$ is $p(x)dx = \mathbf{\frac{1}{8}dx}$. The expected value $E(x)$ is the integral $\mathbf{\int_1^9 \frac{x}{8}\,dx}$. It equals **5.**

6 $v_{\text{ave}} = \frac{1}{2\pi}\int_{-\pi}^{\pi}(\sin x)^9 dx = \mathbf{0}$ (odd function over symmetric interval $-\pi$ to π). This equals $(\sin c)^9$ at $c = -\pi$ and **0** and π.

8 $2\int_1^5 x\,dx = x^2]_1^5 = \mathbf{24}$. Remember to reverse sign in the integral from 5 to 1.

20 **Property 6** proves both (a) and (b) because $v(x) \le |v(x)|$ and also $-v(x) \le |v(x)|$. So their integrals maintain these inequalities.

32 The square has area 1. The area under $y = \sqrt{x}$ is $\int_0^1 \sqrt{x}\,dx = \mathbf{\frac{2}{3}}$.

38 Average size is $\frac{G}{n}$. The chance of an individual belonging to group 1 is $\frac{x_1}{G}$. The expected size is sum of size times probability: $E(x) = \sum \frac{x_i^2}{G}$. This exceeds $\frac{G}{n}$ by the Schwarz inequality: $(1x_1 + \cdots + 1x_n)^2 \le (1^2 + \cdots + 1^2)(x_1^2 + \cdots + x_n^2)$ is the same as $G^2 \le n \sum x_i^2$.

40 This formula for $f(x)$ jumps from 9 to -9. The correct formula (with continuous f) is x^2 then $18 - x^2$. Then $f(4) - f(0) = \mathbf{2}$, which is $\int_0^4 v(x)dx$.

5.7 The Fundamental Theorem and Its Applications (page 219)

The Fundamental Theorem is almost anticlimactic at this point! It has all been built into the theory since page 1. The applications are where most of the challenge and interest lie. You must see the difference between Parts 1 and 2 – derivative of integral and then integral of derivative. In Problems 1-5, find $F'(x)$.

1. $F(x) = \int_{\pi/4}^{x} \sin 4t\,dt$. This is a direct application of the Fundamental Theorem (Part 1 on page 214). Here $v(t) = \sin 4t$. The theorem says that $F'(x) = \sin 4x$. As a check, pretend you don't believe the Fundamental Theorem. Since $-\frac{1}{4}\cos 4t$ is an antiderivative of $\sin 4t$, the integral of v is $F(x) = -\frac{1}{4}\cos 4x + \frac{1}{4}\cos \pi$. The derivative of F is $F'(x) = \sin 4x$. *The integral of $v(t)$ from a to x has derivative $v(x)$.*

2. $F(x) = \int_x^5 \frac{1}{t}dt$. Now x is the lower limit rather than the upper limit. We expect a minus sign.

   - To "fix" this write $F(x) = -\int_5^x \frac{1}{t}dt$. The Fundamental Theorem gives $F'(x) = -\frac{1}{x}$. Notice that *we do not have to know this integral* $F(x)$, in order to know that its derivative is $-\frac{1}{x}$.

3. $F(x) = \int_2^{6x} t^3 dt$. The derivative of $F(x)$ is ***not*** $(6x)^3$.

   - When the limit of integration is a function of x, we need the chain rule. Set $u = 6x$ and $F(x) = \int_2^u t^3 dt$. $\frac{dF}{dx} = \frac{dF}{du}\frac{du}{dx} = u^3 \cdot 6 = (6x)^3 \cdot 6$. *There is an extra 6 from the derivative of the limit!*

4. $F(x) = \int_{2x}^{8x} \sin^{-1} u\,du$. Both limits $a(x)$ and $b(x)$ depend on x.

   - Equation (6) on page 215 is $F'(x) = v(b(x))\frac{db}{dx} - v(a(x))\frac{da}{dx}$. The chain rule applies at both of the limits $b(x) = 8x$ and $a(x) = 2x$. Don't let it bother you that the dummy variable is u and not t. The rule gives $F'(x) = 8\sin^{-1} 8x - 2\sin^{-1} 2x$. The extra 8 and 2 are from derivatives of limits.

5. $F(x) = \frac{1}{x}\int_0^x \tan t\,dt$.

   - $F(x)$ is a product, so this calls for the product rule. (In fact $F(x)$ is the average value of $\tan x$.) Its derivative is $F'(x) = \frac{1}{x}\tan x - \frac{1}{x^2}\int_0^x \tan t\,dt$.

Problems 6 and 7 ask you to set up integrals involving regions other than vertical rectangles.

6. Find the area A bounded by the curve $x = 4 - y^2$, and the two axes.

- Think of the area being approximated by thin *horizontal* rectangles. Each rectangle has *height* Δy and length x. The last rectangle is at $y = 2$ (where the region ends).

  The area is $A \approx \sum x\ \Delta y$, where the sum is taken over the $\frac{2}{\Delta y}$ rectangles. As we let the number of rectangles become infinite, the sum becomes an integral. $A \approx \sum x\ \Delta y$ becomes $A = \int_0^2 x\ dy$. The $\approx$ has become $=$, the $\sum$ has become $\int$, and Δy has become dy. The limits of integration, 0 and 2, are the lowest and highest values of y in the area.

The antiderivative is not $\frac{1}{2}x^2$ because we have "dy" and not "dx." Before integrating we must replace x by $4 - y^2$. Then $A = \int_0^2 (4 - y^2)dy = [4y - \frac{1}{3}y^3]_0^2 = \frac{16}{3}$.

7. Find the volume V of a cone of radius 3 feet and height 6 feet.

- Think of the cone being approximated by a stack of thin disks. Each disk (a very short cylinder) has volume $\pi r^2 \Delta h$. The disk radius decreases with h to approximate the cone. The total volume of the disks is $\sum \pi r^2 \Delta h \approx V$.

  Increasing the number of disks to infinity gives $V = \int_0^6 \pi r^2 dh$. The height variable h goes from 0 to 6. Before integrating we must replace r^2 with an expression using h. By similar triangles, $\frac{r}{h} = \frac{3}{6}$ and $r = \frac{1}{2}h$. Then $V = \int_0^6 \pi(\frac{h}{2})^2 dh = \frac{\pi}{4}[\frac{1}{3}h^3]_0^6 = 18\pi$ cubic feet.

Read-throughs and selected even-numbered solutions :

The area $f(x) = \int_a^x v(t)\, dt$ is a function of **x**. By Part **1** of the Fundamental Theorem, its derivative is $\mathbf{v(x)}$. In the proof, a small change Δx produces the area of a thin **rectangle.** This area Δf is approximately $\mathbf{\Delta x}$ times $\mathbf{v(x)}$. So the derivative of $\int_a^x t^2\, dt$ is $\mathbf{x^2}$.

The integral $\int_x^b t^2\, dt$ has derivative $\mathbf{-x^2}$. The minus sign is because **x is the lower limit.** When both limits $a(x)$ and $b(x)$ depend on x, the formula for df/dx becomes $\mathbf{v(b(x))\frac{db}{dx}}$ minus $\mathbf{v(a(x))\frac{da}{dx}}$. In the example $\int_2^{3x} t\, dt$, the derivative is **9x**.

By Part **2** of the Fundamental Theorem, the integral of df/dx is $\mathbf{f(x) + C}$. In the special case when $df/dx = 0$, this says that **the integral is constant.** From this special case we conclude: If $dA/dx = dB/dx$ then $A(x) = \mathbf{B(x) + C}$. If an antiderivative of $1/x$ is $\ln x$ (whatever that is), then automatically $\int_a^b dx/x = \mathbf{\ln b - \ln a}$.

The square $0 \le x \le s, 0 \le y \le s$ has area $A = \mathbf{s^2}$. If s is increased by Δs, the extra area has the shape of **an L.** That area ΔA is approximately $\mathbf{2s\ \Delta s}$. So $dA/ds = \mathbf{2s}$.

6 $\frac{d}{dx}\int_{-x}^{x/2} v(u)du = \frac{1}{2}v(\frac{x}{2}) - (-1)v(-x) = \mathbf{\frac{1}{2}v(\frac{x}{2}) + v(-x)}$

10 $\frac{d}{dx}(\frac{1}{2}\int_x^{x+2} x^3 dx) = \mathbf{\frac{1}{2}(x+2)^3 - \frac{1}{2}x^3}$ **18** $\frac{d}{dx}\int_{a(x)}^{b(x)} 5dt = 5\frac{db}{dx} - 5\frac{da}{dx}$.

30 When the side s is increased, only *two* strips are added to the square (on the right side and top). So $dA = 2s\ ds$ which agrees with $A = s^2$.

34 $\int x\ dy = \int_0^1 \sqrt{y}\ dy = \frac{2}{3}y^{3/2}]_0^1 = \frac{2}{3}$.

38 The triangle ends at the line $x + y = 1$ or $r\cos\theta + r\sin\theta = 1$. The area is $\frac{1}{2}$, by geometry. So the area integral $\int_{\theta=0}^{\pi/2} \frac{1}{2}r^2 d\theta = \frac{1}{2}$: Substitute $r = \frac{1}{\cos\theta + \sin\theta}$.

40 Rings have area $2\pi r\ dr$, and $\int_2^3 2\pi r\ dr = \pi r^2]_2^3 = 5\pi$. Strips are difficult because they go in and out of the ring (see Figure 14.5b on page 528).

5.8 Numerical Integration (page 226)

1. Estimate $\int_0^1 \sin\frac{\pi x}{2}dx$ using each method: **L, R, M, T, S**. Take $n = 4$ intervals of length $\Delta x = \frac{1}{4}$.

   - The intervals end at $x = 0, \frac{1}{4}, \frac{1}{2}, \frac{3}{4}, 1$. The values of $y(x) = \sin\frac{\pi x}{2}$ at those endpoints are $y_0 = \sin 0 = 0, y_1 = \sin\frac{\pi}{8} = .3827, y_2 = .7071, y_3 = .9239$, and $y_4 = \sin\frac{\pi}{2} = 1$. Now for the five methods:

$$\begin{aligned} \text{Left rectangular area} \quad L_4 &= \Delta x(y_0 + \cdots + y_3) = \tfrac{1}{4}(0 + .3827 + .7071 + .9239) = .5034. \\ \text{Right rectangular area} \quad R_4 &= \tfrac{1}{4}(.3827 + .7071 + .9239 + 1) = .7534. \\ \text{Trapezoidal rule} \quad T_4 &= \tfrac{1}{2}R_4 + \tfrac{1}{2}L_4 = \tfrac{1}{4}(\tfrac{1}{2}\cdot 0 + .3827 + .7071 + 0.9239 + \tfrac{1}{2}\cdot 1) = 0.6284. \end{aligned}$$

To apply the midpoint rule, we need the values of $y = \sin\frac{\pi x}{2}$ at the midpoints $x = \frac{1}{8}, \frac{3}{8}, \frac{5}{8}$, and $\frac{7}{8}$. The first midpoint $\frac{1}{8}$ is halfway between 0 and $\frac{1}{4}$, and the calculator gives $y_{1/2} = \sin(\frac{\pi}{2}(\frac{1}{8})) = 0.1951$. The other midpoint values are $y_{3/2} = \sin(\frac{\pi}{2}(\frac{3}{8})) = 0.5556, y_{5/2} = 0.8315, y_{7/2} = 0.9808$. Now we're ready to find M_4:

$$M_4 = \Delta x(y_{1/2} + y_{3/2} + y_{5/2} + y_{7/2}) = \frac{1}{4}(0.1951 + 0.5556 + 0.8315 + 0.9808) = 0.6408.$$

Simpson's rule gives S_4 as $\frac{1}{3}T_4 + \frac{2}{3}M_4$ or directly from the $1-4-2-4-2-4-2-4-1$ pattern:

$$\begin{aligned} S_4 &= \tfrac{1}{6}\Delta x[y_0 + 4y_{1/2} + 2y_1 + 4y_{3/2} + 2y_2 + 4y_{5/2} + 2y_3 + 4y_{7/2} + y_4] \\ &= \tfrac{1}{6}(\tfrac{1}{4})[0 + 4(0.1951) + 2(0.3827) + 4(0.5556) + 2(0.7071) \\ &\quad +4(0.8315) + 2(0.9239) + 4(0.9808) + 1] = 0.6366. \end{aligned}$$

2. Find $\int_0^1 \sin\frac{\pi x}{2}dx$ by integration. Give the predicted and actual errors for each method in Problem 1.

   - The integral is $\int_0^1 \sin\frac{\pi x}{2}dx = \frac{2}{\pi} \approx 0.6366$ (accurate to 4 places). The predicted error for R_4 is $\frac{1}{2}\Delta x(y_n - y_0) = \frac{1}{2}(\frac{1}{4})(1-0) = 0.125$. The actual error is $0.7534 - 0.6366 = 0.1168$. The predicted error for L_4 is -0.125 (*just change sign!*) and the actual error is -0.1322.

     For T_4, the predicted error is $\frac{1}{12}(\Delta x)^2(y'(b) - y'(a)) = \frac{1}{12}(\frac{1}{4})^2(0 - \frac{\pi}{2}) \approx -0.0082$. The factor $\frac{\pi}{2}$ is $y' = \frac{\pi}{2}\cos\frac{\pi x}{2}$ at $x = 0$. *This prediction is excellent:* $0.6284 - 0.6366 = -0.0082$.

     The predicted error for M_4 is $-\frac{1}{24}(\Delta x)^2[y'(b) - y'(a)]$. This is half the predicted error for T_4, or 0.0041. The actual error is $0.6408 - 0.6366 = 0.0042$.

For Simpson's rule S_4, the predicted error in Figure 5.20 is $\frac{1}{2880}(\Delta x)^4(y'''(b) - y'''(a)) = \frac{1}{2880}(\frac{1}{4})^4(-\frac{\pi^3}{2^3}) = -0.000021$. That is to say, Simpson's error is predicted to be 0 to four places. The actual error to four places is $0.6366 - 0.6366 = 0.0000$.

3. (This is 5.8.6) Compute π to 6 places as $4\int_0^1 \frac{dx}{1+x^2}$ using any rule for numerical integration.

   - Check first that $4\int_0^1 \frac{dx}{1+x^2} = 4\tan^{-1}x]_0^1 = 4(\frac{\pi}{4}) = \pi$. To obtain 6-place accuracy, we want error $< 5\times 10^{-7}$. Examine the error predictions for R_n, T_n, and S_n to see which Δx will be necessary.

     The predicted error for R_n is $\frac{1}{2}\Delta x[y(b) - y(a)] = \frac{1}{2}(\Delta x)[\frac{4}{2} - \frac{4}{1}] = -\Delta x$. This is $-\frac{1}{n}$, since $\Delta x = \frac{1}{n}$ for an interval of length 1 cut into n pieces. Ignore the negative sign since the size of the error is what interests us. To obtain $\frac{1}{n} < 5\cdot 10^{-7}$ we need $n > \frac{1}{5}\cdot 10^7$. Over 2 million rectangles will be necessary!

Go up an order of accuracy to T_n. Its error is estimated by $\frac{1}{12}(\Delta x)^2(y'(1) - y'(0)) = \frac{1}{12}(\frac{1}{n})^2(-2)$. We want $\frac{1}{6n^2} < 5 \cdot 10^{-7}$ or $n^2 > \frac{1}{30} \cdot 10^7$. This means that $n > 577$ trapezoids will be necessary.

Simpson offers a better hope for reducing the work. Simpson's error is about $\frac{1}{2880}(\Delta x)^4(y'''(1) - y'''(0))$. Differentiate $\frac{4}{1+x^2}$ three times to find $y'''(1) = 12$ and $y'''(0) = 0$. Simpson's error is about $\frac{1}{240n^4}$. To make the calculation accurate to 6 places, we need $\frac{1}{240n^4} < 5 \times 10^{-7}$ or $n^4 > \frac{1}{1200} \times 10^7$ or $n > 9$. Much better than 2 million!

Read-throughs and selected even-numbered solutions :

To integrate $y(x)$, divide $[a, b]$ into n pieces of length $\Delta x = \mathbf{(b-a)/n}$. R_n and L_n place a **rectangle** over each piece, using the height at the right or **left** endpoint: $R_n = \Delta x(y_1 + \cdots + y_n)$ and $L_n = \mathbf{\Delta x(y_0 + \cdots + y_{n-1})}$. These are **first**-order methods, because they are incorrect for $y = \mathbf{x}$. The total error on $[0, 1]$ is approximately $\mathbf{\frac{\Delta x}{2}(y(1) - y(0))}$. For $y = \cos \pi x$ this leading term is $\mathbf{-\Delta x}$. For $y = \cos 2\pi x$ the error is very small because $[0, 1]$ is a complete **period.**

A much better method is $T_n = \frac{1}{2}R_n + \mathbf{\frac{1}{2}L_n} = \Delta x[\frac{1}{2}y_0 + \mathbf{1}y_1 + \cdots + \mathbf{\frac{1}{2}}y_n]$. This **trapezoidal** rule is **second**-order because the error for $y = x$ is **zero.** The error for $y = x^2$ from a to b is $\mathbf{\frac{1}{6}(\Delta x)^2(b-a)}$. The **midpoint** rule is twice as accurate, using $M_n = \Delta x[\mathbf{y_{\frac{1}{2}} + \cdots + y_{n-\frac{1}{2}}}]$.

Simpson's method is $S_n = \frac{2}{3}M_n + \mathbf{\frac{1}{3}T_n}$. It is **fourth**-order, because the powers $\mathbf{1, x, x^2, x^3}$ are integrated correctly. The coefficients of $y_0, y_{1/2}, y_1$ are $\mathbf{\frac{1}{6}, \frac{4}{6}, \frac{1}{6}}$ times Δx. Over three intervals the weights are $\Delta x/6$ times $1 - 4 - \mathbf{2} - \mathbf{4} - \mathbf{2} - \mathbf{4} - \mathbf{1}$. Gauss uses **two** points in each interval, separated by $\Delta x/\sqrt{3}$. For a method of order p the error is nearly proportional to $\mathbf{(\Delta x)^p}$.

2 The trapezoidal error has a factor $(\Delta x)^2$. It is reduced by 4 when Δx is cut in half. The error in Simpson's rule is proportional to $(\Delta x)^4$ and is reduced by 16.

8 The trapezoidal rule for $\int_0^{2\pi} \frac{dx}{3+\sin x} = \frac{\pi}{\sqrt{2}} = 2.221441$ gives $\frac{2\pi}{3} \approx 2.09$ (two intervals), $\frac{7\pi}{9} \approx 2.221$ (three intervals), $\frac{17\pi}{24} \approx 2.225$ (four intervals is worse??), and 7 digits for T_5. Curious that $M_n = T_n$ for odd n.

10 The midpoint rule is exact for 1 and x. For $y = x^2$ the integral from 0 to Δx is $\frac{1}{3}(\Delta x)^3$ and the rule gives $(\Delta x)(\frac{\Delta x}{2})^2$. This error $\frac{1}{4}(\Delta x)^3 - \frac{1}{3}(\Delta x)^3 = -\frac{1}{12}(\Delta x)^3$ does equal $-\frac{(\Delta x)^2}{24}(y'(\Delta x) - y'(0))$.

14 Correct answer $\frac{2}{3}$. $T_1 = .5, T_{10} \approx .66051, T_{100} \approx .66646$. $M_1 \approx .707, M_{10} \approx .66838, M_{100} \approx .66673$. What is the rate of decrease of the error?

18 The trapezoidal rule $T_4 = \frac{\pi}{8}(\frac{1}{2} + \cos^2 \frac{\pi}{8} + \cos^2 \frac{\pi}{4} + \cos^2 \frac{3\pi}{8} + 0)$ gives the correct answer $\frac{\pi}{4}$.

22 Any of these stopping points should give the integral as $0.886227 \cdots$ Extra correct digits depend on the computer design.

5 Chapter Review Problems

Review Problems

R1 Draw any up and down curve between $x = 0$ and $x = 3$, starting at $v(0) = -1$. Aligned below it draw the integral $f(x)$ that gives the area from 0 to x.

R2 Find the derivatives of $f(x) = \int_x^{2x} v(t)dt$ and $f(x) = \int_{1/x}^{1+x} v(t)dt$.

R3 Use sketches to illustrate these approximations to $\int_0^1 x^4\, dx$ with $\Delta x = 1$: Left rectangle, Right rectangle, Trapezoidal, Midpoint, and Simpson. Find the errors.

R4 What is the difference between a *definite* integral and an *indefinite* integral?

R5 Sketch a function $v(x)$ on $[a, b]$. Show graphically how the average value relates to $\int_a^b v(x)dx$.

R6 Show how a step function has no antiderivative but has a definite integral.

R7 Compute $\int_{-1}^{4}(5 + |x|)dx$ from a graph (and from the area of a triangle).

R8 1000 lottery tickets are sold. One person will win \$250, ten will win free \$1 tickets to the next lottery. The others lose. Is the expected value of a ticket equal to 26¢?

Drill Problems (Find an antiderivative by substitution or otherwise)

D1 $\int \cos^3 x \sin x\, dx$

D2 $\int x\sqrt{3x^2 - 7}\, dx$

D3 $\int \frac{(x+2)dx}{(x^2+4x+3)^2}$

D4 $\int (x^4 - 20x)^7(x^3 - 5)dx$

D5 $\int \frac{\cos x}{\sqrt{1-\sin x}}dx$

D6 $\int \frac{x\, dx}{x^4+1}$ (try $u = x^2$)

D7 $\int \frac{dx}{x^2\sqrt{x^4-1}}$ (try $u = x^2$)

D8 $\int \frac{x^3\, dx}{\sqrt{x^2+25}}$

D9 $\int \frac{dx}{1+9x^2}$

D10 $\int x^2 \cos(x^3)dx$

D11 $\int (x - 4)\sqrt[3]{5 - 2x}\, dx$

D12 $\int \frac{\sin^{-1} x\, dx}{\sqrt{1-x^2}}$

Questions **D13** – **D18** are definite integrals. Use the Fundamental Theorem of Calculus to check answers.

D13 $\int_0^1 x\sqrt{3x}\, dx$ Ans $\frac{2\sqrt{3}}{5}$

D14 $\int_{-\pi/2}^{\pi/2} \sec^2(\frac{x}{2})dx$ Ans 4

D15 $\int_{-1}^{1} \frac{dx}{1+x^2}$ Ans $\pi/2$

D16 $\int_0^{2\pi} \sin^3 2x\, \cos 2x\, dx$ Ans 0

D17 $\int_2^1 x\sqrt{2 - x}\, dx$ Ans $-\frac{14}{15}$

D18 $\int_0^{\pi/4} \sec^2 x\, \tan x\, dx$ Ans $\frac{1}{2}$

Use numerical methods to estimate **D19** – **D22**. You should get more accurate answers than these:

D19 $\int_1^{10} \frac{dx}{x}$ (close to 2.3)

D20 $\int_{1/2}^{2} \sin \frac{1}{t}dt$ (close to 1.1)

D21 $\int_0^{\pi/3} \sec^3 x\, dx$ (close to 2.4)

D22 $\int_0^5 (x^2 + 4x + 3)dx$ (close to 107)

D23 Find the average value of $f(x) = \sec^2 x$ on the interval $[0, \frac{\pi}{4}]$. Ans $\frac{4}{\pi}$

D24 If the average value of $f(x) = x^3 + kx - 2$ on [0,2] is 6, find k. Ans 6

CHAPTER 6 EXPONENTIALS AND LOGARITHMS

6.1 An Overview (page 234)

The laws of logarithms which are highlighted on pages 229 and 230 apply just as well to "natural logs." Thus $\ln yz = \ln y + \ln z$ and $b = e^{\ln b}$. Also important :

$$b^x = e^{x \ln b} \quad \text{and} \quad \ln x^a = a \ln x \quad \text{and} \quad \ln 1 = 0.$$

Problems 1 – 4 review the rules for logarithms. Don't use your calculator. Find the exponent or power.

1. $\log_7 \frac{1}{49}$ 2. $\log_{12} 72 + \log_{12} 2$ 3. $\log_{10} 6 \cdot \log_6 x$ 4. $\log_{0.5} 8$

- To find $\log_7 \frac{1}{49}$,ask yourself "seven to *what power* is $\frac{1}{49}$?" Since $\frac{1}{49} = 7^{-2}$, the power is $\log_7 \frac{1}{49} = -2$.
- $\log_{12} 72 + \log_{12} 2 = \log_{12}(72 \cdot 2) = \log_{12} 144$. Since the bases are the same (everything is base 12), the log of the product equals the sum of the logs. To find $\log_{12} 144$, ask $12^{\text{what power}} = 144$. The power is 2, so $\log_{12} 144 = 2$.
- Follow the change of base formula $\log_a x = (\log_a b)(\log_b x)$. Here $a = 10$ and $b = 6$. The answer is $\log_{10} x$.
- To find $\log_{0.5} 8$, ask $\frac{1}{2}^{\text{what power}} = 8$. Since $\frac{1}{2} = 2^{-1}$ and $8 = 2^3$, the power is -3. Therefore $\log_{0.5} 8 = -3$.

5. Solve $\log_x 10 = 2$. (This is Problem 6.1.6c) The unknown is the base x.

- The statement $\log_x 10 = 2$ means exactly the same as $x^2 = 10$. Therefore $x = \sqrt{10}$. We can't choose $x = -\sqrt{10}$ since bases must be positive.

6. Draw the graphs for $y = 6^x$ and $y = 5 \cdot 6^x$ on semilog paper (preferably homemade).

- The x axis is scaled normally. The y axis is scaled so that $6^0 = 1$, $6^1 = 6$, $6^2 = 36$, and $6^3 = 216$ are one unit apart. The axes cross at (0,1), not at (0,0) as on regular paper. Both graphs are straight lines. The line $y = 10 \cdot 6^x$ crosses the vertical axis when $x = 0$ and $y = 10$.

7. What are the equations of the functions represented in the right graph?

- This is base 10 semilog paper, so both lines graph functions $y = A \cdot 10^{x \log b}$, where A is the intercept on the vertical axis and $\log b$ is the slope. One graph has $A = 1$ and the slope is $\frac{1}{4}$, so $y = 10^{x/4}$. The intercept on the second graph is 300 and the slope is $\frac{2}{3}$, so $y = 300 \cdot 10^{-2x/3}$.

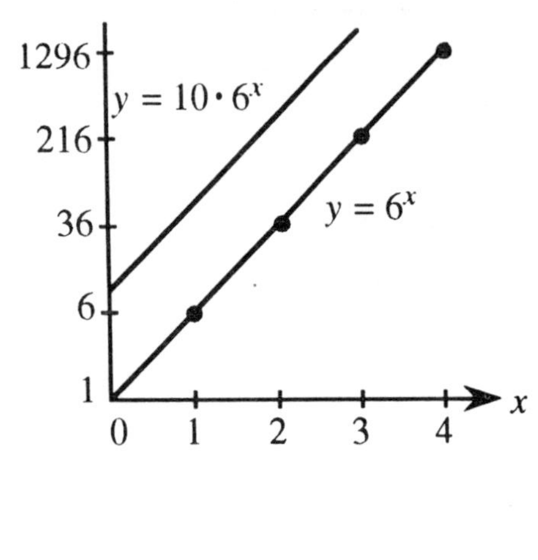

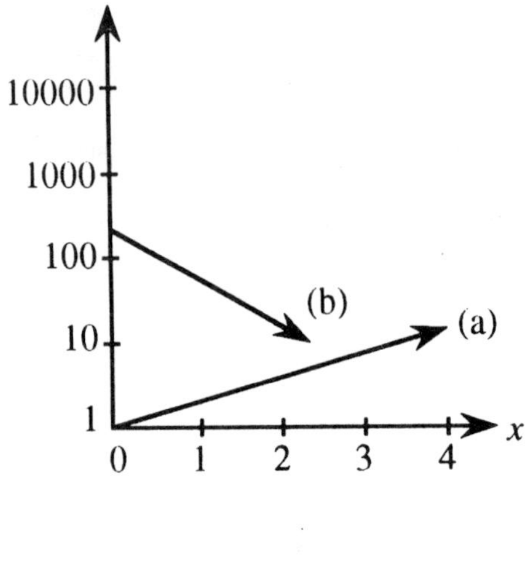

Read-throughs and selected even-numbered solutions :

In $10^4 = 10{,}000$, the exponent 4 is the **logarithm** of 10,000. The base is $b = \mathbf{10}$. The logarithm of 10^m times 10^n is $\mathbf{m+n}$. The logarithm of $10^m/10^n$ is $\mathbf{m-n}$. The logarithm of $10{,}000^x$ is $\mathbf{4x}$. If $y = b^x$ then $x = \mathbf{\log_b y}$. Here x is any number, and y is always **positive**.

A base change gives $b = \mathbf{a^{\log_a b}}$ and $b^x = \mathbf{a^{x \log_a b}}$. Then 8^5 is 2^{15}. In other words $\log_2 y$ is $\mathbf{\log_2 8}$ times $\log_8 y$. When $y = 2$ it follows that $\log_2 8$ times $\log_8 2$ equals **1**.

On ordinary paper the graph of $y = \mathbf{mx+b}$ is a straight line. Its slope is $\mathbf{m}$. On semilog paper the graph of $y = \mathbf{Ab^x}$ is a straight line. Its slope is $\mathbf{\log b}$. On log-log paper the graph of $y = \mathbf{Ax^k}$ is a straight line. Its slope is $\mathbf{k}$.

The slope of $y = b^x$ is $dy/dx = \mathbf{cb^x}$, where c depends on b. The number c is the limit as $h \to 0$ of $\mathbf{\frac{b^h-1}{h}}$. Since $x = \log_b y$ is the inverse, $(dx/dy)(dy/dx) = \mathbf{1}$. Knowing $dy/dx = cb^x$ yields $dx/dy = \mathbf{1/cb^x}$. Substituting b^x for y, the slope of $\log_b y$ is $\mathbf{1/cy}$. With a change of letters, the slope of $\log_b x$ is $\mathbf{1/cx}$.

6 (a) 7 (b) 3 (c) $\sqrt{10}$ (d) $\frac{1}{4}$ (e) $\sqrt{8}$ (f) 5

12 $y = \log_{10} x$ is a straight line on "inverse" semilog paper: y axis normal, x axis scaled logarithmically (so $x = 1, 10, 100$ are equally spaced). Any equation $y = \log_b x + C$ will have a straight line graph.

14 $y = 10^{1-x}$ drops from 10 to 1 to .1 with slope -1 on semilog paper; $y = \frac{1}{2}\sqrt{10}^x$ increases with slope $\frac{1}{2}$ from $y = \frac{1}{2}$ at $x = 0$ to $y = 5$ at $x = 2$.

16 If 440/second is the frequency of middle A, then the next A is **880/second.** The 12 steps from A to A are approximately multiples of $2^{1/12}$. So 7 steps multiplies by $2^{7/12} \approx 1.5$ to give $(1.5)\,(440) = 660$. The seventh note from A is **E.**

22 The slope of $y = 10^x$ is $\frac{dy}{dx} = c10^x$ (later we find that $c = \ln 10$). At $x = 0$ and $x = 1$ the slope is c and $10c$. So the tangent lines are $y - 1 = c(x-0)$ and $y - 10 = 10c(x-1)$.

6.2 The Exponential e^x (page 241)

Problems 1 – 8 use the facts that $\frac{d}{dx}e^u = e^u\frac{du}{dx}$ and $\frac{d}{dx}\ln u = \frac{1}{u}\cdot\frac{du}{dx}$. If the base in the problem is not e, convert to base e. Use the change of base formulas $b^u = e^{(\ln b)u}$ and $\log_b u = \frac{\ln u}{\ln b}$. (And remember that $\ln b$ is just a constant.) In each problem, find dy/dx:

1. $y = \ln 3x$ • Take $u = 3x$ to get $\frac{dy}{dx} = \frac{1}{u}\cdot\frac{du}{dx} = \frac{1}{3x}\cdot 3 = \frac{1}{x}$. This is the same derivative as for $y = \ln x$. *Why?* The answer lies in the laws of logarithms: $\ln 3x = \ln 3 + \ln x$. Since $\ln 3$ is a constant, its derivative is zero. Because $\ln 3x$ and $\ln x$ differ only by a constant, they have the same derivative.

2. $y = \ln\cos 3x$. Assume $\cos 3x$ is positive so $\ln\cos 3x$ is defined.

   • Take $u = \cos 3x$. Then $\frac{du}{dx} = -3\sin 3x$. The answer is $\frac{dy}{dx} = \frac{1}{\cos 3x}(-3\sin 3x) = -3\tan 3x$. When we find a derivative we also find an integral: $-\int 3\tan 3x = \ln\cos 3x + C$.

3. $y = \ln(\ln x^2)$.

- Take $u = \ln x^2$. Then $\frac{du}{dx} = \frac{1}{x^2} \cdot 2x = \frac{2}{x}$. This means

$$\frac{dy}{dx} = (\frac{1}{\ln x^2})(\frac{2}{x}) = \frac{2}{x \ln x^2} = \frac{1}{x \ln x}.$$

Surprise to the author: This is also the derivative of $\ln(\ln x)$. Why does $\ln(\ln x^2)$ have the same derivative?

3. $y = \log_{10} \sqrt{x^2+5}$ • First change the base from 10 to e, by dividing by $\ln 10$. Now you are differentiating $\ln u$ instead of $\log u$: $y = \frac{1}{\ln 10} \ln \sqrt{x^2+5}$. For square roots, it is worthwhile to use the law that $\ln u^{1/2} = \frac{1}{2} \ln u$. Then $\ln \sqrt{x^2+5} = \frac{1}{2} \ln(x^2+5)$. [THIS IS NOT $\frac{1}{2}(\ln x^2 + \ln 5)$] This function is now

$$y = \frac{\ln(x^2+5)}{2 \ln 10} \quad \text{and} \quad \frac{dy}{dx} = (\frac{1}{2 \ln 10})(\frac{1}{x^2+5})(2x) = \frac{1}{\ln 10} \frac{x}{x^2+5}.$$

4. $y = \ln \frac{(x^4-8)^5}{(x^6+5x)\cos x}$ • Here again the laws of logarithms allow you to make things easier. Multiplication of numbers is addition of logs. Division is subtraction. Powers of u become multiples of $\ln u$: $y = 5\ln(x^4-8) - \ln(x^6+5x) - \ln \cos x$. Now dy/dx is long but easy:

$$\frac{dy}{dx} = \frac{5}{x^4-8}(4x^3) - \frac{6x^5+5}{x^6+5x} - \frac{(-\sin x)}{\cos x} = \frac{20x^3}{x^4-8} - \frac{6x^5+5}{x^6+5x} + \tan x.$$

5. $y = e^{\tan x}$ • dy/dx is $e^u du/dx = e^{\tan x}(\sec^2 x)$.

6. $y = \sin(e^{2x})$ • Set $u = e^{2x}$. Then $du/dx = 2e^{2x}$. Using the chain rule,

$$\frac{d}{dx}(\sin u) = (\cos u)(\frac{du}{dx}) = (\cos e^{2x})(2e^{2x}).$$

7. $y = 10^{x^2}$ • First change the base from 10 to e: $y = (e^{\ln 10})^{x^2} = e^{x^2 \ln 10}$. Let $u = x^2 \ln 10$. Then $\frac{du}{dx} = 2x \ln 10$. (Remember $\ln 10$ is a constant, you don't need the product rule.) We have

$$\frac{dy}{dx} = e^u \frac{du}{dx} = e^{x^2 \ln 10}(2x \ln 10).$$

8. $y = x^{-1/x}$ (This is Problem 6.2.18)

- First change to base e: $y = (e^{\ln x})^{-1/x}$. Since the exponent is $u = -\frac{1}{x} \ln x$, we need the product rule to get $du/dx = -\frac{1}{x}(\frac{1}{x}) + \frac{1}{x^2} \ln x$. Therefore

$$\frac{dy}{dx} = e^u \frac{du}{dx} = x^{-1/x} \cdot \frac{1}{x^2}(\ln x - 1).$$

Problems 9 – 14 use the definition $e = \lim_{h \to 0}(1+h)^{1/h}$. By substituting $h = \frac{1}{n}$ this becomes $e = \lim_{n \to \infty}(1+\frac{1}{n})^n$. Evaluate these limits as $n \to \infty$:

9. $\lim(1+\frac{1}{n})^{6n}$ 10. $\lim(1+\frac{1}{6n})^{6n}$ 11. $\lim(1+\frac{1}{2n})^{3n}$ 12. $\lim(1+\frac{r}{n})^n$ (r is constant) 13. $\lim(\frac{n+8}{n})^n$

- Rewrite Problem 9 as $\lim((1+\frac{1}{n})^n)^6$. Since $(1+\frac{1}{n})^n$ goes to e the answer is $e^6 \approx 403$. The calculator shows $(1+\frac{1}{1000})^{6000} \approx 402$.
- Problem 10 is different because $6n$ is both inside and outside the parentheses. If you let $k = 6n$, and note $k \to \infty$ as $n \to \infty$, this becomes $\lim_{k \to \infty}(1+\frac{1}{k})^k = e$. The idea here is : If we have $\lim_{\square \to \infty}(1+\frac{1}{\square})^{\square}$ and all the boxes *are the same*, the limit is e.

- Question 11 can be rewritten as $\lim(1+\frac{1}{2n})^{2n\cdot\frac{3}{2}} = e^{3/2}$. (The box is $\square = 2n$).
- In Question 12 write $n = mr$. Then $\frac{r}{n} = \frac{1}{m}$ and we have $\lim(1+\frac{1}{m})^{mr} = e^r$.
- In 13 write $(\frac{n+8}{n}) = 1 + \frac{8}{n}$. This is Problem 12 with $r = 8$: $\lim_{n\to\infty}(1+\frac{8}{n})^n = e^8$.

14. (This is 6.2.21) Find the limit of $(\frac{11}{10})^{10}, (\frac{101}{100})^{100}, (\frac{1001}{1000})^{1000}, \cdots$. Then find the limit of $(\frac{10}{11})^{10}, (\frac{100}{101})^{100}, (\frac{1000}{1001})^{1000}, \cdots$ and the limit of $(\frac{10}{11})^{11}, (\frac{100}{101})^{101}, (\frac{1000}{1001})^{1001}, \cdots$.

- The terms of the first sequence are $(\frac{n+1}{n})^n = (1+\frac{1}{n})^n$ where $n = 10, 100, 1000, \cdots$. The limit is e. The terms of the second sequence are the reciprocals of those of the first. So the second limit is $\frac{1}{e}$. The terms of the third are each $(\frac{n}{n+1})$ times those of the second. Since $\frac{n}{n+1} \to 1$ as $n \to \infty$, the third limit is again $\frac{1}{e}$.

The third sequence can also be written $(\frac{n-1}{n})^n$ or $(1-\frac{1}{n})^n$. Its limit is e^{-1}. See Problem 12 with $r = -1$.

Exercises 6.2.27 and 6.2.45 – 6.2.54 give plenty of practice in integrating exponential functions. Usually the trick is to locate $e^u du$. Problems 15 – 17 are three models.

15. (This is 6.2.32) Find an antiderivative for $v(x) = \frac{1}{e^x} + \frac{1}{x^e}$.

- The first term is e^{-x}. Its antiderivative is $-e^{-x}$. The second term is just x^n with $n = -e$. Its antiderivative is $\frac{1}{1-e}x^{1-e}$. The answer is $f(x) = -e^{-x} + \frac{1}{1-e}x^{1-e} + C$.

16. Find an antiderivative for $v(x) = 3^{-2x}$. You may change to base e.

- The change produces $e^{-2x\ln 3}$. The coefficient of x in the exponent is $-2\ln 3$. An antiderivative is $f(x) = \frac{-1}{2\ln 3}e^{-2x\ln 3}$ or $\frac{-1}{2\ln 3}3^{-2x}$.
  We need -2 and $\ln 3$ in the denominator, the same way that we needed $n+1$ when integrating x^n.

17. (This is 6.2.52) Find $\int_0^3 e^{(1+x^2)}x\,dx$. Set $u = 1+x^2$ and $du = 2x\,dx$. The integral is $\frac{1}{2}\int_{u(0)}^{u(3)} e^u du$. The new limits of integration are $u(0) = 1+0^2 = 1$ and $u(3) = 1+3^2 = 10$. Now $\frac{1}{2}\int_1^{10} e^u du = \frac{1}{2}e^u]_1^{10} = \frac{1}{2}(e^{10} - e)$. *This is not the same as* $\frac{1}{2}e^9$!

Read-throughs and selected even-numbered solutions :

The number e is approximately **2.78**. It is the limit of $(1+h)$ to the power **1/h**. This gives 1.01^{100} when $h = .01$. An equivalent form is $e = \lim(\mathbf{1} + \frac{\mathbf{1}}{\mathbf{n}})^{\mathbf{n}}$.

When the base is $b = e$, the constant c in Section 6.1 is **1**. Therefore the derivative of $y = e^x$ is $dy/dx = \mathbf{e^x}$. The derivative of $x = \log_e y$ is $dx/dy = \mathbf{1/y}$. The slopes at $x = 0$ and $y = 1$ are both **1**. The notation for $\log_e y$ is **ln y**, which is the **natural** logarithm of y.

The constant c in the slope of b^x is $c =$ **ln b**. The function b^x can be rewritten as $\mathbf{e^{x \ln b}}$. Its derivative is $(\mathbf{\ln b})\mathbf{e^{x\ln b}} = (\mathbf{\ln b})\mathbf{b^x}$. The derivative of $e^{u(x)}$ is $\mathbf{e^{u(x)}\frac{du}{dx}}$. The derivative of $e^{\sin x}$ is $\mathbf{e^{\sin x}\cos x}$. The derivative of e^{cx} brings down a factor **c**.

The integral of e^x is $\mathbf{e^x + C}$. The integral of e^{cx} is $\mathbf{\frac{1}{c}e^{cx} + C}$. The integral of $e^{u(x)}du/dx$ is $\mathbf{e^{u(x)} + C}$. In general the integral of $e^{u(x)}$ by itself is **impossible** to find.

18 $x^{-1/x} = e^{-(\ln x)/x}$ has derivative $(-\frac{1}{x^2} + \frac{\ln x}{x^2})e^{-(\ln x)/x} = (\frac{\ln x - 1}{x^2})x^{-1/x}$

20 $(1+\frac{1}{n})^{2n} \to \mathbf{e^2} \approx 7.7$ and $(1+\frac{1}{n})^{\sqrt{n}} \to \mathbf{1}$. Note that $(1+\frac{1}{n})^{\sqrt{n}}$ is squeezed between 1 and $e^{1/\sqrt{n}}$ which approaches 1.

28 $(e^{3x})(e^{7x}) = e^{10x}$ which is the derivative of $\mathbf{\frac{1}{10}e^{10x}}$

42 $x^{1/x} = e^{(\ln x)/x}$ has slope $e^{(\ln x)/x}(\frac{1}{x^2} - \frac{\ln x}{x^2}) = \mathbf{x^{1/x}(\frac{1-\ln x}{x^2})}$. This slope is zero at $\mathbf{x = e}$, when $\ln x = 1$. The second derivative is *negative* so the maximum of $x^{1/x}$ is $e^{1/e}$. Check: $\frac{d}{dx}e^{(\ln x)/x}(\frac{1-\ln x}{x^2}) = e^{(\ln x)/x}[(\frac{1-\ln x}{x^2})^2 + \frac{(-2-1+2\ln x)}{x^3}] = -\frac{1}{e^3}e^{1/e}$ at $x = e$.

44 $x^e = e^x$ at $x = e$. This is the only point where $x^e e^{-x} = 1$ because the derivative is $x^e(-e^{-x}) + ex^{e-1}e^{-x} = (\frac{e}{x} - 1)x^e e^{-x}$. This derivative is positive for $x < e$ and negative for $x > e$. So the function $x^e e^{-x}$ increases to 1 at $x = e$ and then decreases: it never equals 1 again.

58 The asymptotes of $(1+\frac{1}{x})^x = (\frac{x+1}{x})^x = (\frac{x}{x+1})^{-x}$ are $x = -1$ (from the last formula) and $y = e$ (from the first formula).

62 $\lim \frac{x^6}{e^x} = \lim \frac{6x^5}{e^x} = \lim \frac{30x^4}{e^x} = \lim \frac{120x^3}{e^x} = \lim \frac{360x^2}{e^x} = \lim \frac{720x}{e^x} = \lim \frac{720}{e^x} = 0$.

6.3 Growth and Decay in Science and Economics (page 250)

The applications in this section begin to suggest the power of the mathematics you are learning. Concentrate on understanding how to *use* $y = y_0e^{ct}$ and $y = y_0e^{ct} + \frac{s}{c}(e^{ct} - 1)$ as you work the examples.

In Problems 1 and 2, solve the differential equations starting from $y_0 = 1$ and $y_0 = -1$. Draw both solutions on the same graph.

1. $\frac{dy}{dt} = \frac{1}{3}y$ (pure exponential) 2. $\frac{dy}{dt} = \frac{1}{3}y + 0.8$ (exponential with source term)

- Problem 1 says that the rate of change is proportional to y. There are no other complicating terms. Use the exponential law y_0e^{ct} with $c = \frac{1}{3}$ and $y_0 = \pm 1$. The graphs of $y = \pm e^{t/3}$ are at left below.
- Problem 2 changes Problem 1 into $\frac{dy}{dt} = cy + s$. We have $c = \frac{1}{3}$ and $s = 0.8$. Its solution is $y = y_0e^{t/3} + \frac{0.8}{\frac{1}{3}}(e^{t/3} - 1) = y_0e^{t/3} + 2.4(e^{t/3} - 1)$. Study the graphs to see the effect of y_0 and s. With a graphing calculator you can carry these studies further. See what happens if s is very large, or if s is negative. Exercise 6.3.36 is also good for comparing the effects of various c's and s's.

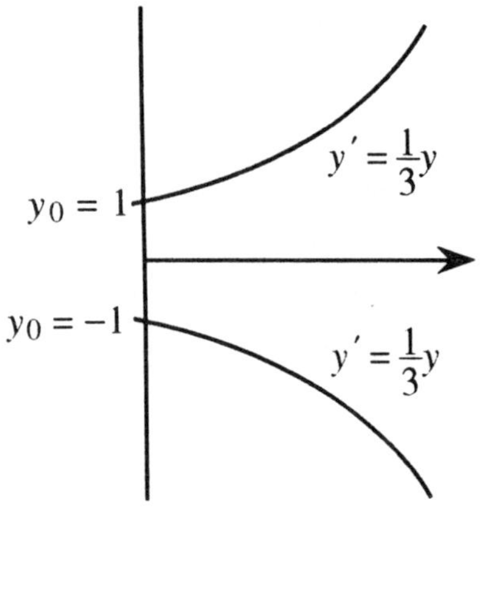

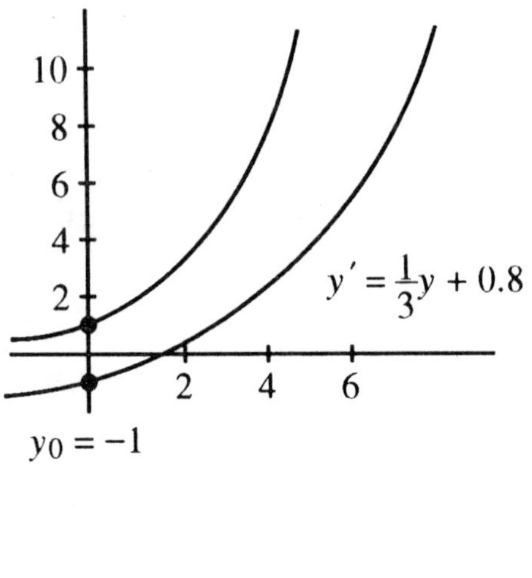

3. (This is 6.3.5) Start from $y_0 = 10$. If $\frac{dy}{dt} = 4y$, at what time does y increase to 100?

- The solution is $y = y_0 e^{ct} = 10e^{4t}$. Set $y = 100$ and solve for t:

$$100 = 10e^{4t} \text{ gives } 10 = e^{4t} \quad \text{and} \quad \ln 10 = 4t. \text{ Then } t = \frac{1}{4}\ln 10.$$

4. Problem 6.3.6 looks the same as the last question, but the right side is $4t$ instead of $4y$. Note that $\frac{dy}{dt} = 4t$ is *not* exponential growth. The slope $\frac{dy}{dt}$ is proportional to t and the solution is simply $y = 2t^2 + C$. Start at $y_0 = C = 10$. Setting $y = 100$ gives $100 = 2t^2 + 10$ and $t = \sqrt{45}$.

   Problems 5 – 10 involve $y = y_0 e^{ct}$.

5. Write the equation describing a bacterial colony growing exponentially. Start with 100 bacteria and end with 10^6 after 30 hours.

- Right away we know $y_0 = 100$ and $y = 100e^{ct}$. We don't yet know c, but at $t = 30$ we have $y = 10^6 = 100e^{30c}$. Taking logarithms of $10^4 = e^{30c}$ gives $4\ln 10 = 30\,c$ or $\frac{4\ln 10}{30} = c$. The equation is $y = 100e^{(\frac{4\ln 10}{30}t)}$. More concisely, since $e^{\ln 10} = 10$ this is $y = 100 \cdot 10^{4t/30}$.

6. The number of cases of a disease increases by 2% a year. If there were 10,000 cases in 1992, how many will there be in 1995?

- The direct approach is to multiply by 1.02 after every year. After three years $(1.02)^3 10,000 \approx 10,612$.
- We can also use $y = 10^4 e^{ct}$. The 2% increase means $c = \ln(1.02)$. After three years (1992 to 1995) we set $t = 3$: $y = 10^4 e^{3(\ln 1.02)} = 10,612$.

  This is *not the same* as $y = 10^4 e^{.02t}$. That is *continuous* growth at 2%. It is continuous compounding, and $e^{.02} = 1.0202\cdots$ is a little different from 1.02.

7. How would Problem 6 change if the number of cases *decreases* by 2%?

- A 2% decrease changes the multiplier to .98. Then $c = \ln(.98)$. In 3 years there would be 9,411 cases.

8. (This is 6.3.15) The population of Cairo grew exponentially from 5 million to 10 million in 20 years. Find the equation for Cairo's population. When was $y = 8$ million?

- Starting from $y_0 = 5$ million $= 5 \cdot 10^6$ the population is $y = 5 \cdot 10^6 e^{ct}$. The doubling time $\frac{\ln 2}{c}$ is 20 years. We deduce that $c = \frac{\ln 2}{20} = .035$ and $y = 5 \cdot 10^6 e^{.035t}$. This reaches 8 million $= 8 \cdot 10^6$ when $\frac{8}{5} = e^{.035t}$. Then $t = \frac{\ln \frac{8}{5}}{.035} \approx 13.6$ years.

9. If $y = 4500$ at $t = 4$ and $y = 90$ at $t = 10$, what was y at $t = 0$? (We are assuming exponential decay.)

- The first part says that $y = 4500e^{c(t-4)}$. The t in the basic formula is replaced by $(t - 4)$. [The "shifted" formula is $y = y_T e^{c(t-T)}$.] Note that $y = 4500$ when $t = 4$, as required. Since $y = 90$ when $t = 10$, we have $90 = 4500e^{6c}$ and $e^{6c} = \frac{90}{4500} = .02$. This means $6c = \ln .02$ and $c = \frac{1}{6}\ln .02$. Finally, set $t = 0$ to get the amount at that time: $y = 4500e^{(\frac{1}{6}\ln .02)(0-4)} \approx 61074$.

10. (Problem 6.3.13) How old is a skull containing $\frac{1}{5}$ as much radiocarbon as a modern skull?

- Information about radioactive dating is on pages 243-245. Since the half-life of carbon 14 is 5568 years, the amount left at time t is $y_0 e^{ct}$ with exponent $c = \frac{\ln 1/2}{5568} = \frac{-\ln 2}{5568}$. We do not know the initial amount y_0. But we can use $y_0 = 1$ (100% at the start) and $y = \frac{1}{5} = 0.2$ at the unknown age t. Then

$$0.2 = e^{\frac{-\ln 2}{5568}t} \text{ yields } t = \frac{(\ln 0.2)5568}{-\ln 2} = 31,425 \text{ years.}$$

11. (Problem 6.3.37) What value $y =$ constant solves $\frac{dy}{dt} = 4 - y$? Show that $y(t) = Ae^{-t} + 4$ is also a solution. Find $y(1)$ and y_∞ if $y_0 = 3$.

- If y is constant, then $\frac{dy}{dt} = 0$. Therefore $y - 4 = 0$. The steady state y_∞ is the constant $y = 4$.
- A non-constant solution is $y(t) = Ae^{-t} + 4$. Check: $\frac{dy}{dt} = -Ae^{-t}$ equals $4 - y = 4 - (Ae^{-t} + 4)$.
- If we know $y(0) = A + 4 = 3$, then $A = -1$. In this case $y(t) = -1e^{-t} + 4$ gives $y(1) = 4 - \frac{1}{e}$.
- To find y_∞, let $t \to \infty$. Then $y = -e^{-t} + 4$ goes to $y_\infty = 4$, the expected steady state.

12. (Problem 6.3.46) (a) To have \$50,000 for college tuition in 20 years, what gift y_0 should a grandparent make now? Assume $c = 10\%$. (b) What continuous deposit should a parent make during 20 years to save \$50,000? (c) If the parent saves $s =$ \$1000 per year, when does the account reach \$50,000?

- Part (a) is a question about the present value y_0, if the gift is worth \$50,000 in 20 years. The formula $y = y_0e^{ct}$ turns into $y_0 = ye^{-ct} = (50,000)e^{-0.1(20)} = \$6767$.
- Part (b) is different because there is a continuous deposit instead of one lump sum. In the formula $y_0e^{ct} + \frac{s}{c}(e^{ct} - 1)$ we know $y_0 = 0$ and $c = 10\% = 0.1$. We want to choose s so that $y = 50,000$ when $t = 20$. Therefore $50,000 = \frac{s}{0.1}(e^{(0.1)20} - 1)$. This gives $s = 782.59$. The parents should continuously deposit \$782.59 per year for 20 years.
- Part (c) asks how long it would take to accumulate \$50,000 if the deposit is $s =$ \$1000 per year.

$$50,000 = \frac{1000}{0.1}(e^{0.1t} - 1) \text{ leads to } 5 = e^{0.1t} - 1 \text{ and } t = \frac{\ln 6}{0.1} = 17.9 \text{ years.}$$

This method takes 17.9 years to accumulate the tuition. The smaller deposit $s =$ \$782.59 took 20 years.

13. (Problem 6.3.50) For how long can you withdraw \$500/year after depositing \$5000 at a continuous rate of 8%? At time t you run dry: and $y(t) = 0$.

- This situation uses both terms of the formula $y = y_0e^{ct} + \frac{s}{c}(e^{ct} - 1)$. There is an initial value $y_0 = 5000$ and a *sink* (negative source) of $s = -500$/year. With $c = .08$ we find the time t when $y = 0$:

$$\text{Multiply } 0 = 5000e^{.08t} - \frac{500}{.08}(e^{.08t} - 1) \text{ by } .08 \text{ to get } 0 = 400e^{.08t} - 500(e^{.08t} - 1).$$

Then $e^{.08t} = \frac{500}{100} = 5$ and $t = \frac{\ln 5}{.08} = 20.1$. You have 20 years of income.

14. Your Thanksgiving turkey is at 40°F when it goes into a 350° oven at 10 o'clock. At noon the meat thermometer reads 110°. When will the turkey be done (195°)?

- Newton's law of cooling applies even though the turkey is warming. Its temperature is approaching $y_\infty = 350°$ from $y_0 = 40°$. Using method 3 (page 250) we have $(y - 350) = (40 - 350)e^{ct}$. The value of c varies from turkey to turkey. To find c for your particular turkey, substitute $y = 110$ when $t = 2$:

$$110 - 350 = -310e^{2c} \Rightarrow \frac{-240}{-310} = e^{2c} \Rightarrow \ln\frac{24}{31} = 2c \Rightarrow c = \frac{1}{2}\ln\frac{24}{31} = -.128.$$

The equation for y is $350 - 310e^{-.128t}$. The turkey is done when $y = 195$:

$$195 = 350 - 310e^{-.128t} \text{ or } \frac{195-350}{-310} = e^{-.128t} \text{ or } -.128\,t = \ln\frac{-155}{-310} = \ln\frac{1}{2}.$$

This gives $t = 5.4$ hours. You can start making gravy at 3 : 24.

Read-throughs and selected even-numbered solutions :

If $y' = cy$ then $y(t) = \mathbf{y_0 e^{ct}}$. If $dy/dt = 7y$ and $y_0 = 4$ then $y(t) = \mathbf{4e^{7t}}$. This solution reaches 8 at $t = \mathbf{\frac{\ln 2}{7}}$. If the doubling time is T then $c = \mathbf{\frac{\ln 2}{T}}$. If $y' = 3y$ and $y(1) = 9$ then y_0 was $\mathbf{9e^{-3}}$. When c is negative, the solution approaches **zero** as $t \to \infty$.

The constant solution to $dy/dt = y + 6$ is $y = \mathbf{-6}$. The general solution is $y = Ae^t - 6$. If $y_0 = 4$ then $A = \mathbf{10}$. The solution of $dy/dt = cy + s$ starting from y_0 is $y = Ae^{ct} + B = \mathbf{(y_0 + \frac{s}{c})e^{ct} - \frac{s}{c}}$. The output from the source is $\mathbf{\frac{s}{c}(e^{ct} - 1)}$. An input at time T grows by the factor $\mathbf{e^{c(t-T)}}$ at time t.

At $c = 10\%$, the interest in time dt is $dy = \mathbf{.01\,y\,dt}$. This equation yields $y(t) = \mathbf{y_0 e^{.01t}}$. With a source term instead of y_0, a continuous deposit of $s = 4000$/year yields $y = \mathbf{40,000(e - 1)}$ after ten years. The deposit required to produce 10,000 in 10 years is $s = \mathbf{yc/(e^{ct} - 1) = 1000/(e - 1)}$. An income of 4000/year forever (!) comes from $y_0 = \mathbf{40,000}$. The deposit to give 4000/year for 20 years is $y_0 = \mathbf{40,000(1 - e^{-2})}$. The payment rate s to clear a loan of 10,000 in 10 years is $\mathbf{1000e/(e - 1)}$ per year.

The solution to $y' = -3y + s$ approaches $y_\infty = \mathbf{s/3}$.

12 To multiply again by 10 takes ten more hours, a total of **20 hours**. If $e^{10c} = 10$ (and $e^{20c} = 100$) then $10c = \ln 10$ and $\mathbf{c = \frac{\ln 10}{10} \approx .23}$.

16 $8e^{.01t} = 6e^{.014t}$ gives $\frac{8}{6} = e^{.004t}$ and $t = \frac{1}{.004}\ln\frac{8}{6} = 250\ln\frac{4}{3} = \mathbf{72\ years}$.

24 Go from 4 mg back down to 1 mg in T hours. Then $e^{-.01T} = \frac{1}{4}$ and $-.01T = \ln\frac{1}{4}$ and $T = \frac{\ln\frac{1}{4}}{-.01} = 139$ hours (not so realistic).

28 Given $mv = mv - v\Delta m + m\Delta v - (\Delta m)\Delta v + \Delta m(v - 7)$; cancel terms to leave $m\Delta v - (\Delta m)\Delta v = 7\Delta m$; divide by Δm and approach the limit $\mathbf{m\frac{dv}{dm} = 7}$. Then $v = 7\ln m + C$. At $t = 0$ this is $20 = 7\ln 4 + C$ so that $v = 7\ln m + 20 - 7\ln 4 = \mathbf{7\ln\frac{m}{4} + 20}$.

36 (a) $\frac{dy}{dt} = 3y + 6$ gives $y \to \infty$ (b) $\frac{dy}{dt} = -3y + 6$ gives $y \to 2$ (c) $\frac{dy}{dt} = -3y - 6$ gives $y \to -2$ (d) $\frac{dy}{dt} = 3y - 6$ gives $y \to -\infty$.

42 \$1000 changes by (\$1000) $(-.04dt)$, a decrease of $40dt$ dollars in time dt. The printing rate should be $s = 40$.

48 The deposit of $4dT$ grows with factor c from time T to time t, and reaches $e^{c(t-T)}4dT$. With $t = 2$ add deposits from $T = 0$ to $T = 1$: $\int_0^1 e^{c(2-T)}4dT = [\frac{4e^{c(2-T)}}{-c}]_0^1 = \mathbf{\frac{4e^c - 4e^{2c}}{-c}}$.

58 If $\frac{dy}{dt} = -y + 7$ then $\frac{dy}{dt}$ is zero at $y_\infty = \mathbf{7}$ (this is $-\frac{s}{c} = \frac{7}{1}$). The derivative of $y - y_\infty$ is $\frac{dy}{dt}$, so the derivative of $y - 7$ is $-(y - 7)$. The decay rate is $c = -1$, and $\mathbf{y - 7 = e^{-t}(y_0 - 7)}$.

60 All solutions to $\frac{dy}{dt} = c(y - 12)$ converge to $\mathbf{y = 12}$ provided c is **negative**.

66 (a) The white coffee cools to $y_\infty + (y_0 - y_\infty)e^{ct} = \mathbf{20 + 40e^{ct}}$. (b) The black coffee cools to $\mathbf{20 + 50e^{ct}}$. The milk warms to $\mathbf{20 - 10e^{ct}}$. The mixture $\frac{5(\text{black coffee}) + 1(\text{milk})}{6}$ has $20 + \frac{250-10}{6}e^{ct} = \mathbf{20 + 40e^{ct}}$. So it doesn't matter when you add the milk!

6.4 Logarithms (page 258)

This short section is packed with important information and techniques – how to differentiate and integrate logarithms, logarithms as areas, approximation of logarithms, and logarithmic differentiation (**LD**). The examples cover each of these topics:

Derivatives The rule for $y = \ln u$ is $\frac{dy}{dx} = \frac{1}{u}\frac{du}{dx}$. With a different base b, the rule for $y = \log_b u = \frac{\ln u}{\ln b}$ is $\frac{dy}{dx} = \frac{1}{u \ln b}\frac{du}{dx}$. Find $\frac{dy}{dx}$ in Problems 1 – 4.

1. $y = \ln(5 - x)$. • $u = 5 - x$ so $\frac{dy}{dx} = (\frac{1}{5-x})(-1) = \frac{1}{x-5}$.

2. $y = \log_{10}(\sin x)$. • Change to base e with $y = \frac{\ln(\sin x)}{\ln 10}$. Now $\frac{dy}{dx} = \frac{1}{\ln 10} \cdot \frac{1}{\sin x} \cdot \cos x$.

3. $y = (\ln x)^3$. • This is $y = u^3$, so $\frac{dy}{dx} = 3u^2\frac{du}{dx} = 3(\ln x)^2\frac{1}{x}$.

4. $y = \tan x \ln \sin x$. • The product rule gives

$$\frac{dy}{dx} = \tan x \cdot \frac{1}{\sin x} \cdot \cos x + \sec^2 x(\ln \sin x) = 1 + \sec^2 x(\ln \sin x).$$

5. (This is 6.4.53) Find $\lim_{x \to 0} \frac{\log_b(1+x)}{x}$.

   • This limit takes the form $\frac{0}{0}$, so turn to l'Hôpital's rule (Section 3.8). The derivative of $\log_b(1 + x)$ is $(\frac{1}{\ln b})(\frac{1}{1+x})$. The derivative of x is 1. The ratio is $\frac{1}{(\ln b)(1+x)}$ which approaches $\frac{1}{\ln b}$.

Logarithms as areas

6. (This is 6.4.56) Estimate the area under $y = \frac{1}{x}$ for $4 \le x \le 8$ by four trapezoids. What is the exact area?

   • Each trapezoid has base $\Delta x = 1$, so four trapezoids take us from $x = 4$ to $x = 8$. With $y = \frac{1}{x}$ the sides of the trapezoids are the heights $y_0, y_1, y_2, y_3, y_4 = \frac{1}{4}, \frac{1}{5}, \frac{1}{6}, \frac{1}{7}, \frac{1}{8}$. The total trapezoidal area is

$$\Delta x(\frac{1}{2}y_0 + y_1 + y_2 + y_3 + \frac{1}{2}y_4) = 1(\frac{1}{8} + \frac{1}{5} + \frac{1}{6} + \frac{1}{7} + \frac{1}{16}) = 0.6970.$$

   To get the exact area we integrate $\int_4^8 \frac{1}{x}dx = \ln 8 - \ln 4 = \ln \frac{8}{4} = \ln 2 \approx 0.6931$.
   It is interesting to compare with the trapezoidal area from $x = 1$ to $x = 2$. *The exact area* $\int_1^2 \frac{1}{x}dx$ *is still* $\ln 2$. Now $\Delta x = \frac{1}{4}$ and the heights are $\frac{1}{1}, \frac{1}{\frac{5}{4}}, \frac{1}{\frac{6}{4}}, \frac{1}{\frac{7}{4}}, \frac{1}{2}$. The total trapezoidal area comes from the same rule:

$$\frac{1}{4}(\frac{1}{2} \cdot \frac{1}{1} + \frac{4}{5} + \frac{4}{6} + \frac{4}{7} + \frac{1}{2} \cdot \frac{1}{2}) = (\frac{1}{8} + \frac{1}{5} + \frac{1}{6} + \frac{1}{7} + \frac{1}{16}) = 0.6970 \text{ as before.}$$

The sum is not changed! This is another way to see why $\ln 8 - \ln 4$ is equal to $\ln 2 - \ln 1$. The area stays the same when we integrate $\frac{1}{x}$ from any a to $2a$.

Questions 7 and 8 are about approximations going as far as the x^3 term:

$$\ln(1 + x) \approx x - \frac{x^2}{2} + \frac{x^3}{3} \quad \text{and} \quad e^x \approx 1 + x + \frac{x^2}{2} + \frac{x^3}{6}.$$

7. Approximate $\ln(.98)$ by choosing $x = -.02$. Then $1 + x = .98$.

- $\ln(1-.02) \approx (-.02) - \frac{(-.02)^2}{2} + \frac{(-.02)^3}{3} = -.0202026667$.
  The calculator gives $\ln .98 = -.0202027073$. Somebody is wrong by $4 \cdot 10^{-8}$.

8. Find a quadratic approximation (this means x^2 terms) near $x = 0$ for $y = 2^x$.

- 2^x is the same as $e^{x \ln 2}$. Put $x \ln 2$ into the series. The approximation is $1 + x\ln 2 + \frac{(\ln 2)^2}{2}x^2$.

Integration The basic rule is $\int \frac{du}{u} = \ln|u| + C$. Why not just $\ln u + C$? Go back to the definition of $\ln u$ = area under the curve $y = \frac{1}{x}$ from $x = 1$ to $x = u$. Here u must be positive since we cannot cross $x = 0$, where $\frac{1}{x}$ blows up. However if u stays negative, there is something we can do. Write $\int \frac{du}{u} = \int \frac{-du}{-u}$. The denominator $-u$ is positive and the numerator is its derivative! In that case, $\int \frac{-du}{-u} = \ln(-u) + C$. The expression $\int \frac{du}{u} = \ln|u| + C$ covers both cases. When you *know* u is positive, as in $\ln(x^2+1)$, leave off the absolute value sign.

For definite integrals, the limits of integration should tell you whether u is negative or positive. Here are two examples with $u = \sin x$:

$$\int_{\pi/4}^{\pi/2} \frac{\cos x}{\sin x}dx = \ln(\sin x)]_{\pi/4}^{\pi/2} \qquad \int_{\pi/2}^{3\pi/4} \frac{\cos x}{\sin x}dx = \ln|\sin x|]_{\pi/2}^{3\pi/4}.$$

The integral $\int_0^\pi \frac{\cos x}{\sin x}dx$ is illegal. It starts and ends with $u = \sin x = 0$

9. Integrate $\int \frac{x\,dx}{1-x^2}$.

- Let $1 - x^2$ equal u. Then $du = -2x\,dx$. The integral becomes $-\frac{1}{2}\int \frac{du}{u} = -\frac{1}{2}\ln|u| + C$. This is $-\frac{1}{2}\ln|1-x^2| + C$. Avoid $x = \pm 1$ where $u = 0$.

10. Integrate $\int \frac{x\,dx}{1-x}$. This is *not* $\int \frac{du}{u}$. But we can write $\frac{x}{1-x}$ as $-1 + \frac{1}{1-x}$:

- $\int \frac{x\,dx}{1-x} = \int(-1 + \frac{1}{1-x})dx = -x - \ln|1-x| + C$.

11. (This is 6.4.18) Integrate $\int_2^e \frac{dx}{x(\ln x)^2}$.

- A sneaky one, not $\frac{du}{u}$. Set $u = \ln x$ and $du = \frac{dx}{x}$:

$$\int_{\ln 2}^{1} \frac{du}{u^2} = -\frac{1}{u}\Big]_{\ln 2}^{1} = -1 + \frac{1}{\ln 2}.$$

Logarithmic differentiation (**LD**) greatly simplifies derivatives of powers and products, and quotients. To find the derivative of $x^{1/x}$, **LD** is the best way to go. (Exponential differentiation in Problem 6.5.70 amounts to the same thing.) The secret is in decomposing the original expression. Here are examples:

12. $y = \frac{(x^2+7)^3}{\sqrt{x^3-9}}(4x^8)$ leads to $\ln y = 3\ln(x^2+7) + \ln 4 + 8\ln x - \frac{1}{2}\ln(x^3-9)$.

Multiplication has become addition. Division has become subtraction. The powers 3, 8, $\frac{1}{2}$ now multiply. This is as far as logarithms can go. *Do not try to separate* $\ln x^2$ *and* $\ln 7$. Take the derivative of $\ln y$:

- $\frac{1}{y}\frac{dy}{dx} = 3\frac{2x}{x^2+7} + 0 + \frac{8}{x} - \frac{3x^2}{2(x^3-9)}$.
  If you substitute back for y then $\frac{dy}{dx} = \frac{(x^2+7)^3 \cdot 4x^8}{\sqrt{x^3-9}}[\frac{6x}{x^2+7} + \frac{8}{x} - \frac{3x^2}{2(x^3-9)}]$.

13. $y = (\sin x)^{x^2}$ has a function $\sin x$ raised to a functional power x^2. **LD** is necessary.

- First take logarithms: $\ln y = x^2 \ln \sin x$. Now take the derivative of both sides. Notice especially the left side: $\frac{1}{y}\frac{dy}{dx} = x^2\frac{\cos x}{\sin x} + 2x\ln\sin x$. Multiply by y to find $\frac{dy}{dx}$.

14. Find the tangent line $y^2(2-x) = x^3$ at the point (1,1). **ID** and **LD** are useful but not necessary.

- We need to know the slope dy/dx at (1,1). Taking logarithms gives

$$\ln y^2 + \ln(2-x) = \ln x^3 \text{ or } 2\ln y + \ln(2-x) = 3\ln x.$$

Now take the x derivative of both sides: $\frac{2}{y}\frac{dy}{dx} + \frac{-1}{2-x} = \frac{3}{x}$. Plug in $x=1, y=1$ to get $2\frac{dy}{dx} + \frac{-1}{1} = 3$ or $\frac{dy}{dx} = 2$. The tangent line through (1,1) with slope 2 is $y-1 = 2(x-1)$.

Read-throughs and selected even-numbered solutions :

The natural logarithm of x is $\int_1^x \frac{dt}{t}$ (or $\int_1^x \frac{dx}{x}$). This definition leads to $\ln xy = \mathbf{\ln x + \ln y}$ and $\ln x^n = \mathbf{n \ln x}$. Then e is the number whose logarithm (area under $1/x$ curve) is **1**. Similarly e^x is now defined as the number whose natural logarithm is **x**. As $x \to \infty, \ln x$ approaches **infinity**. But the ratio $(\ln x)/\sqrt{x}$ approaches **zero**. The domain and range of $\ln x$ are $\mathbf{0 < x < \infty, -\infty < \ln x < \infty}$.

The derivative of $\ln x$ is $\mathbf{\frac{1}{x}}$. The derivative of $\ln(1+x)$ is $\mathbf{\frac{1}{1+x}}$. The tangent approximation to $\ln(1+x)$ at $x=0$ is **x**. The quadratic approximation is $\mathbf{x - \frac{1}{2}x^2}$. The quadratic approximation to e^x is $\mathbf{1 + x + \frac{1}{2}x^2}$.

The derivative of $\ln u(x)$ by the chain rule is $\mathbf{\frac{1}{u(x)}\frac{du}{dx}}$. Thus $(\ln\cos x)' = \mathbf{-\frac{\sin x}{\cos x}} = \mathbf{-\tan x}$. An antiderivative of $\tan x$ is $\mathbf{-\ln\cos x}$. The product $p = xe^{5x}$ has $\ln p = \mathbf{5x + \ln x}$. The derivative of this equation is $\mathbf{p'/p = 5 + \frac{1}{x}}$. Multiplying by p gives $p' = \mathbf{xe^{5x}(5 + \frac{1}{x}) = 5xe^{5x} + e^{5x}}$, which is **LD** or logarithmic differentiation.

The integral of $u'(x)/u(x)$ is $\mathbf{\ln u(x)}$. The integral of $2x/(x^2+4)$ is $\mathbf{\ln(x^2+4)}$. The integral of $1/cx$ is $\mathbf{\frac{\ln x}{c}}$. The integral of $1/(ct+s)$ is $\mathbf{\frac{\ln(ct+s)}{c}}$. The integral of $1/\cos x$, after a trick, is $\mathbf{\ln(\sec x + \tan x)}$. We should write $\ln|x|$ for the antiderivative of $1/x$, since this allows $\mathbf{x < 0}$. Similarly $\int du/u$ should be written $\mathbf{\ln|u|}$.

4 $\frac{x(\frac{1}{x}) - (\ln x)}{x^2} = \mathbf{\frac{1-\ln x}{x^2}}$ **6** Use $(\log_e 10)(\log_{10} x) = \log_e x$. Then $\frac{d}{dx}(\log_{10} x) = \frac{1}{\log_e 10}\cdot\frac{1}{x} = \mathbf{\frac{1}{x \ln 10}}$.

16 $y = \frac{x^3}{x^2+1}$ equals $x - \frac{x}{x^2+1}$. Its integral is $[\frac{1}{2}x^2 - \frac{1}{2}\ln(x^2+1)]_0^2 = \mathbf{2 - \frac{1}{2}\ln 5}$.

20 $\int \frac{\sin x}{\cos x}dx = \int \frac{-du}{u} = -\ln u = -\ln(\cos x)]_0^{\pi/4} = -\ln\frac{1}{\sqrt{2}} + 0 = \mathbf{\frac{1}{2}\ln 2}$.

24 Set $u = \ln\ln x$. By the chain rule $\frac{du}{dx} = \frac{1}{\ln x}\frac{1}{x}$. Our integral is $\int \frac{du}{u} = \ln u = \mathbf{\ln(\ln(\ln x))} + C$.

28 $\ln y = \frac{1}{2}\ln(x^2+1) + \frac{1}{2}\ln(x^2-1)$. Then $\frac{1}{y}\frac{dy}{dx} = \frac{x}{x^2+1} + \frac{x}{x^2-1} = \frac{2x^3}{x^4-1}$. Then $\frac{dy}{dx} = \frac{2x^3 y}{x^4-1} = \frac{2x^3}{\sqrt{x^4-1}}$.

36 $\ln y = -\ln x$ so $\frac{1}{y}\frac{dy}{dx} = \frac{-1}{x}$ and $\frac{dy}{dx} = -\frac{e^{-\ln x}}{x}$. Alternatively we have $y = \frac{1}{x}$ and $\frac{dy}{dx} = \mathbf{-\frac{1}{x^2}}$.

40 $\frac{d}{dx}\ln x = \mathbf{\frac{1}{x}}$. Alternatively use $\frac{1}{x^2}\frac{d}{dx}(x^2) - \frac{1}{x}\frac{d}{dx}(x) = \frac{1}{x}$.

54 Use l'Hôpital's Rule: $\lim_{x\to 0}\frac{b^x \ln b}{1} = \mathbf{\ln b}$. We have redone the derivative of b^x at $x=0$.

62 $\frac{1}{x}\ln\frac{1}{x} = -\frac{\ln x}{x} \to 0$ as $x \to \infty$. This means $\mathbf{y \ln y \to 0}$ as $y = \frac{1}{x} \to 0$. (Emphasize: The factor $y \to 0$ is "stronger" than the factor $\ln y \to -\infty$.)

70 **LD:** $\ln p = x\ln x$ so $\frac{1}{p}\frac{dp}{dx} = 1 + \ln x$ and $\frac{dp}{dx} = p(1+\ln x) = x^x(1+\ln x)$. Now find the same answer by **ED:** $\frac{d}{dx}(e^{x\ln x}) = e^{x\ln x}\frac{d}{dx}(x\ln x) = x^x(1+\ln x)$.

6.5 Separable Equations Including the Logistic Equation (page 266)

Separation of variables works so well (when it works) that there is a big temptation to use it often and wildly. I asked my class to integrate the function $y(x) = \frac{d}{dx}(e^{1+x^2})$ from $x = 0$ to $x = 3$. The point of this question is that you don't have to take the derivative of e^{1+x^2}. When you integrate, that brings back the original function. So the answer is

$$\int_0^3 y(x)dx = [e^{1+x^2}]_0^3 = e^{10} - e.$$

One mistake was to write that answer as e^9. The separation of variables mistake was in $y\ dy$:

from $y = \frac{d}{dx}(e^{1+x^2})$ the class wrote $\int y\ dy = \int \frac{d}{dx}(e^{1+x^2})dx$.

You can't multiply one side by dy and the other side by dx. This mistake leads to $\frac{1}{2}y^2$ which shouldn't appear. Separation of variables starts from $\frac{dy}{dx} = u(y)v(x)$ and does *the same thing to both sides*. Divide by $u(y)$, multiply by dx, and integrate. Then $\int dy/u(y) = \int v(x)dx$. Now a y-integral equals an x-integral.
Solve the differential equations in Problems 1 and 2 by separating variables.

1. $\frac{dy}{dx} = \sqrt{xy}$ with $y_0 = 4$ (which means $y(0) = 4$.)

   - First, move dx to the right side and $\sqrt{y}$ to the left: $\frac{dy}{y^{1/2}} = x^{1/2}dx$. Second, integrate both sides: $2y^{1/2} = \frac{2}{3}x^{3/2} + C$. (This constant C combines the constants for each integral.) Third, solve for $y = (\frac{1}{3}x^{3/2} + C)^2$. Here $C/2$ became C. Half a constant is another constant. This is the general solution. Fourth, use the starting value $y_0 = 4$ to find C:

   $$4 = (\frac{1}{3}(0)^{3/2} + C)^2 \text{ yields } C = \pm 2. \text{ Then } y = (\frac{1}{3}x^{3/2} \pm 2)^2.$$

2. Solve $(x-3)t\ dt + (t^2+1)dx = 0$ with $x = 5$ when $t = 0$.

   - Divide both sides by $(x-3)(t^2+1)$ to separate t from x :

   $$\frac{t\ dt}{t^2+1} + \frac{dx}{x-3} = 0 \text{ or } -\int \frac{t\ dt}{t^2+1} = \int \frac{dx}{x-3}.$$

   Integrating gives $-\frac{1}{2}\ln(t^2+1) = \ln(x-3) + C$ or $(t^2+1)^{-1/2} = e^C(x-3)$. Since $x = 5$ when $t = 0$ we have $1 = 2e^C$. Put $e^C = \frac{1}{2}$ into the solution to find $x - 3 = 2(t^2+1)^{-1/2}$ or $x = 3 + 2(t^2+1)^{-1/2}$.
   Problems 3 – 5 deal with the logistic equation $y' = cy - by^2$.

3. (This is 6.5.15.) Solve $\frac{dz}{dt} = -z + 1$ with $z_0 = 2$. Turned upside down, what is $y = \frac{1}{z}$? Graph y and z.

   - Separation of variables gives $\frac{dz}{-z+1} = dt$ or $-\ln|-z+1| = t + C$. Put in $z = 2$ when $t = 0$ to find $C = 0$. Also notice that $-2+1$ is *negative*. The absolute value is reversing the sign. So we have

   $$-\ln(z-1) = t \quad \text{or} \quad z - 1 = e^{-t} \quad \text{or} \quad z = e^{-t} + 1.$$

   Now $y = \frac{1}{z} = \frac{1}{1+e^{-t}}$. According to Problem 6.3.15, this y solves the logistic equation $y' = y - y^2$.

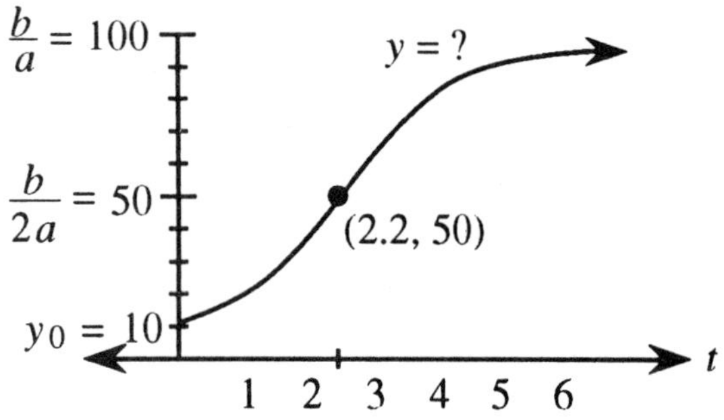

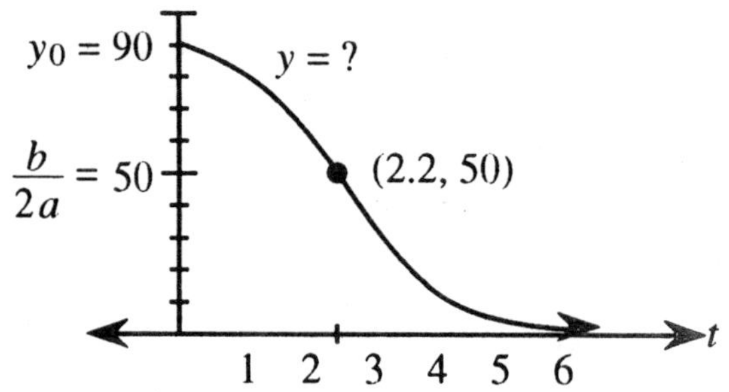

4. Each graph above is an S-curve that solves a logistic equation $y' = \pm y \pm by^2$ with $c = 1$ or $c = -1$. Each has an inflection point at (2.2,50). Find the differential equations and the solutions.

- The first graph shows $y_0 = 10$. The inflection point is at height $\frac{c}{2b} = 50$. Then $c = 1$ and $b = \frac{c}{100} = .01$. The limiting value $y_\infty = \frac{c}{b}$ is twice as high at $y_\infty = 100$. The differential equation is $dy/dt = y - .01y^2$. The solution is given by equation (12) on page 263:

$$y = \frac{c}{b + de^{-ct}} \text{ where } d = \frac{c - by_0}{y_0} = \frac{1 - (.01)(10)}{10} = .09. \text{ Then } y = \frac{1}{.01 - .09e^{-t}}.$$

The second graph must solve the differential equation $\frac{dy}{dt} = -y + by^2$. Its slope is just the opposite of the first. Again we have $\frac{c}{2b} = 50$ and $b = 0.01$. Substitute $c = -1$ and $y_0 = 90$:

$$d = \frac{c - by_0}{y_0} = \frac{-1 + .01(90)}{90} = \frac{-1}{900} \text{ and } y(t) = \frac{-1}{-.01 - \frac{e^t}{900}} = \frac{900}{9 + e^t}.$$

This is a case where *death wins*. Since $y_0 < \frac{c}{b} = 100$ the population dies out before the cooperation term $+by^2$ is strong enough to save it. See Example 6 on page 264 of the text.

5. Change y_0 in Problem 4 to 110. Then $y_0 > \frac{c}{b} = 100$. Find the solution $y(t)$ and graph it.

- As in 4(b) the equation is $\frac{dy}{dt} = -y + .01y^2$. Since y_0 is now 110, the solution has

$$d = \frac{-1 + .01(110)}{110} = \frac{1}{1100} \text{ and } y(t) = \frac{-1}{-.01 + \frac{e^t}{1100}} = \frac{1100}{11 - e^t}.$$

The graph is sketched below. After a sluggish start, the population blows up at $t = \ln 11$.

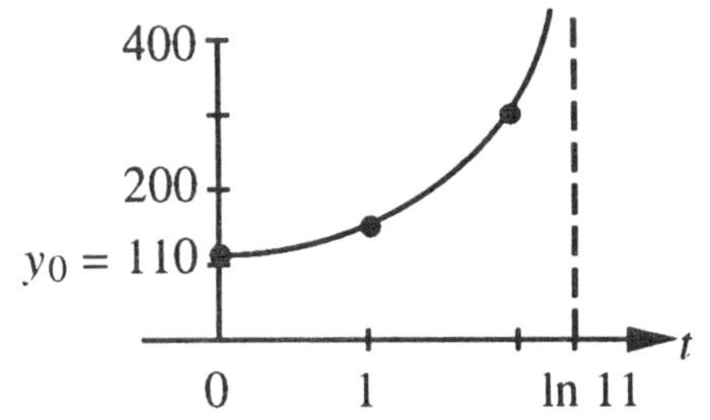

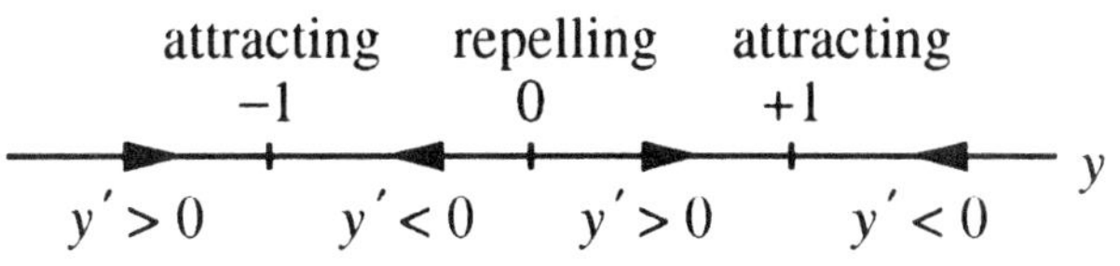

6. Draw a y–line for $y' = y - y^3$. Which steady states are approached from which initial values y_0?

- Factor $y - y^3$ to get $y' = y(1-y)(1+y)$. A steady state has $y' = 0$. This occurs at $y = 0$, 1, and -1. Plot those points on the straight line. They are not all attracting.

  Now consider the sign of $y(1-y)(1+y) = y'$. If y is below $-1, y'$ is positive. (Two factors y and $1+y$ are negative but their product is positive.) If y is between -1 and 0, y' is negative and y decreases. If y is between 0 and 1, all factors are positive and so is y'. Finally, if $y > 1$ then y' is negative.

The signs of y' are $+-+-$. The curved line $f(y)$ is sketched to show those signs. A positive y' means an increasing y. So the solution moves toward -1 and also toward $+1$. It moves away from $y = 0$, because y is increasing on the right of zero and decreasing on the left of zero.

The arrows in the y–line point to the left when y' is negative. The sketch shows that $y = -1$ and $y = +1$ are *stable steady states.* They are attracting, while $y = 0$ is an unstable (or *repelling*) stationary point. The solution approaches -1 from $y_0 < 0$, and it approaches $+1$ from $y_0 > 0$.

Read-throughs and selected even-numbered solutions :

The equations $dy/dt = cy$ and $dy/dt = cy + s$ and $dy/dt = u(y)v(t)$ are called **separable** because we can separate y from t. Integration of $\int dy/y = \int c\, dt$ gives $\mathbf{\ln y = ct + constant}$. Integration of $\int dy/(y + s/c) = \int c\, dt$ gives $\mathbf{\ln(y + \frac{s}{c}) = ct + C}$. The equation $dy/dx = -x/y$ leads to $\mathbf{\int y\, dy = -\int x\, dx}$. Then $y^2 + x^2 =$ **constant** and the solution stays on a circle.

The logistic equation is $dy/dt = \mathbf{cy - by^2}$. The new term $-by^2$ represents **competition** when cy represents growth. Separation gives $\int dy/(cy - by^2) = \int dt$, and the y-integral is $1/c$ times $\ln \mathbf{\frac{y}{c-by}}$. Substituting y_0 at $t = 0$ and taking exponentials produces $y/(c - by) = e^{ct}\mathbf{y_0/(c - by_0)}$. As $t \to \infty$, y approaches $\mathbf{\frac{c}{b}}$. That is the steady state where $cy - by^2 = \mathbf{0}$. The graph of y looks like an **S**, because it has an inflection point at $\mathbf{\frac{1}{2}\frac{c}{b}}$.

In biology and chemistry, concentrations y and z react at a rate proportional to y times **z**. This is the Law of **Mass Action.** In a model equation $dy/dt = c(y)y$, the rate c depends on **y**. The MM equation is

$dy/dt = -\mathbf{cy}/(\mathbf{y}+\mathbf{K})$. Separating variables yields $\int \frac{\mathbf{y+K}}{\mathbf{y}}dy = \int -\mathbf{c}\,\mathbf{dt} = -ct + C$.

6 $\frac{dy}{\tan y} = \cos x dx$ gives $\ln(\sin y) = \sin x + C$. Then $C = \ln(\sin 1)$ at $x = 0$. After taking exponentials $\mathbf{\sin y = (\sin 1)e^{\sin x}}$. No solution after $\sin y$ reaches 1 (at the point where $(\sin 1)e^{\sin x} = 1$).

8 $e^y dy = e^t dt$ so $e^y = e^t + C$. Then $C = e^e - 1$ at $t = 0$. After taking logarithms $y = \mathbf{\ln(e^t + e^e - 1)}$.

10 $\frac{d(\ln y)}{d(\ln x)} = \frac{dy/y}{dx/x} = n$. Therefore $\ln y = n \ln x + C$. Therefore $y = (x^n)(e^C) =$ **constant times $\mathbf{x^n}$**.

16 Equation (14) is $z = \frac{1}{c}(b + \frac{c-by_0}{y_0}e^{-ct})$. Turned upside down this is $y = \frac{c}{b+de^{-ct}}$ with $d = \frac{c-by_0}{y_0}$.

20 $y' = y + y^2$ has $c = 1$ and $b = -1$ with $y_0 = 1$. Then $y(t) = \mathbf{\frac{1}{-1+2e^{-t}}}$ by formula (12). The denominator is zero and y blows up when $2e^{-t} = 1$ or $\mathbf{t = \ln 2}$.

26 At the middle of the S-curve $y = \frac{c}{2b}$ and $\frac{dy}{dt} = c(\frac{c}{2b}) - b(\frac{c}{2b})^2 = \frac{c^2}{4b}$. If b and c are multiplied by 10 then so is this slope $\frac{c^2}{4b}$, which becomes **steeper.**

28 If $\frac{cy}{y+K} = d$ then $cy = dy + dK$ and $\mathbf{y = \frac{dK}{c-d}}$. At this steady state the maintenance dose replaces the aspirin being eliminated.

30 The rate $R = \frac{cy}{y+K}$ is a **decreasing** function of K because $\frac{dR}{dK} = \frac{-cy}{(y+K)^2}$.

34 $\frac{d[A]}{dt} = -r[A][B] = -r[A](b_0 - \frac{n}{m}(a_0 - [A]))$. The changes $a_0 - [A]$ and $b_0 - [B]$ are in the proportion m to n; we solved for $[B]$.

6.6 Powers Instead of Exponentials (page 276)

1. Write down a power series for $y(x)$ whose derivative is $\frac{1}{2}y(x)$. Assume that $y(0) = 1$.

- *First method*: Look for $y = a_0 + a_1x + a_2x^2 + \cdots + a_nx^n + \cdots$, and choose the a's so that $y' = \frac{1}{2}y$. Start with $a_0 = 1$ so that $y(0) = 1$. Then take the derivative of each term:

$$y' = 0 + a_1 + 2a_2x + 3a_3x^2 + 4a_4x^3 + \cdots + na_nx^{n-1} + \cdots$$

Matching this series with $\frac{1}{2}y$ gives $a_1 = \frac{1}{2}a_0$ and $2a_2 = \frac{1}{2}a_1$. Therefore $a_1 = \frac{1}{2}$ and $a_2 = \frac{1}{8}$. Similarly $3a_3$ matches $\frac{1}{2}a_2$ and na_n matches $\frac{1}{2}a_{n-1}$. The pattern continues with $a_3 = \frac{1}{3}\cdot\frac{1}{2}\cdot a_2$ and $a_4 = \frac{1}{4}\cdot\frac{1}{2}\cdot a_3$. The typical term is $a_n = \frac{1}{n!2^n}$:

$$\text{The series is } y(x) = 1 + \frac{x}{1\cdot 2} + \frac{x^2}{2\cdot 2^2} + \frac{x^3}{3!2^3} + \cdots + \frac{x^n}{n!2^n} + \cdots.$$

- *Second method*: We already know the solution to $y' = \frac{1}{2}y$. It is $y_0e^{\frac{1}{2}x}$. Starting from $y_0 = 1$, the solution is $y = e^{\frac{1}{2}x}$. We also know the exponential series $e^x = 1 + x + \frac{x^2}{2!} + \frac{x^3}{3!} + \cdots + \frac{x^n}{n!} + \cdots$. So just substitute the new exponent $\frac{1}{2}x$ in place of x:

$$y = e^{\frac{1}{2}x} = 1 + \frac{1}{2}x + \frac{1}{2!}(\frac{x}{2})^2 + \frac{1}{3!}(\frac{x}{2})^3 + \cdots + \frac{1}{n!}(\frac{x}{2})^n + \cdots = \text{same answer.}$$

2. (This is 6.6.19) Solve the difference equation $y(t+1) = 3y(t) + 1$ with $y_0 = 0$.

- Follow equations 8 and 9 on page 271. In this problem $a = 3$ and $s = 1$. Each step multiplies the previous y by 3 and adds 1. From $y_0 = 0$ we have $y_1 = 1$ and $y_2 = 4$. Then $y_3 = 13$ and $y_4 = 40$. The solution is

$$y(t) = 3^t \cdot 0 + 1\frac{(3^t - 1)}{3 - 1} \quad \text{or} \quad y(t) = \frac{3^t - 1}{2}.$$

3. If prices rose $\frac{3}{10}\%$ in the last month, what is the equivalent annual rate of inflation?

- *The answer is not* 12 times $\frac{3}{10} = 3.6\%$. The monthly increases are ***compounded***. A \$1 price at the beginning of the year would be $(1 + .003)^{12} \approx 1.0366$ at the end of the year. The annual rate of inflation is .0366 or 3.66%.

4. If inflation stays at 4% a year, find the present value that yields a dollar after 10 years.

- Use equation 2 on page 273 with $n = 1$ and $y = 1$. The rate is .04 instead of .05, for 10 years instead of 20. We get $y_0 = (1 + \frac{.04}{1})^{-10}1 = 0.6755$. In a decade a dollar will be worth what 67.55 cents is worth today.

5. Write the difference equation and find the steady state for this situation: Every week 80% of the cereal is sold and 400 more boxes are delivered to the supermarket.

- If $C(t)$ represents the number of cereal boxes after t weeks, the problem states that $C(t + 1) = 0.2C(t) + 400$. The reason for 0.2 is that 80% are sold and 20% are left. The difference equation has $a = 0.2$ and $s = 400$. Since $|a| < 1$, a steady state is approached: $C_\infty = \frac{s}{1-a} = \frac{400}{.8} = 500$. At that steady state, 80% of 500 boxes are sold (that means 400) and they are replaced by 400 new boxes.

Read-throughs and selected even-numbered solutions :

The infinite series for e^x is $\mathbf{1 + x + \frac{1}{2}x^2 + \frac{1}{6}x^3} + \cdots$. Its derivative is $\mathbf{e^x}$. The denominator $n!$ is called **"n factorial"** and is equal to $\mathbf{n(n-1)} \cdots \mathbf{(1)}$. At $x = 1$ the series for e is $\mathbf{1 + 1 + \frac{1}{2} + \frac{1}{6}} + \cdots$.

To match the original definition of e, multiply out $(1 + 1/n)^n = \mathbf{1 + n(\frac{1}{n}) + \frac{n(n-1)}{2}(\frac{1}{n})^2}$ (first three terms). As $n \to \infty$ those terms approach $\mathbf{1 + 1 + \frac{1}{2}}$ in agreement with e. The first three terms of $(1 + x/n)^n$ are $\mathbf{1 + n(\frac{x}{n}) + \frac{n(n-1)}{2}(\frac{x}{n})^2}$. As $n \to \infty$ they approach $\mathbf{1 + x + \frac{1}{2}x^2}$ in agreement with e^x. Thus $(1 + x/n)^n$ approaches $\mathbf{e^x}$. A quicker method computes $\ln(1 + x/n)^n \approx \mathbf{x}$ (first term only) and takes the exponential.

Compound interest (n times in one year at annual rate x) multiplies by $\mathbf{(1 + \frac{x}{n})^n}$. As $n \to \infty$, continuous compounding multiplies by $\mathbf{e^x}$. At $x = 10\%$ with continuous compounding, \$1 grows to $\mathbf{e^{.1}} \approx$ **\$1.105** in a year.

The difference equation $y(t+1) = ay(t)$ yields $y(t) = \mathbf{a^t}$ times y_0. The equation $y(t+1) = ay(t) + s$ is solved by $y = a^t y_0 + s[1 + a + \cdots + a^{t-1}]$. The sum in brackets is $\mathbf{\frac{1-a^t}{1-a}}$ or $\mathbf{\frac{a^t-1}{a-1}}$. When $a = 1.08$ and $y_0 = 0$, annual deposits of $s = 1$ produce $y = \mathbf{\frac{1.08^t-1}{.08}}$ after t years. If $a = \frac{1}{2}$ and $y_0 = 0$, annual deposits of $s = 6$ leave $\mathbf{12(1 - \frac{1}{2^t})}$ after t years, approaching $y_\infty = \mathbf{12}$. The steady equation $y_\infty = ay_\infty + s$ gives $y_\infty = \mathbf{s/(1-a)}$.

When i = interest rate per period, the value of $y_0 = \$1$ after N periods is $y(N) = \mathbf{(1+i)^N}$. The deposit to produce $y(N) = 1$ is $y_0 = \mathbf{(1+i)^{-N}}$. The value of $s = \$1$ deposited after each period grows to $y(N) =$

$\frac{1}{i}((1+i)^N - 1)$. The deposit to reach $y(N) = 1$ is $s = \frac{1}{i}(1 - (1+i)^{-N})$.

Euler's method replaces $y' = cy$ by $\Delta y = cy\Delta t$. Each step multiplies y by $1 + c\Delta t$. Therefore y at $t = 1$ is $(1 + c\Delta t)^{1/\Delta t} y_0$, which converges to $y_0 e^c$ as $\Delta t \to 0$. The error is proportional to Δt, which is too **large** for scientific computing.

4 A larger series is $1 + 1 + \frac{1}{2} + \frac{1}{4} + \frac{1}{8} + \cdots = 3$. This is greater than $1 + 1 + \frac{1}{2} + \frac{1}{6} + \cdots = e$.
8 The exact sum is $e^{-1} \approx .37$ (Problem 6). After five terms $1 - 1 + \frac{1}{2} - \frac{1}{6} + \frac{1}{24} = \frac{9}{24} = .375$.
14 $y(0) = 0, y(1) = 1, y(2) = 3, y(3) = 7$ (and $y(n) = 2^n - 1$). **24** Ask for $\frac{1}{2}y(0) - 6 = y(0)$. Then $y(0) = -12$.
30 The equation $-dP(t+1) + b = cP(t)$ becomes $-2P(t+1) + 8 = P(t)$ or $P(t+1) = -\frac{1}{2}P(t) + 4$. Starting from $P(0) = 0$ the solution is $P(t) = 4[\frac{(-\frac{1}{2})^t - 1}{-\frac{1}{2} - 1}] = \frac{8}{3}(1 - (-\frac{1}{2})^t) \to \frac{8}{3}$.
38 Solve $\$1000 = \$8000\,[\frac{.1}{1-(1.1)^{-n}}]$ for n. Then $1 - (1.1)^{-n} = .8$ or $(1.1)^{-n} = .2$. Thus $1.1^n = 5$ and $n = \frac{\ln 5}{\ln 1.1} \approx$ **17 years**.
40 The interest is $(.05)1000 = \$50$ in the first month. You pay \$60. So your debt is now \$1000 − \$10 = **\$990**. Suppose you owe $y(t)$ after month t, so $y(0) = \$1000$. The next month's interest is $.05y(t)$. You pay \$60. So $y(t+1) = 1.05y(t) - 60$. After 12 months $y(12) = (1.05)^{12}1000 - 60[\frac{(1.05)^{12} - 1}{1.05 - 1}]$. This is also $\frac{60}{.05} + (1000 - \frac{60}{.05})(1.05)^{12} \approx \$841$.
44 Use the loan formula with $.09/n$ not $.09n$: payments $s = 80{,}000\frac{.09/12}{[1-(1+\frac{.09}{12})^{-360}]} \approx \$643.70$. Then 360 payments equal \$231,732.

6.7 Hyperbolic Functions (page 280)

1. Given $\sinh x = \frac{5}{12}$, find the values of $\cosh x, \tanh x, \coth x, \operatorname{sech} x$ and $\operatorname{csch} x$.

- Use the identities on page 278. The one to remember is similar to $\cos^2 x + \sin^2 x = 1$:

$$\cosh^2 x - \sinh^2 x = 1 \text{ gives } \cosh^2 x = 1 + \frac{25}{144} = \frac{169}{144} \text{ and } \cosh x = \frac{13}{12}.$$

Note that $\cosh x$ is always positive. Then $\tanh x = \frac{\sinh x}{\cosh x}$ is $\frac{5/12}{13/12} = \frac{5}{13}$. The others are upside down:

$$\coth x = \frac{1}{\tanh x} = \frac{13}{5} \text{ and } \operatorname{sech} x = \frac{1}{\cosh x} = \frac{12}{13} \text{ and } \operatorname{csch} x = \frac{1}{\sinh x} = \frac{12}{5}.$$

2. Find $\cosh(2\ln 10)$. Substitute $x = 2\ln 10 = \ln 100$ into the definition of $\cosh x$:

- $\cosh(2\ln 10) = \frac{e^{\ln 100} + e^{-\ln 100}}{2} = \frac{100 + \frac{1}{100}}{2} = \frac{100.01}{2} = 50.005.$

3. Find $\frac{dy}{dx}$ when $y = \sinh(4x^3)$. Use the chain rule with $u = 4x^3$ and $\frac{du}{dx} = 12x^2$

- The derivative of $\sinh u(x)$ is $(\cosh u)\frac{du}{dx} = 12x^2\cosh(4x^3)$.

4. Find $\frac{dy}{dx}$ when $y = \ln\tanh 2x$. • Let $y = \ln u$, where $u = \tanh 2x$. Then

$$\frac{dy}{dx} = \frac{1}{u}\frac{du}{dx} = \frac{2\operatorname{sech}^2 2x}{\tanh 2x} = \frac{2}{\sinh 2x \cosh 2x}.$$

5. Find $\frac{dy}{dx}$ when $y = \text{sech}^{-1}6x$. • See equation (3) on page 279. If $u = 6x$ then

$$\frac{dy}{dx} = \frac{-1}{u\sqrt{1-u^2}}\frac{du}{dx} = \frac{-6}{6x\sqrt{1-36x^2}} = \frac{-1}{x\sqrt{1-36x^2}}.$$

6. Find $\int \frac{dx}{\sqrt{x^2+9}}$. • Except for the 9, this looks like $\int \frac{dx}{\sqrt{x^2+1}} = \sinh^{-1} x + C$ on page 279. Factoring out $\sqrt{9}$ leaves $\sqrt{x^2+9} = \sqrt{9}\sqrt{\frac{x^2}{9}+1}$. So the problem has $u = \frac{x}{3}$ and $du = \frac{1}{3}dx$:

$$\int \frac{dx}{3\sqrt{(\frac{x}{3})^2+1}} = \int \frac{du}{\sqrt{u^2+1}} = \sinh^{-1} u + C = \sinh^{-1}(\frac{x}{3} + C).$$

7. Find $\int \cosh^2 x \sinh x\, dx$ (This is 6.7.53.) Remember that $u = \cosh x$ has $\frac{du}{dx} = +\sinh x$:

- The problem is really $\int u^2 du$ with $u = \cosh x$. The answer is $\frac{1}{3}u^3 + C = \frac{1}{3}\cosh^3 x + C$.

8. Find $\int \frac{\sinh x}{1+\cosh x}dx$. (This is 6.7.29.) The top is the derivative of the bottom!

- $\int \frac{du}{u} = \ln|u| + C = \ln(1+\cosh x) + C$.

The absolute value sign is dropped because $1 + \cosh x$ is always positive.

9. (This is Problem 6.7.54) A falling body with friction equal to velocity squared obeys $\frac{dv}{dt} = g - v^2$. (a) Show that $v(t) = \sqrt{g}\tanh\sqrt{g}t$ satisfies the equation. (b) Derive this yourself by integrating $\frac{dv}{g-v^2} = dt$. (c) Integrate $v(t)$ to find the distance $f(t)$.

- (a) The derivative of $\tanh x$ is $\text{sech}^2 x$. The derivative of $v(t) = \sqrt{g}\tanh\sqrt{g}\,t$ has $u = \sqrt{g}\,t$. The chain rule gives $\frac{dv}{dt} = \sqrt{g}(\text{sech}^2 u)\frac{du}{dt} = g\,\text{sech}^2\sqrt{g}t$. Now use the identity $\text{sech}^2 u = 1 - \tanh^2 u$:

$$\frac{dv}{dt} = g(1 - \tanh^2\sqrt{g}t) = g - v^2.$$

- (b) The differential equation is $\frac{dv}{dt} = g - v^2$. Separate variables to find $\frac{dv}{g-v^2} = dt$:

$$\int \frac{dv}{g-v^2} = \int \frac{dv}{g[1-(\frac{v}{\sqrt{g}})^2]} = \frac{1}{\sqrt{g}}\tanh^{-1}\frac{v}{\sqrt{g}} \text{ by equation (2), on page 279.}$$

The integral of dt is $t + C$. Assuming the body falls from rest ($v = 0$ at $t = 0$), we have $C = 0$. Then $t = \frac{1}{\sqrt{g}}\tanh^{-1}\frac{v}{\sqrt{g}}$ turns into $v = \sqrt{g}\tanh\sqrt{g}t$.

- (c) $\int v\,dt = \int \sqrt{g}\tanh\sqrt{g}t\,dt = \ln\cosh\sqrt{g}t + C$.

Read-throughs and selected even-numbered solutions :

Cosh $x = \mathbf{\frac{1}{2}(e^x + e^{-x})}$ and sinh $x = \mathbf{\frac{1}{2}(e^x - e^{-x})}$ and $\cosh^2 x - \sinh^2 x = \mathbf{1}$. Their derivatives are **sinh x** and **cosh x** and **zero**. The point $(x, y) = (\cosh t, \sinh t)$ travels on the hyperbola $\mathbf{x^2 - y^2 = 1}$. A cable hangs in the shape of a catenary $y = \mathbf{a\cosh\frac{x}{a}}$.

The inverse functions $\sinh^{-1} x$ and $\tanh^{-1} x$ are equal to $\ln[x+\sqrt{x^2+1}]$ and $\frac{1}{2}\ln\frac{\mathbf{1+x}}{\mathbf{1-x}}$. Their derivatives are $\mathbf{1/\sqrt{x^2+1}}$ and $\frac{1}{1-x^2}$. So we have two ways to write the anti**derivative**. The parallel to $\cosh x + \sinh x = e^x$ is Euler's formula $\mathbf{\cos x + i \sin x = e^{ix}}$. The formula $\cos x = \frac{1}{2}(e^{ix}+e^{-ix})$ involves **imaginary** exponents. The parallel formula for $\sin x$ is $\mathbf{\frac{1}{2i}(e^{ix} - e^{-ix})}$.

12 $\sinh(\ln x) = \frac{1}{2}(e^{\ln x} - e^{-\ln x}) = \frac{1}{2}(x - \frac{1}{x})$ with derivative $\mathbf{\frac{1}{2}(1 + \frac{1}{x^2})}$.

16 $\frac{1+\tanh x}{1-\tanh x} = e^{2x}$ by the equation following (4). Its derivative is $\mathbf{2e^{2x}}$. More directly the quotient rule gives $\frac{(1-\tanh x)\text{sech}^2 x + (1+\tanh x)\text{sech}^2 x}{(1-\tanh x)^2} = \frac{2\,\text{sech}^2 x}{(1-\tanh x)^2} = \frac{2}{(\cosh x - \sinh x)^2} = \frac{2}{e^{-2x}} = \mathbf{2e^{2x}}$.

18 $\frac{d}{dx}\ln u = \frac{du/dx}{u} = \frac{\text{sech}\, x \tanh x - \text{sech}^2 x}{\text{sech}\, x + \tanh x}$. Because of the minus sign we do not get $\text{sech}\, x$. The integral of $\text{sech}\, x$ is $\sin^{-1}(\tanh x) + C$.

30 $\int \coth x\, dx = \int \frac{\cosh x}{\sinh x} dx = \mathbf{\ln(\sinh x)} + C$. **32** $\sinh x + \cosh x = e^x$ and $\int e^{nx} dx = \mathbf{\frac{1}{n}e^{nx}} + C$.

36 $y = \text{sech} x$ looks like a bell-shaped curve with $y_{\max} = 1$ at $x = 0$. The x axis is the asymptote. But note that y decays like $2e^{-x}$ and not like e^{-x^2}.

40 $\frac{1}{2}\ln(\frac{1+x}{1-x})$ approaches $+\infty$ as $x \to 1$ and $-\infty$ as $x \to -1$. The function is *odd* (so is the tanh function). The graph is an **S** curve rotated by 90°.

44 The x derivative of $x = \sinh y$ is $1 = \cosh y \frac{dy}{dx}$. Then $\frac{dy}{dx} = \frac{1}{\cosh y} = \frac{1}{\sqrt{1+\sinh^2 y}} = \mathbf{\frac{1}{\sqrt{1+x^2}}}$ = slope of $\sinh^{-1} x$.

50 Not hyperbolic! Just $\int (x^2+1)^{-1/2} x\, dx = \mathbf{(x^2+1)^{1/2}} + C$.

58 $\cos ix = \frac{1}{2}(e^{i(ix)} + e^{-i(ix)}) = \frac{1}{2}(e^{-x} + e^x) = \cosh x$. Then $\cos i = \cosh 1 = \mathbf{\frac{e+e^{-1}}{2}}$ (real!).

6 Chapter Review Problems

Graph Problems (Sketch the graphs and locate maxima, minima, and inflection points)

G1 $y = x \ln x$

G2 $y = e^{-x^2}$

G3 $y = e^{-x^3}$

G4 $y = x^2 - 72\ln x$

G5 $y = x^6 e^{-x}$

G6 $y = e^{\ln x}$ (watch the domain)

G7 Sketch $\ln 3$ as an area under a curve. Approximate the area using four trapezoids.

G8 Sketch $y = \ln x$ and $y = \ln\frac{1}{x}$. Also sketch $y = e^x$ and $y = e^{-1/x}$.

G9 Sketch $y = 2 + e^x$ and $y = e^{x+2}$ and $y = 2e^x$ on the same axes.

Review Problems

R1 Give an example of a linear differential equation and a nonlinear differential equation. If possible find their solutions starting from $y(0) = A$.

R2 Give examples of differential equations that can and cannot be solved by separation of variables.

R3 In exponential growth, the rate of change of y is directly proportional to ______. In exponential decay, dy/dt is proportional to ______. The difference is that ______.

R4 What is a steady state? Give an example for $\frac{dy}{dt} = y + 3$.

R5 Show from the definition that $d(\cosh x) = \sinh x\, dx$ and $d(\text{sech}\ x) = \text{sech}\ x \tanh x\, dx$.

R6 A particle moves along the curve $y = \cosh x$ with $dx/dt = 2$. Find dy/dt when $x = 1$.

R7 A chemical is decomposing with a half-life of 3 hours. Starting with 120 grams how much remains after 3 hours and how much after 9 hours?

R8 A radioactive substance decays with a half-life of 10 hours. Starting with 100 grams, show that the average during the first 10 hours is $100/\ln 2$ grams.

R9 How much money must be deposited now at 6% interest (compounded continuously) to build a nest egg of $40,000 in 15 years?

R10 Show that a continuous deposit of $1645 per year at 6% interest yields more than $40,000 after 15 years.

Drill Problems (Find dy/dx in **D1** to **D 12**.)

D1 $y = e^{\cos x}$

D2 $e^y + e^{-y} = 2x$

D3 $\sin x = e^y$

D4 $y = \pi^x + \pi^{-x}$

D5 $y = \frac{e^x}{x}$

D6 $y = \sec e^x$

D7 $y = \ln \frac{x-2}{x+2}$

D8 $y^2 = \ln(x^2 + y^2)$

D9 $y = \frac{\sqrt{x^2+5}(2x-3)^2}{\sqrt[3]{x^4(x+1)}}$ (use **LD**)

D10 $y = x^{\cos x}$

D11 $y = \ln(\tanh x^2)$

D12 $y = \cosh x \sinh x$

Find the integral in **D13** to **D20**.

D13 $\int 5^x \, dx$

D14 $\int x \, e^{x^2+1} \, dx$

D15 $\int \frac{e^{\sqrt{x}}}{\sqrt{x}} \, dx$

D16 $\int \frac{e^x}{5+e^x} \, dx$

D17 $\int \frac{\cos x}{4+\sin x} \, dx$

D18 $\int \sinh x \cosh x \, dx$

D19 $\int \tanh^2 x \, \text{sech}^2 x \, dx$

D20 $\int \frac{dx}{x \ln \frac{1}{x}}$

Solve the differential equations **D21** *to* **D26**

D21 $y' = -4y$ with $y(0) = 2$

D22 $\frac{dy}{dt} = 2 - 3y$ with $y_0 = 1$

D23 $\frac{dy}{dt} = t^2\sqrt{y}$ with $y_0 = 9$

D24 $\frac{dy}{dt} = 2ty^2$ with $y_0 = 1$

D25 $\frac{dy}{dx} = e^{xy}$ with $y_0 = 10$

D26 $\frac{dy}{dt} = y - 2y^2$ with $y_0 = 100$

Solutions $y = 2e^{-4t}$ $\quad y = \frac{1}{3}e^{-3t} + \frac{2}{3}$ $\quad y = (\frac{t^3}{6} + 3)^2$ $\quad y = \frac{-1}{t^2-1}$ $\quad y = -\ln|e^{-10} - e^x|$ $\quad y = \frac{1}{2-1.99e^{-t}}$

D27 If a population grows continuously at 2% a year, what is its percentage growth after 20 years?

CHAPTER 7 TECHNIQUES OF INTEGRATION

7.1 Integration by Parts (page 287)

Integration by parts aims to exchange a difficult problem for a possibly longer but probably easier one. It is up to you to make the problem easier! The key lies in choosing "u" and "dv" in the formula $\int u\,dv = uv - \int v\,du$. Try to pick u so that du is simple (or at least no worse than u). For $u = x$ or x^2 the derivative 1 or $2x$ is simpler. For $u = \sin x$ or $\cos x$ or e^x it is no worse. On the other hand, choose "dv" to have a nice integral. Good choices are $dv = \sin x\,dx$ or $\cos x\,dx$ or $e^x\,dx$.

Of course the selection of u also decides dv (since $u\,dv$ is the given integration problem). Notice that $u = \ln x$ is a good choice because $du = \frac{1}{x}dx$ is simpler. On the other hand, $\ln x\,dx$ is usually a poor choice for dv, because its integral $x \ln x - x$ is more complicated. Here are more suggestions:

Good choices for u: $\ln x$, inverse trig functions, $x^n, \cos x, \sin x, e^x$ or e^{cx}.

These are just suggestions. It's a free country. Integrate $1 - 6$ by parts:

1. $\int xe^{-x}dx$.

   - Pick $u = x$ because $\frac{du}{dx} = 1$ is simpler. Then $dv = e^{-x}dx$ gives $v = -e^{-x}$. Watch all the minus signs:

$$\int u\,dv = \underset{u}{x}\ \underset{v}{(-e^{-x})} - \int \underset{v}{(-e^{-x})}\ \underset{du}{dx} = -xe^{-x} - e^{-x} + C$$

2. $\int x \sec^{-1} x\,dx$.

   - If we choose $u = x$, we are faced with $dv = \sec^{-1} x\,dx$. Its integral is difficult. Better to try $u = \sec^{-1} x$, so that $du = \frac{dx}{|x|\sqrt{x^2-1}}$. Is that simpler? It leaves $dv = x\,dx$, so that $v = \frac{1}{2}x^2$. Our integral is now $uv - \int v\,du$:

$$(\sec^{-1} x)(\tfrac{1}{2}x^2) - \int \tfrac{1}{2}x^2 \cdot \frac{dx}{|x|\sqrt{x^2-1}} = \tfrac{1}{2}x^2 \sec^{-1} x \pm \tfrac{1}{2}\int \frac{x\,dx}{\sqrt{x^2-1}}$$
$$= \tfrac{1}{2}x^2 \sec^{-1} x \pm \tfrac{1}{2}\sqrt{x^2-1} + C.$$

   The $\pm$ sign comes from $|x|$; plus if $x > 0$ and minus if $x < 0$.

3. $\int e^x \sin x\,dx$. (Problem 7.1.9) This example requires *two* integrations by parts. First choose $u = e^x$ and $dv = \sin x\,dx$. This makes $du = e^x\,dx$ and $v = -\cos x$. The first integration by parts is $\int e^x \sin x\,dx = -e^x \cos x + \int e^x \cos x\,dx$. The new integral on the right is no simpler than the old one on the left. For the new one, $dv = \cos x\,dx$ brings back $v = \sin x$:

$$\int e^x \cos x\,dx = e^x \sin x - \int e^x \sin x\,dx.$$

   Are we back where we started? Not quite. Put the second into the first:

$$\int e^x \sin x\,dx = -e^x \cos x + e^x \sin x - \int e^x \sin x\,dx.$$

   The integrals are now the same. Move the one on the right side to the left side, *and divide by* 2:

$$\int e^x \sin x\,dx = \frac{1}{2}e^x(\sin x - \cos x) + C.$$

4. $\int x^2 \ln x\, dx$ (Problem 7.1.6). The function $\ln x$ (if it appears) is almost always the choice for u. Then $du = \frac{dx}{x}$. This leaves $dv = x^2 dx$ and $v = \frac{1}{3}x^3$. Therefore

$$\int x^2 \ln x\, dx = \frac{1}{3}x^3 \ln x - \int \frac{1}{3}x^2 dx = \frac{1}{3}x^3 \ln x - \frac{1}{9}x^3 + C.$$

5. $\int \frac{x^3 dx}{\sqrt{x^2+1}}$.

- Generally we choose u for a nice derivative, and dv is what's left. In this case it pays for dv to have a nice integral. We don't know $\int \frac{x^3}{\sqrt{x^2+1}} dx$ but we do know $\int \frac{x}{\sqrt{x^2+1}} dx = \sqrt{x^2+1}$. This leaves $u = x^2$ with $du = 2x\, dx$:

$$\begin{aligned} \int \frac{x^3 dx}{\sqrt{x^2+1}} &= x^2\sqrt{x^2+1} - \int 2x\sqrt{x^2+1}\, dx \\ &= x^2\sqrt{x^2+1} - \tfrac{2}{3}(x^2+1)^{3/2} + C. \end{aligned}$$

Note Integration by parts is not the only way to do this problem. You can directly substitute $u = x^2 + 1$ and $du = 2x\, dx$. Then x^2 is $u - 1$ and $x\, dx$ is $\frac{1}{2}du$. The integral is

$$\begin{aligned} \frac{1}{2}\int \frac{u-1}{\sqrt{u}} du &= \frac{1}{2}\int (u^{1/2} - u^{-1/2}) du = \frac{1}{3}u^{3/2} - u^{1/2} + C \\ &= \frac{1}{3}(x^2+1)^{3/2} - (x^2+1)^{1/2} + C \quad \text{(same answer in disguise).} \end{aligned}$$

6. Derive a reduction formula for $\int (\ln x)^n dx$.

- A reduction formula gives this integral in terms of an integral of $(\ln x)^{n-1}$. Let $u = (\ln x)^n$ so that $du = n(\ln x)^{n-1}(\frac{1}{x})dx$. Then $dv = dx$ gives $v = x$. This cancels the $\frac{1}{x}$ in du :

$$\int (\ln x)^n dx = x(\ln x)^n - \int n(\ln x)^{n-1} dx.$$

6′. Find a similar reduction from $\int x^n e^x\, dx$ to $\int x^{n-1} e^x\, dx$.

7. Use this reduction formula as often as necessary to find $\int (\ln x)^3 dx$.

- Start with $n = 3$ to get $\int (\ln x)^3 dx = x(\ln x)^3 - 3\int (\ln x)^2 dx$. Now use the formula with $n = 2$. The last integral is $x(\ln x)^2 - 2\int \ln x\, dx$. Finally $\int \ln x\, dx$ comes from $n = 1$: $\int \ln x\, dx = x(\ln x) - \int (\ln x)^0 dx = x(\ln x) - x$. Substitute everything back:

$$\begin{aligned} \int (\ln x)^3 dx &= x(\ln x)^3 - 3[x(\ln x)^2 - 2[x \ln x - x]] + C \\ &= x(\ln x)^3 - 3x(\ln x)^2 + 6x \ln x - 6x + C. \end{aligned}$$

Problems 8 and 9 are about the step function $U(x)$ and its derivative the delta function $\delta(x)$.

8. Find $\int_{-2}^{6} (x^2 - 8)\delta(x) dx$

- Since $\delta(x) = 0$ everywhere except at $x = 0$, we are only interested in $v(x) = x^2 - 8$ at $x = 0$. At that point $v(0) = -8$. We separate the problem into two parts:

$$\int_{-2}^{2} (x^2 - 8)\delta(x) dx + \int_{2}^{6} (x^2 - 8)\delta(x) dx = -8 + 0.$$

The first integral is just like 7B, picking out $v(0)$. The second integral is zero since $\delta(x) = 0$ in the interval [2,6]. The answer is -8.

9. (This is 7.1.54) Find the area under the graph of $\frac{\Delta U}{\Delta x} = [U(x+\Delta x) - U(x)]/\Delta x$.

- For the sake of this discussion let Δx be positive. The step function has $U(x) = 1$ if $x \geq 0$. In that case $U(x+\Delta x) = 1$ also. Subtraction $U(x+\Delta x) - U(x)$ leaves zero. The only time $U(x+\Delta x)$ is different from $U(x)$ is when $x + \Delta x \geq 0$ and $x < 0$. In that case

$$U(x+\Delta x) - U(x) = 1 - 0 = 1 \text{ and } \frac{U(x+\Delta x)-U(x)}{\Delta x} = \frac{1}{\Delta x}.$$

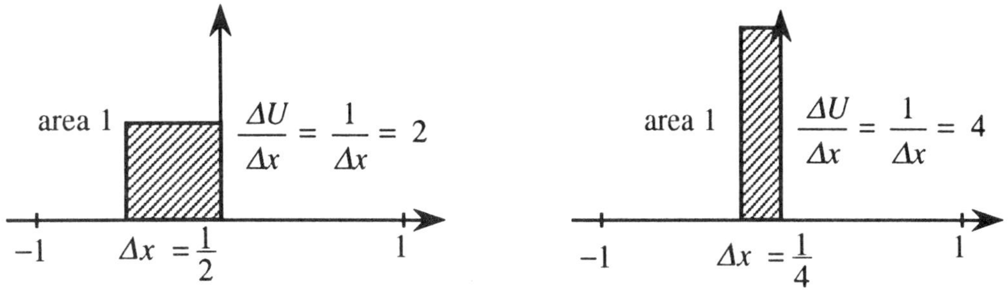

The sketches show the small interval $-\Delta x \leq x < 0$ where this happens. The base of the rectangle is Δx but the height is $\frac{1}{\Delta x}$. ***The area stays constant at 1.***

The limit of $\frac{U(x+\Delta x)-U(x)}{\Delta x}$ is the slope of the step function. This is the delta function $U'(x) = \delta(x)$. Certainly $\delta(x) = 0$ except at $x = 0$. But the integral of the delta function across the spike at $x = 0$ is 1. (The area hasn't changed as $\Delta x \to 0$.) A strange function.

Read-throughs and selected even-numbered solutions :

Integration by parts is the reverse of the **product** rule. It changes $\int u\, dv$ into **uv** minus $\int \mathbf{v\, du}$. In case $u = x$ and $dv = e^{2x}dx$, it changes $\int xe^{2x}dx$ to $\mathbf{\frac{1}{2}xe^{2x}}$ minus $\int \mathbf{\frac{1}{2}\, e^{2x}dx}$. The definite integral $\int_0^2 xe^{2x}dx$ becomes $\mathbf{\frac{3}{4}e^4}$ minus $\mathbf{\frac{1}{4}}$. In choosing u and dv, the **derivative** of u and the **integral** of dv/dx should be as simple as possible. Normally $\ln x$ goes into **u** and e^x goes into **v**. Prime candidates are $u = x$ or x^2 and $v = \sin x$ or $\mathbf{\cos x}$ or $\mathbf{e^x}$. When $u = x^2$ we need **two** integrations by parts. For $\int \sin^{-1} x\, dx$, the choice $dv = dx$ leads to $\mathbf{x \sin^{-1} x}$ minus $\int \mathbf{x\, dx/\sqrt{1-x^2}}$.

If U is the unit step function, $dU/dx = \delta$ is the unit **delta** function. The integral from $-A$ to A is $U(A) - U(-A) = \mathbf{1}$. The integral of $v(x)\delta(x)$ equals $\mathbf{v(0)}$. The integral $\int_{-1}^{1} \cos x\, \delta(x)dx$ equals **1**. In engineering, the balance of forces $-dv/dx = f$ is multiplied by a displacement $u(x)$ and integrated to give a balance of **work**.

14 $\int \cos(\ln x)dx = uv - \int v du = \cos(\ln x)x + \int x \sin(\ln x)\frac{1}{x}dx$. Cancel x with $\frac{1}{x}$. Integrate by parts again to get $\cos(\ln x)x + \sin(\ln x)x - \int x\cos(\ln x)\frac{1}{x}dx$. Move the last integral to the left and divide by 2. The answer is $\frac{x}{2}(\cos(\ln x) + \sin(\ln x)) + C$.

18 $uv - \int v\, du = \cos^{-1}(2x)x + \int x\frac{2\, dx}{\sqrt{1-(2x)^2}} = x\cos^{-1}(2x) - \frac{1}{2}(1-4x^2)^{1/2} + C$.

22 $uv - \int v\, du = x^3(-\cos x) + \int(\cos x)3x^2dx =$ (use Problem 5) $= -x^3\cos x + 3x^2\sin x + 6x\cos x - 6\sin x + C$.

28 $\int_0^1 e^{\sqrt{x}}dx = \int_{u=0}^{1} e^u(2u\, du) = 2e^u(u-1)]_0^1 = \mathbf{2}$. **38** $\int x^n \sin x\, dx = -x^n\cos x + n\int x^{n-1}\cos x\, dx$.

44 (a) $e^0 = \mathbf{1}$; (b) $\mathbf{v(0)}$ (c) **0** (limits do not enclose zero).

46 $\int_{-1}^{1}\delta(2x)dx = \int_{u=-2}^{2}\delta(u)\frac{du}{2} = \mathbf{\frac{1}{2}}$. Apparently $\delta(2x)$ equals $\frac{1}{2}\delta(x)$; both are zero for $x \neq 0$.

48 $\int_0^1 \delta(x-\frac{1}{2})dx = \int_{-1/2}^{1/2}\delta(u)du = \mathbf{1}$; $\int_0^1 e^x\delta(x-\frac{1}{2})dx = \int_{-1/2}^{1/2}e^{u+\frac{1}{2}}\delta(u)du = \mathbf{e^{1/2}}$; $\delta(x)\delta(x-\frac{1}{2}) = \mathbf{0}$.

60 $A = \int_1^e \ln x\, dx = [x\ \ln x - x]_1^e = \mathbf{1}$ is the area under $y = \ln x$. $B = \int_0^1 e^y dy = \mathbf{e-1}$ is the area to the left of $y = \ln x$. Together the area of the rectangle is $1 + (e-1) = e$.

7.2 Trigonometric Integrals (page 293)

This section integrates powers and products of sines and cosines and tangents and secants. We are constantly using $\sin^2 x = 1 - \cos^2 x$. Starting with $\int \sin^3 x\, dx$, we convert it to $\int (1 - \cos^2 x) \sin x\, dx$. Are we unhappy about that one remaining $\sin x$? *Not at all.* It will be part of du, when we set $u = \cos x$. Odd powers are actually easier than even powers, because the extra term goes into du. For even powers use the double-angle formula in Problem 2 below.

1. $\int (\sin x)^{-3/2} (\cos x)^3 dx$ is a product of sines and cosines.

   - The angles x are the same and *the power 3 is odd.* ($-\frac{3}{2}$ is neither even nor odd.) Change all but one of the cosines to sines by $\cos^2 x = 1 - \sin^2 x$. The problem is now

   $$\int (\sin x)^{-3/2}(1 - \sin^2 x) \cos x\, dx = \int (u^{-3/2} - u^{1/2}) du.$$

   Here $u = \sin x$ and $du = \cos x\, dx$. The answer is

   $$-2u^{-1/2} - \frac{2}{3}u^{3/2} + C = -2(\sin x)^{-1/2} - \frac{2}{3}(\sin x)^{3/2} + C.$$

2. $\int \sin^4 3x \cos^2 3x\, dx$ has even powers 4 and 2, with the same angle $3x$.

   - *Use the double-angle method.* Replace $\sin^2 3x$ with $\frac{1}{2}(1 - \cos 6x)$ and $\cos^2 3x$ with $\frac{1}{2}(1 + \cos 6x)$. The problem is now

   $$\int \frac{(1-\cos 6x)^2}{4} \frac{(1+\cos 6x)}{2} dx = \frac{1}{8} \int (1 - 2\cos 6x + \cos^2 6x)(1 + \cos 6x) dx$$
   $$= \frac{1}{8} \int (1 - \cos 6x - \cos^2 6x + \cos^3 6x) dx.$$

   The integrals of the first two terms are x and $\frac{1}{6} \sin 6x$. The third integral is another double angle:

   $$\int \cos^2 6x\, dx = \int \frac{1}{2}(1 + \cos 12x) dx = \frac{1}{2}x + \frac{1}{24} \sin 12x.$$

   For $\int \cos^3 6x\, dx$, with an odd power, change $\cos^2$ to $1 - \sin^2$:

   $$\int \cos^3 6x\, dx = \int (1 - \sin^2 6x) \cos 6x\, dx = \int (1 - u^2) \frac{du}{6} = \frac{1}{6} \sin 6x - \frac{1}{18} \sin^3 6x.$$

   Putting all these together, the final solution is

   $$\frac{1}{8}[x - \frac{1}{6} \sin 6x - (\frac{1}{2}x + \frac{1}{24} \sin 12x) + \frac{1}{6} \sin 6x - \frac{1}{18} \sin^3 6x] = \frac{1}{16}x - \frac{1}{192} \sin 12x - \frac{1}{144} \sin^3 6x + C.$$

3. $\int \sin 10x \cos 4x\, dx$ has different angles $10x$ and $4x$. Use the identity $\sin 10x \cos 4x = \frac{1}{2} \sin(10 + 4)x + \frac{1}{2} \sin(10 - 4)x$. Now integrate:

   $$\int (\frac{1}{2} \sin 14x + \frac{1}{2} \sin 6x) dx = -\frac{1}{28} \cos 14x - \frac{1}{12} \cos 6x + C.$$

4. $\int \cos x \cos 4x \cos 8x \, dx$ has three different angles!

- Use the identity $\cos 4x \cos 8x = \frac{1}{2}\cos(4+8)x + \frac{1}{2}\cos(4-8)x$. The integral is now $\frac{1}{2}\int(\cos x \cos 12x + \cos x \cos 4x)dx$. Apply the $\cos px \cos qx$ identity twice more to get

$$\frac{1}{2}\int(\frac{1}{2}\cos 13x + \frac{1}{2}\cos 11x + \frac{1}{2}\cos 5x + \frac{1}{2}\cos 3x)dx = \frac{1}{4}(\frac{\sin 13x}{13} + \frac{\sin 11x}{11} + \frac{\sin 5x}{5} + \frac{\sin 3x}{3}) + C.$$

5. $\int \tan^5 x \ \sec^4 x \ dx$. Here are three ways to deal with tangents and secants.

- *First*: Remember $d(\tan x) = \sec^2 x \ dx$ and convert the other $\sec^2 x$ to $1 + \tan^2 x$. The problem is

$$\int \tan^5 x(1+\tan^2 x)\sec^2 x \ dx = \int (u^5 + u^7)du.$$

- *Second*: Remember $d(\sec x) = \sec x \tan x \ dx$ and convert $\tan^4 x$ to $(\sec^2 x - 1)^2$. The integral is

$$\int (\sec^2 x - 1)^2 \sec^3 x \sec x \tan x \ dx = \int (u^2-1)^2 u^3 du = \int (u^7 - 2u^5 + u^3)du.$$

- *Third*: Convert $\tan^5 x \sec^4 x$ to sines and cosines as $\frac{\sin^5 x}{\cos^9 x}$. Eventually take $u = \cos x$:

$$\begin{aligned} \int \frac{(1-\cos^2 x)^2}{\cos^9 x} \sin x \ dx &= \int (\cos^{-9} x - 2\cos^{-7} x + \cos^{-5} x) \sin x \ dx \\ &= \int (-u^{-9} + 2u^{-7} - u^{-5}) du. \end{aligned}$$

6. Use the substitution $u = \tan\frac{x}{2}$ in the text equation (11) to find $\int_0^{\pi/4} \frac{dx}{1-\sin x}$.

- The substitutions are $\sin x = \frac{2u}{1+u^2}$ and $dx = \frac{2du}{1+u^2}$. This gives

$$\int \frac{1}{1-\sin x} dx = \int \frac{1}{1-\frac{2u}{1+u^2}} \cdot \frac{2du}{1+u^2} = \int \frac{2du}{(1+u^2)-2u} = \int \frac{2du}{(1-u)^2} = \frac{2}{1-u}.$$

The definite integral is from $x = 0$ to $x = \frac{\pi}{4}$. Then $u = \tan\frac{x}{2}$ goes from 0 to $\tan\frac{\pi}{8}$. The answer is $\frac{2}{1-\tan\frac{\pi}{8}} - \frac{2}{1} \approx 1.41$.

7. Problem 7.2.26 asks for $\int_0^\pi \sin 3x \sin 5x \ dx$. First write $\sin 3x \sin 5x$ in terms of $\cos 8x$ and $\cos 2x$.

- The formula for $\sin px \sin qx$ gives

$$\int_0^\pi (-\frac{1}{2}\cos 8x + \frac{1}{2}\cos 2x)dx = [-\frac{1}{16}\sin 8x + \frac{1}{4}\sin 2x]_0^\pi = 0.$$

8. Problem 7.2.33 is the Fourier sine series $A\sin x + B\sin 2x + C\sin 3x + \cdots$ that adds to x. Find A.

- Multiply both sides of $x = A\sin x + B\sin 2x + C\sin 3x + \cdots$ by $\sin x$. *Integrate from* 0 *to* π:

$$\int_0^\pi x\sin x \ dx = \int_0^\pi A\sin^2 x \ dx + \int_0^\pi B\sin 2x \ \sin x \ dx + \int_0^\pi C\sin 3x \sin x \ dx + \cdots.$$

All of the definite integrals on the right are zero, except for $\int_0^\pi A\sin^2 x \ dx$. For example the integral of $\sin 2x \sin x$ is $[-\frac{1}{6}\sin 3x + \frac{1}{2}\sin x]_0^\pi = 0$. The only nonzero terms are $\int_0^\pi x\sin x \ dx = \int_0^\pi A\sin^2 x \ dx$. Integrate $x\sin x$ by parts to find one side of this equation for A :

$$\int_0^\pi x\sin x \ dx = [-x\cos x]_0^\pi + \int_0^\pi \cos x \ dx = [-x\cos x + \sin x]_0^\pi = \pi.$$

On the other side $\int_0^\pi A\sin^2 x \ dx = \frac{A}{2}[x - \sin x\cos x]_0^\pi = \frac{A\pi}{2}$. Then $\frac{A\pi}{2} = \pi$ and $A = 2$.

- You should *memorize* those integrals $\int_0^\pi \sin^2 x dx = \int_0^\pi \cos^2 x\, dx = \frac{\pi}{2}$. They say that the average value of $\sin^2 x$ is $\frac{1}{2}$, and the average value of $\cos^2 x$ is $\frac{1}{2}$.
- You would find B by multiplying the Fourier series by $\sin 2x$ instead of $\sin x$. This leads in the same way to $\int_0^\pi x \sin 2x\, dx = \int_0^\pi B \sin^2 2x\, dx = B\frac{\pi}{2}$ because all other integrals are zero.

9. When a sine and a cosine are added, the resulting wave is best expressed as a single cosine: $a\cos x + b\sin x = \sqrt{a^2+b^2}\cos(x-\alpha)$. Show that this is correct and find the angle α (Problem 7.2.56).

- Expand $\cos(x-\alpha)$ into $\cos x\cos\alpha + \sin x \sin\alpha$. Choose α so that $\cos\alpha = \frac{a}{\sqrt{a^2+b^2}}$ and $\sin\alpha = \frac{b}{\sqrt{a^2+b^2}}$. Our formula becomes correct. The reason for $\sqrt{a^2+b^2}$ is to ensure that $\cos^2\alpha + \sin^2\alpha = \frac{a^2+b^2}{a^2+b^2} = 1$. Dividing $\sin\alpha$ by $\cos\alpha$ gives $\tan\alpha = \frac{b}{a}$ or $\alpha = \tan^{-1}\frac{b}{a}$. Thus $3\cos x + 4\sin x = 5\cos(x - \tan^{-1}\frac{4}{3})$.

10. Use the previous answer (Problem 9) to find $\int \frac{dx}{\sqrt{3}\cos x + \sin x}$.

- With $a = \sqrt{3}$ and $b = 1$ we have $\sqrt{a^2+b^2} = \sqrt{3+1} = 2$ and $\alpha = \tan^{-1}\frac{1}{\sqrt{3}} = \frac{\pi}{6}$. Therefore

$$\int \frac{dx}{\sqrt{3}\cos x + \sin x} = \int \frac{dx}{2\cos(x-\pi/6)} = \frac{1}{2}\int \sec(x-\frac{\pi}{6})dx = \frac{1}{2}\ln|\sec(x-\frac{\pi}{6}) + \tan(x-\frac{\pi}{6})| + C.$$

The figure shows the waves $\sqrt{3}\cos x$ and $\sin x$ adding to $2\cos(x-\frac{\pi}{6})$.

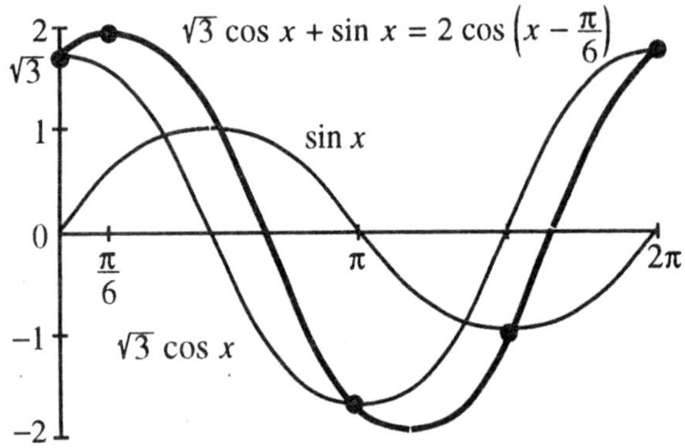

11. What is the distance from the equator to latitude 45° on a Mercator world map? From 45° to 70°?

- The distance north is the integral of $\sec x$, multiplied by the radius R of the earth (on your map). See Figure 7.3 in the text. The equator is at 0°. The distance to 45° $= \frac{\pi}{4}$ radians is

$$R\int_0^{\pi/4} \sec x\, dx = R\ln(\sec x + \tan x)_0^{\pi/4} = R\ln(\sqrt{2}+1) - R\ln 1 \approx 0.88R.$$

The distance from 45° to 70° is almost the same: $R\ln|\sec x + \tan x|\Big|_{45^\circ}^{70^\circ} \approx 0.85R$.

Read-throughs and selected even-numbered solutions :

To integrate $\sin^4 x \cos^3 x$, replace $\cos^2 x$ by $\mathbf{1 - \sin^2 x}$. Then $(\sin^4 x - \sin^6 x)\cos x\, dx$ is $(\mathbf{u^4 - u^6})du$. In terms of $u = \sin x$ the integral is $\mathbf{\frac{1}{5}u^5 - \frac{1}{7}u^7}$. This idea works for $\sin^m x \cos^n x$ if m or n is **odd**.

If both m and n are even, one method is integration by **parts**. For $\int \sin^4 x\, dx$, split off $dv = \sin x\, dx$.

Then $-\int v\,du$ is $\int \mathbf{3\sin^2 x\cos^2 x}$. Replacing $\cos^2 x$ by $\mathbf{1-\sin^2 x}$ creates a new $\sin^4 x\,dx$ that combines with the original one. The result is a *reduction* to $\int \sin^2 x\,dx$, which is known to equal $\mathbf{\frac{1}{2}(x-\sin x\cos x)}$.

The second method uses the double-angle formula $\sin^2 x = \mathbf{\frac{1}{2}(1-\cos 2x)}$. Then $\sin^4 x$ involves $\cos^2 \mathbf{2x}$. Another doubling comes from $\cos^2 2x = \mathbf{\frac{1}{2}(1+\cos 4x)}$. The integral contains the sine of $\mathbf{4x}$.

To integrate $\sin 6x\cos 4x$, rewrite it as $\frac{1}{2}\sin 10x + \mathbf{\frac{1}{2}\sin 2x}$. The integral is $\mathbf{-\frac{1}{20}\cos 10x - \frac{1}{4}\cos 2x}$. The definite integral from 0 to 2π is **zero.** The product $\cos px\ \cos qx$ is written as $\frac{1}{2}\cos(p+q)x + \mathbf{\frac{1}{2}\cos(p-q)x}$. Its integral is also zero, except if $\mathbf{p = q}$ when the answer is π.

With $u=\tan x$, the integral of $\tan^9 x\ \sec^2 x$ is $\mathbf{\frac{1}{10}\tan^{10} x}$. Similarly $\int \sec^9 x(\sec x\ \tan x\ dx) = \mathbf{\frac{1}{10}\sec^{10} x}$. For the combination $\tan^m x\ \sec^n x$ we apply the identity $\tan^2 x = \mathbf{1+\sec^2 x}$. After reduction we may need $\int \tan x\ dx = \mathbf{-\ln\cos x}$ and $\int \sec x\ dx = \mathbf{\ln(\sec x+\tan x)}$.

6 $\int \sin^3 x\cos^3 x\,dx = \int \sin^3 x(1-\sin^2 x)\cos x\,dx = \mathbf{\frac{1}{4}\sin^4 x - \frac{1}{6}\sin^6 x} + C$

10 $\int \sin^2 ax\cos ax\,dx = \mathbf{\frac{\sin^3 ax}{3a}} + C$ and $\int \sin ax\cos ax\,dx = \mathbf{\frac{\sin^2 ax}{2a}} + C$

16 $\int \sin^2 x\cos^2 2x\,dx = \int \frac{1-\cos 2x}{2}\cos^2 2x\,dx = \int\left(\frac{1+\cos 4x}{4} - \frac{\cos 2x}{2}(1-\sin^2 2x)\right)dx =$

$\mathbf{\frac{x}{4} + \frac{\sin 4x}{16} - \frac{\sin 2x}{4} + \frac{\sin^3 2x}{12}} + C$. This is a hard one.

18 Equation (7) gives $\int_0^{\pi/2}\cos^n x\,dx = \left[\frac{\cos^{n-1}x\sin x}{n}\right]_0^{\pi/2} + \frac{n-1}{n}\int_0^{\pi/2}\cos^{n-2}x\,dx$. The integrated term is zero because $\cos\frac{\pi}{2} = 0$ and $\sin 0 = 0$. The exception is $n=1$, when the integral is $[\sin x]_0^{\pi/2} = 1$.

26 $\int_0^\pi \sin 3x\sin 5x\,dx = \int_0^\pi \frac{-\cos 8x+\cos 2x}{2}dx = \left[\mathbf{\frac{-\sin 8x}{16} + \frac{\sin 2x}{4}}\right]_0^\pi = \mathbf{0}$.

30 $\int_0^{2\pi}\sin x\sin 2x\sin 3x\,dx = \int_0^{2\pi}\sin 2x\left(\frac{-\cos 4x+\cos 2x}{2}\right)dx = \int_0^{2\pi}\sin 2x\left(\frac{1-2\cos^2 2x+\cos 2x}{2}\right)dx =$

$\left[\mathbf{-\frac{\cos 2x}{4} + \frac{\cos^3 2x}{6} - \frac{\cos^2 2x}{8}}\right]_0^{2\pi} = \mathbf{0}$. Note: The integral has other forms.

32 $\int_0^\pi x\cos x\,dx = [x\sin x]_0^\pi - \int_0^\pi \sin x\,dx = [\mathbf{x\sin x+\cos x}]_0^\pi = \mathbf{-2}$.

34 $\int_0^\pi 1\sin 3x\,dx = \int_0^\pi(A\sin x + B\sin 2x + C\sin 3x + \cdots)\sin 3x\,dx$ reduces to $\left[-\frac{\cos 3x}{3}\right]_0^\pi = 0+0+C\int_0^\pi \sin^2 3x\,dx$. Then $\frac{2}{3} = C(\frac{\pi}{2})$ and $\mathbf{C = \frac{4}{3\pi}}$.

44 First by substituting for $\tan^2 x$: $\int \tan^2 x\sec x\,dx = \int \sec^3 x\,dx - \int \sec x\,dx$. Use Problem 62 to integrate $\sec^3 x$: final answer $\mathbf{\frac{1}{2}(\sec x\tan x - \ln|\sec x+\tan x|)} + C$. Second method from line 1 of Example 11: $\int \tan^2 x\sec x\,dx = \sec x\tan x - \int \sec^3 x\,dx$. Same final answer.

52 This should have an asterisk! $\int \frac{\sin^6 x}{\cos^3 x}dx = \int \frac{(1-\cos^2 x)^3}{\cos^3 x}dx = \int(\sec^3 x - 3\sec x + 3\cos x - \cos^3 x)dx =$ use Example 11 = Problem 62 for $\int \sec^3 x\,dx$ and change $\int \cos^3 x\,dx$ to $\int(1-\sin^2 x)\cos x\,dx$.

Final answer $\mathbf{\frac{\sec x\tan x}{2} - \frac{5}{2}\ln|\sec x+\tan x| + 2\sin x + \frac{\sin^3 x}{3}} + C$.

54 $\mathbf{A = 2}$: $2\cos(x+\frac{\pi}{3}) = 2\cos x\cos\frac{\pi}{3} - 2\sin x\sin\frac{\pi}{3} = \cos x - \sqrt{3}\sin x$. Therefore $\int \frac{dx}{(\cos x-\sqrt{3}\sin x)^2} =$

$\int \frac{dx}{4\cos^2(x+\frac{\pi}{3})} = \mathbf{\frac{1}{4}\tan(x+\frac{\pi}{3})} + C$.

58 When lengths are scaled by $\sec x$, area is scaled by $\mathbf{\sec^2 x}$. The area from the equator to latitude x is then proportional to $\int \sec^2 x\,dx = \tan x$.

7.3 Trigonometric Substitutions (page 299)

The substitutions may be easier to remember from these right triangles:

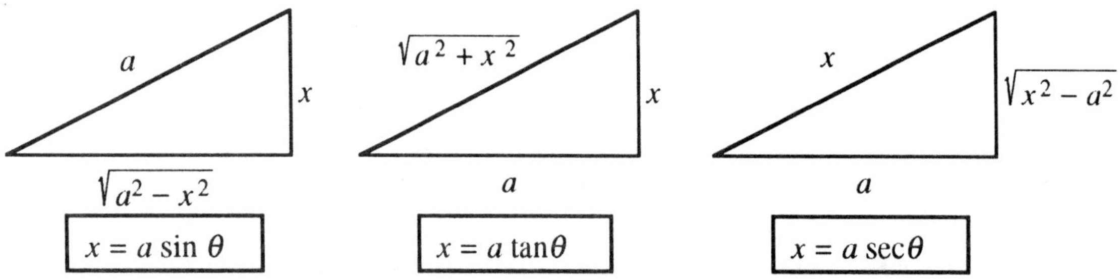

Each triangle obeys Pythagoras. The squares of the legs add to the square of the hypotenuse. The first triangle has $\sin\theta = \frac{\text{opposite}}{\text{hypotenuse}} = \frac{x}{a}$. Thus $x = a\sin\theta$ and $dx = a\cos\theta d\theta$.

Use these triangles in Problems 1–3 or use the table of substitutions in the text.

1. $\int_1^4 \frac{dx}{x^2\sqrt{x^2+9}}$ has a plus sign in the square root (*second triangle*).

   - Choose the second triangle with $a = 3$. Then $x = 3\tan\theta$ and $dx = 3\sec^2\theta d\theta$ and $\sqrt{x^2+9} = 3\sec\theta$. Substitute and then write $\sec\theta = \frac{1}{\cos\theta}$ and $\tan\theta = \frac{\sin\theta}{\cos\theta}$:

   $$\int \frac{dx}{x^2\sqrt{x^2+9}} = \int \frac{3\sec^2\theta d\theta}{(9\tan^2\theta)(3\sec\theta)} = \int \frac{\cos\theta d\theta}{9\sin^2\theta} = \frac{-1}{9\sin\theta}.$$

   The integral was $\frac{1}{9}u^{-2}du$ with $u = \sin\theta$. The original limits of integration are $x = 1$ and $x = 4$. Instead of converting them to θ, we convert $\sin\theta$ back to x. The second triangle above shows

   $$\frac{1}{\sin\theta} = \frac{\text{hypotenuse}}{\text{opposite}} = \frac{\sqrt{x^2+9}}{x} \text{ and then } \left[\frac{-\sqrt{x^2+9}}{9x}\right]_1^4 = \frac{-5}{36} + \frac{\sqrt{10}}{9} \approx 0.212.$$

2. $\int \sqrt{100 - x^2}\, dx$ contains the square root of $a^2 - x^2$ with $a = 10$.

   - Choose the *first* triangle: $x = 10\sin\theta$ and $\sqrt{100 - x^2} = 10\cos\theta$ and $dx = 10\cos\theta d\theta$:

   $$\int \sqrt{100 - x^2}\, dx = \int (10\cos\theta)10\cos\theta d\theta = 100\int \cos^2\theta d\theta = 100 \cdot \frac{1}{2}(\theta + \sin\theta\cos\theta) + C.$$

   Returning to x this is $50(\sin^{-1}\frac{x}{10} + \frac{x}{10}\cdot\frac{\sqrt{100-x^2}}{10}) = 50\sin^{-1}\frac{x}{10} + \frac{1}{2}x\sqrt{100 - x^2}$.

3. $\int \frac{dx}{x^2\sqrt{9x^2-25}}$ does not exactly contain $x^2 - a^2$. But try the third triangle.

   - Factor $\sqrt{9} = 3$ from the square root to leave $\sqrt{x^2 - \frac{25}{9}}$. Then $a^2 = \frac{25}{9}$ and $a = \frac{5}{3}$. The third triangle has $x = \frac{5}{3}\sec\theta$ and $dx = \frac{5}{3}\sec\theta\tan\theta d\theta$ and $\sqrt{9x^2 - 25} = 5\tan\theta$. The problem is now

   $$\int \frac{\frac{5}{3}\sec\theta\tan\theta d\theta}{(\frac{5}{3})^2\sec^2\theta(5\tan\theta)} = \frac{3}{25}\int \cos\theta d\theta = \frac{3}{25}\sin\theta + C.$$

   The third triangle converts $\frac{3}{25}\sin\theta$ back to $\frac{\sqrt{9x^2-25}}{25x}$.

4. For $\int \frac{x^3dx}{\sqrt{1-x^2}}$ the substitution $x = \sin\theta$ will work. But try $u = 1 - x^2$.

- Then $du = -2x\,dx$ and $x^2 = 1 - u$. The problem becomes $-\frac{1}{2}\int \frac{(1-u)du}{u^{1/2}} = \frac{1}{2}\int (u^{-1/2} - u^{-1/2})du$. In this case the old way is simpler than the new.

Problems 5 and 6 require *completing the square* before a trig substitution.

5. $\int \frac{dx}{\sqrt{x^2-8x+6}}$ requires us to complete $(x-4)^2$. We need $4^2 = 16$ so add and subtract 10 :

$$x^2 - 8x + 6 = (x^2 - 8x + 16) - 10 = (x-4)^2 - 10.$$

This has the form $u^2 - a^2$ with $u = x - 4$ and $a = \sqrt{10}$. Finally set $u = \sqrt{10}\sec\theta$:

$$\int \frac{dx}{\sqrt{x^2-8x+6}} = \int \frac{du}{\sqrt{u^2-10}} = \int \frac{\sqrt{10}\sec\theta\tan\theta d\theta}{\sqrt{10}\tan\theta} = \int \sec\theta d\theta.$$

6. $\int \frac{\sqrt{2x-x^2}}{x}dx$ requires us to complete $2x - x^2$ (watch the minus sign):

$$2x - x^2 = -(x^2 - 2x) = -(x^2 - 2x + 1) + 1 = 1 - (x-1)^2.$$

This is $1 - u^2$ with $u = x - 1$ and $x = 1 + u$. The trig substitution is $u = \sin\theta$:

$$\int \frac{\sqrt{2x-x^2}}{x} = \int \frac{\sqrt{1-u^2}}{1+u}du = \int \frac{\cos^2\theta d\theta}{1+\sin\theta} = \int \frac{(1-\sin^2\theta)}{1+\sin\theta)}d\theta = \int (1-\sin\theta)d\theta.$$

Read-throughs and selected even-numbered solutions :

The function $\sqrt{1-x^2}$ suggests the substitution $x = \mathbf{\sin\theta}$. The square root becomes $\mathbf{\cos\theta}$ and dx changes to $\mathbf{\cos\theta\, d\theta}$. The integral $\int(1-x^2)^{3/2}dx$ becomes $\int \mathbf{\cos^4}\theta\, d\theta$. The interval $\frac{1}{2} \le x \le 1$ changes to $\mathbf{\frac{\pi}{6}} \le \theta \le \mathbf{\frac{\pi}{2}}$.

For $\sqrt{a^2 - x^2}$ the substitution is $x = \mathbf{a\sin\theta}$ with $dx = \mathbf{a\cos\theta\, d\theta}$. For $x^2 - a^2$ we use $x = \mathbf{a\sec\theta}$ with $dx = \mathbf{a\sec\theta\tan\theta}$. (**Insert:** For $x^2 + a^2$ use $x = a\ \tan\theta$). Then $\int dx/(1+x^2)$ becomes $\int\ d\theta$, because $1 + \tan^2\theta = \mathbf{sec^2}\theta$. The answer is $\theta = \tan^{-1}x$. We already knew that $\mathbf{\frac{1}{1+x^2}}$ is the derivative of $\tan^{-1}x$.

The quadratic $x^2 + 2bx + c$ contains a **linear** term $2bx$. To remove it we **complete** the square. This gives $(x+b)^2 + C$ with $C = \mathbf{c - b^2}$. The example $x^2 + 4x + 9$ becomes $\mathbf{(x+2)^2 + 5}$. Then $u = x + 2$. In case x^2 enters with a minus sign, $-x^2 + 4x + 9$ becomes $\mathbf{-(x-2)^2 + 13}$. When the quadratic contains $4x^2$, start by factoring out **4**.

2 $x = a\sec\theta, x^2 - a^2 = a^2\tan^2\theta, \int \frac{dx}{\sqrt{x^2-a^2}} = \int \frac{a\sec\theta\tan\theta d\theta}{a\tan\theta} = \ln|\sec\theta + \tan\theta| = \mathbf{\ln|\frac{x}{a} + \sqrt{\frac{x^2}{a^2} - 1}|} + C$

4 $x = \frac{1}{3}\tan\theta, 1 + 9x^2 = \sec^2\theta, \int \frac{dx}{1+9x^2} = \int \frac{\frac{1}{3}\sec^2\theta d\theta}{\sec^2\theta} = \frac{\theta}{3} = \mathbf{\frac{1}{3}\tan^{-1}3x} + C.$

12 Write $\sqrt{x^6 - x^8} = x^3\sqrt{1-x^2}$ and set $x = \sin\theta$: $\int \sqrt{x^6-x^8}dx = \int \sin^3\theta\cos\theta(\cos\theta d\theta) =$
$\int \sin\theta(\cos^2\theta - \cos^4\theta)d\theta = -\frac{\cos^3\theta}{3} + \frac{\cos^5\theta}{5} = \mathbf{-\frac{1}{3}(1-x^2)^{3/2} + \frac{1}{5}(1-x^2)^{5/2}} + C$

14 $x = \sin\theta, \int \frac{dx}{(1-x^2)^{3/2}} = \int \frac{\cos\theta d\theta}{\cos^3\theta} = \tan\theta + C = \mathbf{\frac{x}{\sqrt{1-x^2}}} + C.$

32 First use geometry: $\int_{1/2}^{1}\sqrt{1-x^2}dx =$ half the area of the unit circle beyond $x = \frac{1}{2}$ which breaks into

$\frac{1}{2}(120°$ wedge minus $120°$ triangle$) = \frac{1}{2}(\frac{\pi}{3} - \frac{1}{2} \cdot \frac{1}{2} \cdot 2\sqrt{1-(\frac{1}{2})^2}) = \mathbf{\frac{\pi}{6} - \frac{\sqrt{3}}{8}}$.

Check by integration: $\int_{1/2}^{1} \sqrt{1-x^2}dx = [\frac{1}{2}(x\sqrt{1-x^2} + \sin^{-1} x)]_{1/2}^{1} = \frac{1}{2}(\frac{\pi}{2} - \frac{1}{2}\frac{\sqrt{3}}{2} - \frac{\pi}{6}) = \frac{\pi}{6} - \frac{\sqrt{3}}{8}$.

34 $\int \frac{dx}{\cos x} = \int \sec x \, dx = \mathbf{\ln|\sec x + \tan x|} + C; \int \frac{dx}{1+\cos x}(\frac{1-\cos x}{1-\cos x}) = \int \frac{dx}{\sin^2 x} - \int \frac{\cos x \, dx}{\sin^2 x} = \int \csc^2 x \, dx - \int \frac{du}{u^2} = -\cot x + \frac{1}{\sin x} = \mathbf{\frac{1-\cos x}{\sin x}} + C; \int \frac{dx}{\sqrt{1+\cos x}} = \int \frac{dx}{\sqrt{2}\cos\frac{x}{2}} = \mathbf{\sqrt{2}\ln|\sec\frac{x}{2} + \tan\frac{x}{2}|} + C$

40 $x = \cosh\theta : \int \frac{\sqrt{x^2-1}}{x^2}dx = \int \frac{\sinh\theta}{\cosh^2\theta}\sinh\theta d\theta = \int \tanh^2\theta d\theta = \int(1-\operatorname{sech}^2\theta)d\theta = \theta - \tanh\theta = \mathbf{\cosh^{-1}x - \frac{\sqrt{x^2-1}}{x}} + C$

44 $-x^2 + 2x + 8 = -(x-1)^2 + 9$

50 $\int \frac{dx}{\sqrt{9-(x-1)^2}} = \int \frac{du}{\sqrt{9-u^2}}$. Set $u = 3\sin\theta : \int \frac{\cos\theta d\theta}{\cos\theta} = \theta = \sin^{-1}\frac{u}{3} = \mathbf{\sin^{-1}\frac{x-1}{3}} + C;$

$\int \frac{dx}{10-x^2} = \mathbf{\frac{1}{2\sqrt{10}} \ln \frac{x-\sqrt{10}}{x+\sqrt{10}}} + C; \int \frac{dx}{(x+2)^2-16} = \int \frac{du}{u^2-16} = \frac{1}{8}\ln\frac{2u-8}{2u+8} = \mathbf{\frac{1}{8}\ln\frac{x-2}{x+6}} + C$

52 (a) $u = x-2$ (b) $u = x+1$ (c) $u = x-5$ (d) $u = x - \frac{1}{4}$

7.4 Partial Fractions (page 304)

This method applies to ratios $\frac{P(x)}{Q(x)}$, where P and Q are polynomials. The goal is to split the ratio into pieces that are easier to integrate. We begin by comparing this method with substitutions , on some basic problems where both methods give the answer.

1. $\int \frac{2x}{x^2-1}dx$. The substitution $u = x^2 - 1$ produces $\int \frac{du}{u} = \ln|u| = \ln|x^2-1|$.

   - Partial fractions breaks up this problem into smaller pieces:

   $$\frac{2x}{x^2-1} = \frac{2x}{(x+1)(x-1)} \text{ splits into } \frac{A}{x+1} + \frac{B}{x-1} = \frac{1}{x+1} + \frac{1}{x-1}.$$

   Now integrate the pieces to get $\ln|x+1| + \ln|x-1|$. This equals $\ln|x^2-1|$, the answer from substitution. We review how to find the numbers A and B starting from $\frac{2x}{x^2-1}$.

   - First, factor $x^2 - 1$ to get the denominators $x+1$ and $x-1$. Second, cover up $(x-1)$ and set $x = 1$:

   $$\frac{2x}{(x+1)(x-1)} = \frac{A}{x+1} + \frac{B}{x-1} \text{ becomes } \frac{2x}{(x+1)} = \frac{2}{2} = B. \text{ Thus } B = 1.$$

   Third, cover up $(x+1)$ and set $x = -1$ to find A:

   $$\text{At } x = -1 \text{ we get } A = \frac{2x}{(x-1)} = \frac{-2}{-2} = 1.$$

   That is it. Both methods are good. Use substitution or partial fractions.

2. $\int \frac{1}{x^2-1}dx$. The substitution $x = \sec\theta$ gives $\int \frac{4\sec\theta\tan\theta}{\tan^2\theta}d\theta = \int \frac{4}{\sin\theta}d\theta$.

   - The integral of $\frac{1}{\sin\theta}$ is not good. This time partial fractions look better:

   $$\frac{4}{x^2-1} = \frac{4}{(x+1)(x-1)} \text{ splits into } \frac{A}{x+1} + \frac{B}{x-1} = \frac{-2}{x+1} + \frac{2}{x-1}.$$

   The integral is $-2\ln|x+1| + 2\ln|x-1| = 2\ln|\frac{x-1}{x+1}|$. Remember the cover-up:

$$x = 1 \text{ gives } B = \frac{4}{(x+1)} = 2. \qquad x = -1 \text{ gives } A = \frac{4}{(x-1)} = -2.$$

3. $\int \frac{2x+4}{x^2-1}dx$ is the sum of the previous two integrals. Add A's and B's:

$$\frac{2x+4}{x^2-1} = \frac{A}{x+1} + \frac{B}{x-1} = \frac{-1}{x+1} + \frac{3}{x-1}.$$

In practice I would find $A = -1$ and $B = 3$ by the usual cover-up:

$$x = 1 \text{ gives } B = \frac{2x+4}{(x+1)} = \frac{6}{2}. \qquad x = -1 \text{ gives } A = \frac{2x+4}{(x-1)} = \frac{2}{-2}.$$

The integral is immediately $-\ln|x+1| + 3\ln|x-1|$. In this problem partial fractions is much better than substitutions. This case is $\frac{\text{linear}}{\text{quadratic}} = \frac{\text{degree 1}}{\text{degree 2}}$. *That is where partial fractions work best.*
The text solves the logistic equation by partial fractions. Here are more difficult ratios $\frac{P(x)}{Q(x)}$.

- It is the algebra, not the calculus, that can make $\frac{P(x)}{Q(x)}$ difficult. A reminder about division of polynomials may be helpful. If the degree of $P(x)$ is greater than or equal to the degree of $Q(x)$, *you first divide Q into P*. The example $\frac{x^3}{x^2+2x+1}$ requires long division:

$$\begin{array}{rrl} & x & \text{divide } x^2 \text{ into } x^3 \text{ to get } x \\ x^2+2x+1 & \sqrt{x^3} & \\ & x^3+2x^2+x & \text{multiply } x^2+2x+1 \text{ by } x \\ \hline & -2x^2-x & \text{subtract from } x^3 \end{array}$$

The first part of the division gives x. If we stop there, division leaves $\frac{x^3}{x^2+2x+1} = x + \frac{-2x^2-x}{x^2+2x+1}$. This new fraction is $\frac{\text{degree 2}}{\text{degree 2}}$. So the division has to continue one more step:

$$\begin{array}{rrl} & x-2 & \text{divide } x^2 \text{ into } -2x^2 \text{ to get } -2 \\ x^2+2x+1 & \sqrt{x^3} & \\ & x^3+2x^2+x & \\ \hline & -2x^2-x & \\ & -2x^2-4x-2 & \text{multiply } x^2+2x+1 \text{ by } -2 \\ \hline & 3x+2 & \text{subtract to find remainder} \end{array}$$

Now stop. The remainder $3x+2$ has lower degree than x^2+2x+1:

$$\frac{x^3}{x^2+2x+1} = x - 2 + \frac{3x+2}{x^2+2x+1} \text{ is ready for partial fractions.}$$

Factor x^2+2x+1 into $(x+1)^2$. Since $x+1$ is repeated, we look for

$$\frac{3x+2}{(x+1)^2} = \frac{A}{(x+1)} + \frac{B}{(x+1)^2} \quad \text{(notice this form!)}$$

Multiply through by $(x+1)^2$ to get $3x+2 = A(x+1)+B$. Set $x=-1$ to get $B=-1$. Set $x=0$ to get $A+B=2$. This makes $A=3$. The algebra is done and we integrate:

$$\int \frac{x^3}{x^2+2x+1}dx = \int (x-2+\frac{3x+2}{x^2+2x+1})dx = \int (x-2+\frac{3}{x+1} - \frac{1}{(x+1)^2})dx$$
$$= \frac{1}{2}x^2 - 2x + 3\ln|x+1| + (x+1)^{-1} + C.$$

4. $\int \frac{x^2+x}{x^2-4}dx$ also needs long division. The top and bottom have equal degree 2:

$$\begin{array}{rl} & \quad 1 \\ x^2+0x-4 & \sqrt{x^2+x} \\ & \quad x^2+0x-4 \\ \hline & \quad x+4 \end{array}$$

divide x^2 into x^2 to get 1

multiply x^2-4 by 1

subtract to find remainder $x+4$

This says that $\frac{x^2+x}{x^2-4} = 1 + \frac{x+4}{x^2-4} = 1 + \frac{x+4}{(x-2)(x+2)}$. To decompose the remaining fraction, let

$$\frac{x+4}{(x-2)(x+2)} = \frac{A}{x-2} + \frac{B}{x+2}.$$

Multiply by $x-2$ so the problem is $\frac{x+4}{x+2} = A + \frac{B(x-2)}{x+2}$. Set $x=2$ to get $A = \frac{6}{4} = \frac{3}{2}$. Cover up $x+2$ to get (in the mind's eye) $\frac{x+4}{x-2} = \frac{A(x+2)}{(x-2)} + B$. Set $x=-2$ to get $B = -\frac{1}{2}$. All together we have

$$\int \frac{x^2+x}{x^2-4}dx = \int (1 + \frac{\frac{3}{2}}{x-2} - \frac{\frac{1}{2}}{x+2})dx = x + \frac{3}{2}\ln|x-2| - \frac{1}{2}\ln|x+2| + C.$$

5. $\int \frac{7x^2+14x+15}{(x^2+3)(x+7)}dx$ requires no division. Why not? *We have degree 2 over degree 3.* Also x^2+3 cannot be factored further, so there are just two partial fractions:

$$\frac{7x^2+14x+15}{(x^2+3)(x+7)} = \frac{A}{x+7} + \frac{Bx+C}{x^2+3}.$$ Use $Bx+C$ over a quadratic, not just B !

Cover up $x+7$ and set $x=-7$ to get $\frac{260}{52} = A$, or $A=5$. So far we have

$$\frac{7x^2+14x+15}{(x^2+3)(x+7)} = \frac{5}{x+7} + \frac{Bx+C}{x^2+3}.$$

We can set $x=0$ (because zero is easy) to get $\frac{15}{21} = \frac{5}{7} + \frac{C}{3}$, or $C=0$. Then set $x=-1$ to get $\frac{8}{24} = \frac{5}{6} + \frac{-B}{4}$. Thus $B=2$. Our integration problem is $\int(\frac{5}{x+7} + \frac{2x}{x^2+3})dx = 5\ln|x+7| + \ln(x^2+3) + C$.

6. (Problem 7.5.25) By substitution change $\int \frac{1+e^x}{1-e^x}$ to $\int \frac{P(u)}{Q(u)}du$. Then integrate.

- The ratio $\frac{1+e^x}{1-e^x}dx$ does not contain polynomials. Substitute $u=e^x$, $du=e^x\,dx$, and $dx = \frac{du}{u}$ to get $\frac{(1+u)du}{(1-u)u}$. A perfect set-up for partial fractions!

$$\frac{1+u}{u(1-u)} = \frac{A}{u} + \frac{B}{1-u} = \frac{1}{u} + \frac{2}{1-u}.$$

The integral is $\ln\ u - 2\ln|1-u| = x - 2\ln|1-e^x| + C$.

Read-throughs and selected even-numbered solutions :

The idea of **partial** fractions is to express $P(x)/Q(x)$ as a **sum** of simpler terms, each one easy to integrate. To begin, the degree of P should be **less than** the degree of Q. Then Q is split into **linear** factors like $x-5$ (possibly repeated) and quadratic factors like x^2+x+1 (possibly repeated). The quadratic factors have two **complex** roots, and do not allow real linear factors.

A factor like $x-5$ contributes a fraction $A/(\mathbf{x}-\mathbf{5})$. Its integral is $\mathbf{A}\ \mathbf{ln}(\mathbf{x}-\mathbf{5})$. To compute A, cover up $\mathbf{x}-\mathbf{5}$ in the denominator of P/Q. Then set $x=\mathbf{5}$, and the rest of P/Q becomes A. An equivalent method puts all fractions over a common denominator (which is **Q**). Then match the **numerators.** At the same point ($x = 5$) this matching gives A.

A repeated linear factor $(x-5)^2$ contributes not only $A/(x-5)$ but also $B/(\mathbf{x}-\mathbf{5})^{\mathbf{2}}$. A quadratic factor like x^2+x+1 contributes a fraction $(\mathbf{Cx}+\mathbf{D})/(x^2+x+1)$ involving C and D. A repeated quadratic factor or a triple linear factor would bring in $(Ex+F)/(x^2+x+1)^2$ or $G/(x-5)^3$. The conclusion is that any P/Q can be split into partial **fractions**, which can always be integrated.

6 $\frac{1}{x(x-1)(x+1)} = -\frac{1}{x} + \frac{1/2}{x-1} + \frac{1/2}{x+1}$

14 $x + 1\sqrt{x^2+0x+1}$ $\quad \frac{x^2+1}{x+1} = x - 1 + \frac{2}{x+1}$ $\quad$ **16** $\frac{1}{x^2(x-1)} = -\frac{1}{x} - \frac{1}{x^2} + \frac{1}{x-1}$

18 $\frac{x^2}{(x-3)(x+3)} = \frac{A(x+3)+B(x-3)}{(x-3)(x+3)}$ is impossible (no x^2 in the numerator on the right side).
Divide first to rewrite $\frac{x^2}{(x-3)(x+3)} = 1 + \frac{9}{(x-3)(x+3)} =$ (now use partial fractions) $1 + \frac{3/2}{x-3} - \frac{3/2}{x+3}$.

22 Set $u = \sqrt{x}$ so $u^2 = x$ and $2u\,du = dx$. Then $\int \frac{1-\sqrt{x}}{1+\sqrt{x}}dx = \int \frac{1-u}{1+u}2u\,du =$ (*divide* $u+1$ *into* $-2u^2+2u$) $=$ $\int(-2u+4-\frac{4}{u+1})du = -u^2+4u-4\ln(u+1)+C = \mathbf{-x+4\sqrt{x}-4\ln(\sqrt{x}+1)} + C$.

7.5 Improper Integrals (page 309)

An improper integral is really a limit: $\int_0^\infty y(x)dx$ means $\lim_{b\to\infty}\int_0^b y(x)dx$. Usually we just integrate and substitute $b=\infty$. If the integral of $y(x)$ contains e^{-x} then $e^{-\infty}=0$. If the integral contains $\frac{1}{x}$ or $\frac{1}{x+7}$ then $\frac{1}{\infty}=0$. If the integral contains $\tan^{-1}x$ then $\tan^{-1}\infty = \frac{\pi}{2}$. The numbers are often convenient when the upper limit is $b=\infty$.

Similarly $\int_{-\infty}^{\infty} y(x)dx$ is really the sum of two limits. You have to use a and b to keep those limits separate: $\lim_{a\to-\infty}\int_a^0 y(x)dx + \lim_{b\to\infty}\int_0^b y(x)dx$. Normally just integrate $y(x)$ and substitute $a=-\infty$ and $b=\infty$.

EXAMPLE 1 $\int_{-\infty}^{\infty}\frac{dx}{x^2+5} = [\frac{1}{5}\tan^{-1}\frac{x}{5}]_{-\infty}^{\infty} = \frac{1}{5}(\frac{\pi}{2}) - \frac{1}{5}(-\frac{\pi}{2}) = \frac{\pi}{5}$.

Notice the lower limit, where $\tan^{-1}\frac{a}{5}$ approaches $-\frac{\pi}{2}$ as a approaches $-\infty$. Strictly speaking the solution should have separated the limits $a=-\infty$ and $b=\infty$:

$$\int_{-\infty}^{\infty}\frac{dx}{x^2+5} = \lim_{a\to-\infty}[\frac{1}{5}\tan^{-1}\frac{x}{5}]_a^0 + \lim_{b\to\infty}[\frac{1}{5}\tan^{-1}\frac{x}{5}]_0^b.$$

If $y(x)$ blows up *inside* the interval, the integral is really the sum of a left-hand limit and a right-hand limit.

EXAMPLE 2 $\int_{-2}^{3}\frac{dx}{x^3}$ blows up at $x=0$ inside the interval.

- If this was not in a section labeled "improper integrals," would your answer have been $[-\frac{1}{2}x^{-2}]_{-2}^{3} = \frac{1}{2}(\frac{1}{9}-\frac{1}{4}) = \frac{5}{72}$? This is a very easy mistake to make. But since $\frac{1}{x^3}$ is infinite at $x=0$, the integral is improper. Separate it into the part up to $x=0$ and the part beyond $x=0$.

  The integral of $\frac{1}{x^3}$ is $\frac{-1}{2x^2}$ which blows up at $x=0$. Those integrals from -2 to 0 and from 0 to 3 are both infinite. This improper integral *diverges*.

- Notice: The question is whether the *integral* blows up, not whether $y(x)$ blows up. $\int_0^1 \frac{1}{\sqrt{x}}\,dx$ is OK.

  Lots of times you only need to know whether or not the integral converges. This is where the comparison test comes in. Assuming $y(x)$ is positive, try to show that its unknown integral is smaller than a known finite integral (or greater than a known divergent integral).

EXAMPLE 3 $\int_1^\infty \frac{2+\cos x}{x^3}dx$ has $2+\cos x$ between 1 and 3. Therefore $\frac{2+\cos x}{x^3} \le \frac{3}{x^3}$. Since $\int_1^\infty \frac{3}{x^3}dx$ converges to a finite answer, the original integral must converge. You could have started with $\frac{2+\cos x}{x^3} \ge \frac{1}{x^3}$. This is true, but it is not helpful! It only shows that the integral is greater than a convergent integral. The greater one could converge or diverge – this comparison doesn't tell.

EXAMPLE 4 $\int_5^\infty \frac{\sqrt{x}}{x+8}dx$ has $\frac{\sqrt{x}}{x+8} \approx \frac{\sqrt{x}}{x} = \frac{1}{x^{1/2}}$ for large x.
We suspect divergence (the area under $x^{-1/2}$ is infinite). To show a comparison, note $\frac{\sqrt{x}}{x+8} > \frac{\sqrt{x}}{3x}$. This is because 8 is smaller than $2x$ beyond our lower limit $x = 5$. Increasing the denominator to $3x$ makes the fraction smaller. The official reasoning is

$$\int_5^\infty \frac{\sqrt{x}\,dx}{x+8} > \int_5^\infty \frac{\sqrt{x}}{3x}dx = \int_5^\infty \frac{dx}{3x^{1/2}} = \lim_{b\to\infty} \frac{2}{3}x^{1/2}\Big]_5^b = \infty.$$

5. (Problem 7.5.37) What is improper about the area between $y = \sec x$ and $y = \tan x$?

   The area under the secant graph minus the area under the tangent graph is

$$\int_0^{\pi/2} \sec x\; dx - \int_0^{\pi/2} \tan x\; dx = \ln(\sec x + \tan x)]_0^{\pi/2} + \ln(\cos x)]_0^{\pi/2} = \infty - \infty.$$

   The separate areas are infinite! However we can subtract before integrating:

$$\begin{aligned}\int_0^{\pi/2}(\sec x - \tan x)dx &= [\ln(\sec x + \tan x) + \ln(\cos x)]_0^{\pi/2}\\ &= [\ln(\cos x)(\sec x + \tan x)]_0^{\pi/2} = [\ln(1+\sin x)]_0^{\pi/2} = \ln 2 - \ln 1 = \ln 2.\end{aligned}$$

   This is perfectly correct. The difference of areas comes in Section 8.1.

Read-throughs and selected even-numbered solutions :

An improper integral $\int_a^b y(x)dx$ has lower limit $a = -\infty$ or upper limit $b = \infty$ or y becomes **infinite** in the interval $a \le x \le b$. The example $\int_1^\infty dx/x^3$ is improper because $\mathbf{b} = \infty$. We should study the limit of $\int_1^b dx/x^3$ as $\mathbf{b} \to \infty$. In practice we work directly with $-\frac{1}{2}x^{-2}]_1^\infty = \frac{\mathbf{1}}{\mathbf{2}}$. For $p > 1$ the improper integral $\int_p^\infty x^{-p}dx$ is finite. For $p < 1$ the improper integral $\int_0^1 x^{-p}dx$ is finite. For $y = e^{-x}$ the integral from 0 to ∞ is **1**.

Suppose $0 \le u(x) \le v(x)$ for all x. The convergence of $\int$ **v(x) dx** implies the convergence of $\int$ **u(x)dx**. The divergence of $\int u(x)dx$ **implies** the divergence of $\int v(x)dx$. From $-\infty$ to ∞, the integral of $1/(e^x + e^{-x})$ converges by comparison with $\mathbf{1/e^{|x|}}$. Strictly speaking we split $(-\infty, \infty)$ into $(-\infty, 0)$ and $(0, \infty)$. Changing to $1/(e^x - e^{-x})$ gives divergence, because $\mathbf{e^X = e^{-X}}$ **at** $\mathbf{x = 0}$. Also $\int_{-\pi}^{\pi} dx/\sin x$ diverges by comparison with $\int \mathbf{dx/x}$. The regions left and right of zero don't cancel because $\infty - \infty$ is **not zero.**

2 $\int_0^1 \frac{dx}{x^\pi} = [\frac{x^{1-\pi}}{1-\pi}]_0^1$ diverges at $x = 0$: infinite area

8 $\int_{-\infty}^{\infty} \sin x\, dx$ is not defined because $\int_a^b \sin x\, dx = \cos a - \cos b$ does not approach a limit as $b \to \infty$ and $a \to -\infty$

16 $\int_0^\infty \frac{e^x\, dx}{(e^x-1)^p} =$ (set $u = e^x - 1$) $\int_0^\infty \frac{du}{u^p}$ which is infinite: diverges at $u = 0$ if $p \geq 1$, diverges at $u = \infty$ if $p \leq 1$.

18 $\int_0^1 \frac{dx}{x^6+1} < \int_0^1 \frac{dx}{1} = 1$: convergence

24 $\int_0^1 \sqrt{-\ln x}dx < \int_0^{1/e}(-\ln x)dx + \int_{1/e}^1 1dx = [-x\ln x + x]_0^{1/e} + [x]_{1/e}^1 = \frac{1}{e} + 1$: convergence (note $x \ln x \to 0$ as $x \to 0$)

36 $\int_a^b \frac{x\, dx}{1+x^2} = [\frac{1}{2}\ln(1+x^2)]_a^b = \frac{1}{2}\ln(1+b^2) - \frac{1}{2}\ln(1+a^2)$. As $b \to \infty$ or as $a \to -\infty$ (separately!) there is no limiting value. If $a = -b$ then the answer is zero – but we are not allowed to connect a and b.

40 The red area in the right figure has an extra unit square (area 1) compared to the red area on the left.

7 Chapter Review Problems

Review Problems

R1 Why is $\int u(x)v(x)dx$ not equal to $(\int u(x)dx)(\int v(x)dx)$? What formula is correct?

R2 What method of integration would you use for these integrals?

$\int x\cos(2x^2+1)dx \quad \int x\cos(2x+1) \quad \int \cos^2(2x+1)dx \quad \int \cos(2x+1)\sin(2x+1)dx$

$\int \cos^3(2x+1)\sin^5(2x+1)dx \quad \int \cos^4(2x+1)\sin^2(2x+1)dx \quad \int \cos 2x \sin 3x\, dx \quad \int \frac{\cos(2x+1)}{\sin(2x+1)}dx$

R3 Which eight methods will succeed for these eight integrals?

$\int \frac{x}{\sqrt{3+x^2}}dx \quad \int \frac{dx}{\sqrt{3+x^2}} \quad \int \frac{dx}{\sqrt{3-x^2}} \quad \int \frac{dx}{x^2-3}$

$\int \frac{dx}{x^2+2x-3} \quad \int \frac{dx}{\sqrt{x^2+2x}} \quad \int \frac{x\, dx}{x^2-3} \quad \int \frac{x^3 dx}{x^2-3}$

R4 What is an improper integral? Show by example four ways a definite integral can be improper.

R5 Explain with two pictures the comparison tests for convergence and divergence of improper integrals.

Drill Problems

D1 $\int x^2 \ln x \, dx$

D2 $\int e^x \sin 2x \, dx$

D3 $\int x^3\sqrt{1-x^2} \, dx$

D4 $\int \frac{x^3 dx}{x^2+4x+3}$

D5 $\int \frac{\ln x}{\sqrt{x}} \, dx$

D6 $\int \tan^3 2x \sec^2 2x \, dx$

D7 $\int e^{e^x} e^x \, dx$

D8 $\int \frac{dx}{\sqrt{3-2x-x^2}}$

D9 $\int \sin(\ln x) \, dx$

D10 $\int \frac{6x+4}{x^3-4x} \, dx$

D11 $\int \sin^{-1} \sqrt{x} \, dx$

D12 $\int \cos^4 2x \sin^2 2x \, dx$

D13 $\int \frac{dx}{(4-x^2)^{3/2}}$

D14 $\int \frac{9x+36}{x^3+6x^2+9x} \, dx$

Evaluate the improper integrals **D15** to **D20** or show that they diverge.

D15 $\int_0^{\pi/2} \frac{\cos x \, dx}{1-\sin x}$

D16 $\int_1^{\infty} \frac{\ln x}{\sqrt{x}} dx$

D17 $\int_0^{\infty} xe^{-x} dx$

D18 $\int_{-1}^{\infty} x^{-1/3} dx$

D19 $\int_0^{33} \frac{dx}{(1-x)^{2/5}}$

D20 $\int_e^{\infty} \frac{dx}{x(\ln x)^{3/2}}$

CHAPTER 8 APPLICATIONS OF THE INTEGRAL

8.1 Areas and Volumes by Slices (page 318)

1. Find the area of the region enclosed by the curves $y_1 = \frac{1}{4}x^2$ and $y_2 = x + 3$.

   - The first step is to sketch the region. Find the points where the curves intersect. In this case $\frac{1}{4}x^2 = x + 3$ gives $x = 6$ and $x = -2$. (You need these for your limits of integration.) A thin rectangle between the curves has area $(y_2 - y_1)\Delta x$. Because the rectangle is vertical, its thin side is Δx (not Δy). Then you integrate with respect to x. The area integral is $\int_{-2}^{6}(x + 3 - \frac{1}{4}x^2)dx = 10\frac{2}{3}$.

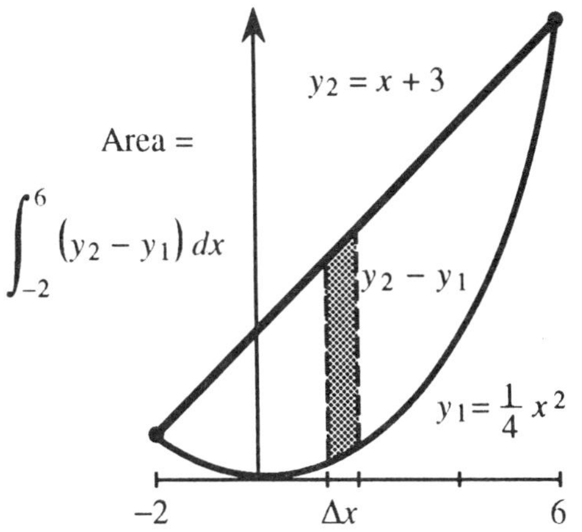

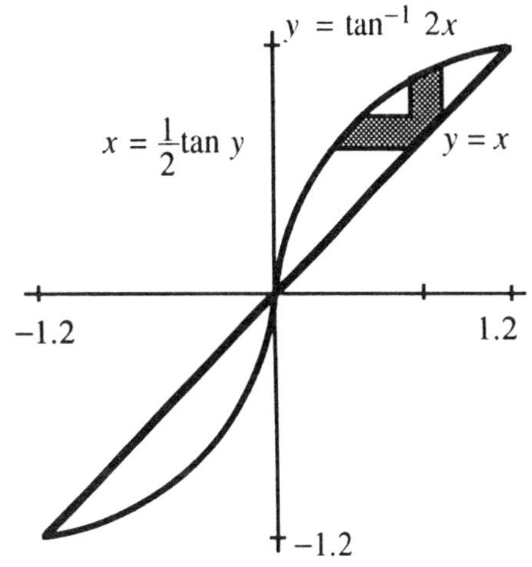

2. Find the area between the graphs of $y = \tan^{-1}(2x)$ and $y = x$.

   - A graphing calculator shows that these curves intersect near (1.2,1.2) and (−1.2, −1.2). The plan is to find the area in the first quadrant and double it. This shortcut is possible because of the symmetry in the graph. We will use horizontal strips.

     The length of the horizontal strip is $x_2 - x_1$, where x_2 is on the straight line and x_1 is on the arctan graph. Therefore $x_2 = y$ and $x_1 = \frac{1}{2}\tan y$. Since the thin part of the rectangle is Δy, the area is an integral with respect to y:

$$\text{Area } = 2 \cdot \int_0^{1.2} (y - \frac{1}{2}\tan y)dy = 2(\frac{1}{2}y^2 + \frac{1}{2}\ln\cos y)]_0^{1.2} \approx 0.42.$$

Area and volume can be found using vertical or horizontal strips. It is good to be familiar with both, because sometimes one method is markedly easier than the other. With vertical strips, Problem 2 would integrate $\tan^{-1}(2x)$. With horizontal strips it was $\frac{1}{2}\tan y$. Those functions are *inverses* (which we expect in switching between x and y).

The techniques for area are also employed for volume – especially for solids of revolution. A sketch is needed to determine the intersection points of the curves. Draw a thin strip – horizontal (Δy) or vertical (Δx) – whichever seems easier. Imagine the path of the strip as it is revolved. If the strip is perpendicular to the axis of revolution, the result is a thin disk or a washer. For a disk use $V = \int \pi y^2 dx$ or $V = \int \pi x^2 dy$. *A washer is a difference of circles.* Integrate $\pi(y_2^2 - y_1^2)$ or $\pi(x_2^2 - x_1^2)$.

If the strip is parallel to the axis, rotation produces a cylindrical shell. The volume of the shell is $\int 2\pi$ (radius) (height) (thickness). For an upright shell this is $\int 2\pi x\ y\ dx$.

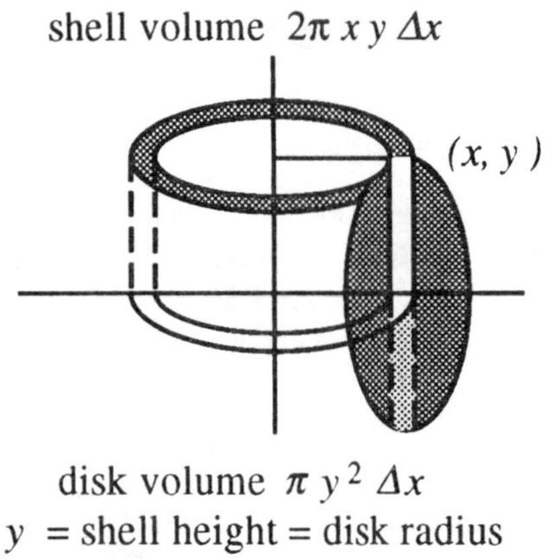

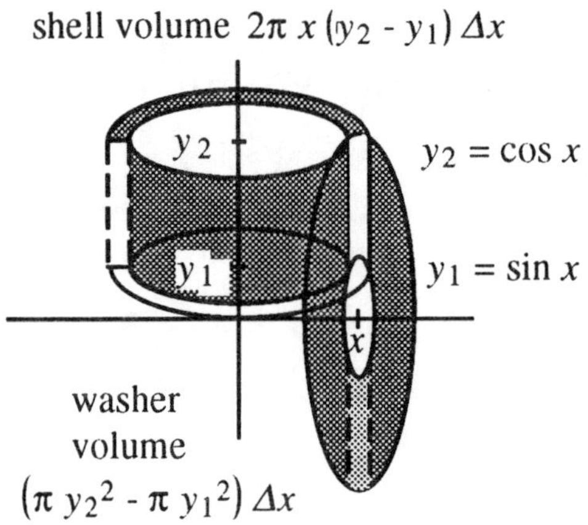

3. (This is 8.1.24) Find the volume when the first-quadrant region bounded by $y = \sin x, y = \cos x$ and $x = 0$ is revolved around (a) the x axis and (b) the y axis.

- (a) $y_1 = \sin x$ intersects $y_2 = \cos x$ at $x = \frac{\pi}{4}, y = \frac{\sqrt{2}}{2}$. A vertical strip is sketched. Imagine it revolving about the x axis. The result is a thin washer whose volume is $(\pi y_2^2 - \pi y_1^2)\Delta x$. This leads to the integral $\pi \int_0^{\pi/4}(\cos^2 x - \sin^2 x)dx$. Since $\cos^2 x - \sin^2 x = \cos 2x$, the total volume is $\pi \int_0^{\pi/4} \cos 2x\, dx = \frac{\pi}{2} \sin 2x]_0^{\pi/4} = \frac{\pi}{2}$.
- (b) When the same region is revolved about the y axis, the same strip carves out a thin shell. Its volume is approximately **circumference** $\times$ **height** $\times$ $\boldsymbol{\Delta x}$ $= 2\pi r(y_2 - y_1)\Delta x$. The radius of the shell is the distance $r = x$ from the y axis to the strip. The total volume is $V = \int_0^{\pi/4} 2\pi x(\cos x - \sin x)dx$. After integration by parts this is

$$[2\pi x(\sin x + \cos x)]_0^{\pi/4} - \int_0^{\pi/4} 2\pi(\sin x + \cos x)dx = 2\pi(\frac{\pi}{2}\sqrt{2} - 1) \approx 0.70.$$

4. Find the volume when the triangle with vertices (1,1), (4,1), and (6,6) is revolved around (a) the x axis and (b) the y axis.

- A horizontal strip has been drawn between two sides of the triangle. The left side is the line $y = x$, so $x_1 = y$. The right side is the line $y = \frac{5}{2}x - 9$, so $x_2 = \frac{2}{5}(y + 9)$. Since the thin side of the strip is Δy, integrals will be with respect to y.

  (a) When the triangle is revolved around the x axis, the strip is parallel to the axis of revolution. This produces a shell of radius y :

$$V = \int 2\pi(\text{radius})\ (\text{height})\ dy = 2\pi \int_1^6 y(\frac{2}{5}(y+9) - y)dy = 40\pi.$$

- (b) When the region is revolved around the y axis, the path of the strip is a washer (disk with hole). The volume is

$$\int_1^6 \pi(x_2^2 - x_1^2)dy = \pi \int_1^6 ((\frac{2}{5}(y+9))^2 - y^2)dy = 55\pi.$$

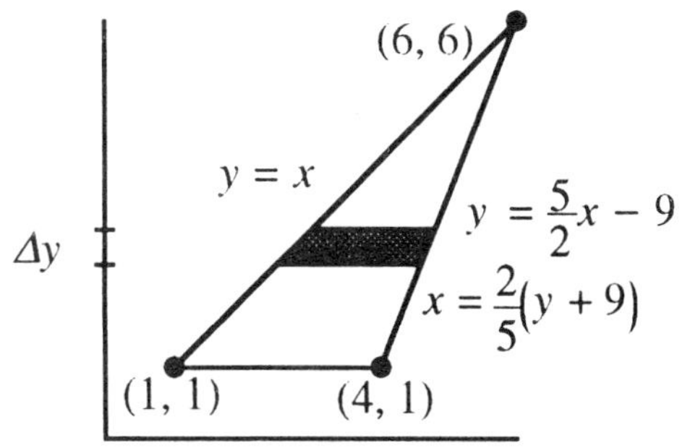

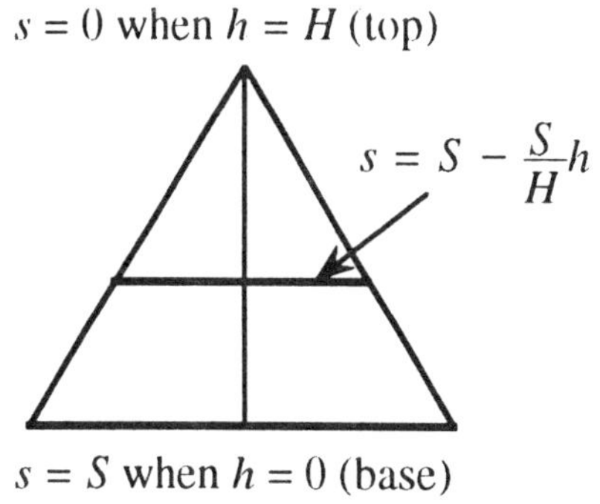

5. Find the volume of a pyramid, when the height is H and the square base has sides of length S. This is not a solid of revolution!

- Imagine the pyramid sliced horizontally. Each slice is like a thin square card. The thickness of each card is Δh. The sides of the square decrease in length from $s = S$ when $h = 0$ to $s = 0$ when $h = H$. *We can write* $s = S - \frac{S}{H}h$. That is probably the hardest step.

  The volume of each card is $s^2\Delta h$ or $(S - \frac{S}{H}h)^2\Delta h$. If we add those volumes and let $\Delta h \to 0$ (the number of cards goes to ∞) the sum becomes the integral

$$V = \int_0^H (S - \frac{S}{H}h)^2 dh = [-\frac{H}{3S}(S - \frac{S}{H}h)^3]_0^H = \frac{1}{3}S^2H.$$

Read-throughs and selected even-numbered solutions :

The area between $y = x^3$ and $y = x^4$ equals the integral of $\mathbf{x^3 - x^4}$. If the region ends where the curves intersect, we find the limits on x by solving $\mathbf{x^3 = x^4}$. Then the area equals $\int_0^1 (\mathbf{x^3 - x^4})\mathbf{dx} = \frac{1}{4} - \frac{1}{5} = \frac{1}{20}$. When the area between $y = \sqrt{x}$ and the y axis is sliced horizontally, the integral to compute is $\int \mathbf{y^2 dy}$.

In three dimensions the volume of a slice is its thickness dx times its **area.** If the cross-sections are squares of side $1 - x$, the volume comes from $\int \mathbf{(1 - x)^2 dx}$. From $x = 0$ to $x = 1$, this gives the volume $\frac{1}{3}$ of a square **pyramid.** If the cross-sections are circles of radius $1 - x$, the volume comes from $\int \pi\mathbf{(1 - x)^2 dx}$. This gives the volume $\frac{\pi}{3}$ of a circular **cone.**

For a solid of revolution, the cross sections are **circles.** Rotating the graph of $y = f(x)$ around the x axis gives a solid volume $\int \pi\mathbf{(f(x))^2 dx}$. Rotating around the y axis leads to $\int \pi\mathbf{(f^{-1}(y))^2 dy}$. Rotating the area between $y = f(x)$ and $y = g(x)$ around the x axis, the slices look like **washers.** Their areas are $\pi\mathbf{(f(x))^2} - \pi\mathbf{(g(x))^2} = \mathbf{A(x)}$ so the volume is $\int \mathbf{A(x)dx}$.

Another method is to cut the solid into thin cylindrical **shells.** Revolving the area under $y = f(x)$ around the y axis, a shell has height $\mathbf{f(x)}$ and thickness dx and volume $\mathbf{2\pi x\, f(x)dx}$. The total volume is $\int \mathbf{2\pi x\, f(x)dx}$.

6 $y = x^{1/5}$ and $y = x^4$ intersect at (0,0) and (1,1): area $= \int_0^1 (x^{1/5} - x^4)dx = \frac{5}{6} - \frac{1}{5} = \mathbf{\frac{19}{30}}$.

12 The region is a curved triangle between $x = -1$ (where $e^{-x} = e$) and $x = 1$ (where $e^x = e$). Vertical strips end at e^{-x} for $x < 0$ and at e^x for $x > 0$: Area $= \int_{-1}^0 (e - e^{-x})dx + \int_0^1 (e - e^x)dx = \mathbf{2}$.

18 Volume $= \int_0^\pi \pi \sin^2 x dx = [\pi(\frac{x - \sin x \cos x}{2})]_0^\pi = \mathbf{\frac{\pi^2}{2}}$.

20 Shells around the y axis have radius x and height $2\sin x$ and volume $(2\pi x)2\sin x dx$. Integrate for the volume of the galaxy: $\int_0^\pi 4\pi x \sin x dx = [4\pi(\sin x - x\cos x)]_0^\pi = \mathbf{8\pi^2}$.

26 The region is a curved triangle, with base between $x = 3, y = 0$ and $x = 9, y = 0$. The top point is where $y = \sqrt{x^2 - 9}$ meets $y = 9 - x$; then $x^2 - 9 = (9 - x)^2$ leads to $x = 5, y = 4$. (a) Around the x axis: Volume $= \int_3^5 \pi(x^2 - 9)dx + \int_5^9 \pi(9 - x)^2 dx = \mathbf{36\pi}$. (b) Around the y axis: Volume $= \int_3^5 2\pi x\sqrt{x^2 - 9}dx + \int_5^9 2\pi x(9 - x)dx = [\frac{2\pi}{3}(x^2 - 9)^{3/2}]_3^5 + [9\pi x^2 - \frac{2\pi x^3}{3}]_5^9 = \frac{2\pi}{3}(64) + 9\pi(9^2 - 5^2) - \frac{2\pi}{3}(9^3 - 5^3) = \mathbf{144\pi}$.

28 The region is a circle of radius 1 with center (2,1). (a) Rotation around the x axis gives a torus with no hole: it is Example 10 with $a = b = 1$ and volume $\mathbf{2\pi^2}$. The integral is $\pi \int_1^3 [(1 + \sqrt{1 - (x-2)^2}) - (1 - \sqrt{1 - (x-2)^2}]dx = 4\pi \int_1^3 \sqrt{1 - (x-2)^2}dx = 4\pi \int_{-1}^1 \sqrt{1 - x^2}dx = 2\pi^2$. (b) Rotation around the y axis also gives a torus. The center now goes around a circle of radius 2 so by Example 10 $V = \mathbf{4\pi^2}$. The volume by shells is $\int_1^3 2\pi x[(1 + \sqrt{1 - (x-2)^2}) - (1 - \sqrt{1 - (x-2)^2})]dx = 4\pi \int_1^3 x\sqrt{1 - (x-2)^2}dx = 4\pi \int_{-1}^1 (x+2)\sqrt{1 - x^2}dx =$ (odd integral is zero) $8\pi \int_{-1}^1 \sqrt{1 - x^2}dx = 4\pi^2$.

34 The area of a semicircle is $\frac{1}{2}\pi r^2$. Here the diameter goes from the base $y = 0$ to the top edge $y = 1 - x$ of the triangle. So the semicircle radius is $r = \frac{1-x}{2}$. The volume by slices is $\int_0^1 \frac{\pi}{2}(\frac{1-x}{2})^2 dx = [-\frac{\pi}{8}\frac{(1-x)^3}{3}]_0^1 = \mathbf{\frac{\pi}{24}}$.

36 The tilted cylinder has circular slices of area πr^2 (at all heights from 0 to h). So the volume is $\int_0^h \pi r^2 dy = \mathbf{\pi r^2 h}$. This equals the volume of an *untilted* cylinder (Cavalieri's principle: same slice areas give same volume).

40 (a) The slices are **rectangles.** (b) The slice area is $\mathbf{2\sqrt{1 - y^2}}$ **times** $\mathbf{y \tan\theta}$. (c) The volume is $\int_0^1 2\sqrt{1 - y^2}y \tan\theta dy = [-\frac{2}{3}(1 - y^2)^{3/2}\tan\theta]_0^1 = \mathbf{\frac{2}{3}\tan\theta}$. (d) Multiply radius by r and volume by $\mathbf{r^3}$.

50 Volume by shells $= \int_0^2 2\pi x(8 - x^3)dx = [8\pi x^2 - \frac{2\pi}{5}x^5]_0^2 = 32\pi - \frac{64\pi}{5} = \mathbf{\frac{96\pi}{5}}$; volume by horizontal disks $= \int_0^8 \pi(y^{1/3})^2 dy = [\frac{3\pi}{5}y^{5/3}]_0^8 = \frac{3\pi}{5}32 = \mathbf{\frac{96\pi}{5}}$.

56 $\int_1^{100} 2\pi x(\frac{1}{x})dx = 2\pi(99) = \mathbf{198\pi}$.

62 Shells around x axis: volume $= \int_{y=0}^1 2\pi y(1)dy + \int_{y=1}^e 2\pi y(1 - \ln y)dy = [\pi y^2]_0^1 + [\pi y^2 - 2\pi\frac{y^2}{2}\ln y + 2\pi\frac{y^2}{4}]_1^e = \pi + \pi e^2 - \pi e^2 + 2\pi\frac{e^2}{4} - \pi + 0 - 2\pi\frac{1}{4} = \mathbf{\frac{\pi}{2}(e^2 - 1)}$. Check disks: $\int_0^1 \pi(e^x)^2 dx = [\pi\frac{e^{2x}}{2}]_0^1 = \frac{\pi}{2}(e^2 - 1)$.

8.2 Length of a Plane Curve (page 324)

1. (This is 8.2.6) Find the length of $y = \frac{x^4}{4} + \frac{1}{8x^2}$ from $x = 1$ to $x = 2$.

   - The length is $\int ds = \int_1^2 \sqrt{1 + (\frac{dy}{dx})^2}\, dx$. The square of $\frac{dy}{dx} = x^3 - \frac{1}{4}x^{-3}$ is $(\frac{dy}{dx})^2 = x^6 - \frac{1}{2} + \frac{1}{16}x^{-6}$. Add 1 and take the square root. The arc length is

$$\int_1^2 \sqrt{x^6 + \frac{1}{2} + \frac{1}{16}x^{-6}}\, dx = \int_1^2 (x^3 + \frac{1}{4}x^{-3})dx = \frac{x^4}{4} - \frac{1}{8}x^{-2}]_1^2 = \frac{123}{32}.$$

   In "real life" the expression $\sqrt{1 + (\frac{dy}{dx})^2}$ almost never simplifies as nicely as this. We found a perfect square under the square root sign. Numerical methods are usually required to find lengths of curves.

2. Find the length of $x = \ln\cos y$ from $y = -\pi/4$ to $y = \pi/4$.

   - It seems simpler to find dx/dy than to rewrite the curve as $y = \cos^{-1} e^x$ and compute dy/dx. When you work with dx/dy, use the form $\int ds = \int_{y_1}^{y_2} \sqrt{(\frac{dx}{dy})^2 + 1}\, dy$. In this case $\frac{dx}{dy} = -\tan y$ and $(\frac{dx}{dy})^2 + 1 = \tan^2 y + 1 = \sec^2 y$. Again this has a good square root. The arc length is

$$\int_{-\pi/4}^{\pi/4} \sec y \, dy = \ln(\sec y + \tan y)]_{-\pi/4}^{\pi/4} = \ln(\frac{\sqrt{2}+1}{\sqrt{2}-1}) \approx 1.76.$$

3. The position of a point is given by $x = e^t \cos t, y = e^t \sin t$. Find the length of the path from $t = 0$ to $t = 6$.

- Since the curve is given in parametric form, it is best to use $ds = \sqrt{(\frac{dx}{dt})^2 + (\frac{dy}{dt})^2}\, dt$. This example has $dx/dt = -e^t \sin t + e^t \cos t$ and $dy/dt = e^t \cos t + e^t \sin t$. When you square and add, the cross terms involving $\sin t \cos t$ cancel each other. This leaves

$$\frac{ds}{dt} = \sqrt{2e^{2t}\sin^2 t + 2e^{2t}\cos^2 t} = \sqrt{2e^{2t}(\sin^2 t + \cos^2 t)} = \sqrt{2}e^t.$$

The arc length is $\int ds = \int_0^6 \sqrt{2}\, e^t dt = \sqrt{2}(e^6 - 1)$.

Read-throughs and selected even-numbered solutions :

The length of a straight segment (Δx across, Δy up) is $\Delta s = \sqrt{(\mathbf{\Delta x})^2 + (\mathbf{\Delta y})^2}$. Between two points on the graph of $y(x), \Delta y$ is approximately dy/dx times $\mathbf{\Delta x}$. The length of that piece is approximately $\sqrt{(\Delta x)^2 + (\mathbf{dy/dx})^2(\mathbf{\Delta x})^2}$. An infinitesimal piece of the curve has length $ds = \sqrt{\mathbf{1 + (dy/dx)^2}}\, \mathbf{dx}$. Then the arc length integral is $\int \mathbf{ds}$.

For $y = 4 - x$ from $x = 0$ to $x = 3$ the arc length is $\int_{\mathbf{0}}^{\mathbf{3}} \sqrt{\mathbf{2}}\, \mathbf{dx} = \mathbf{3\sqrt{2}}$. For $y = x^3$ the arc length integral is $\int \sqrt{\mathbf{1 + 9x^4}}\, \mathbf{dx}$.

The curve $x = \cos t, y = \sin t$ is the same as $\mathbf{x^2 + y^2 = 1}$. The length of a curve given by $x(t), y(t)$ is $\int \sqrt{(\mathbf{dx/dt})^2 + (\mathbf{dy/dt^2})}dt$. For example $x = \cos t, y = \sin t$ from $t = \pi/3$ to $t = \pi/2$ has length $\int_{\pi/\mathbf{3}}^{\pi/\mathbf{2}} \mathbf{dt}$. The speed is $ds/dt = \mathbf{1}$. For the special case $x = t, y = f(t)$ the length formula goes back to $\int \sqrt{\mathbf{1 + (f'(x))^2}}dx$.

4 $y = \frac{1}{3}(x^2 - 2)^{3/2}$ has $\frac{dy}{dx} = x(x^2 - 2)^{1/2}$ and length $= \int_2^4 \sqrt{1 + x^2(x^2 - 2)}dx = \int_2^4 (x^2 - 1)dx = \mathbf{\frac{50}{3}}$.

10 $\frac{dx}{dt} = \cos t - \sin t$ and $\frac{dy}{dt} = -\sin t - \cos t$ and $(\frac{dx}{dt})^2 + (\frac{dy}{dt})^2 = 2$. So length $= \int_0^\pi \sqrt{2}dt = \mathbf{\sqrt{2}\pi}$. The curve is a **half** of a circle of radius $\sqrt{2}$ because $x^2 + y^2 = 2$ and t stops at π.

14 $\frac{dx}{dt} = (1 - \frac{1}{2}\cos 2t)(-\sin t) + \sin 2t \cos t = \frac{3}{2}\sin t \cos 2t$. Note: first rewrite $\sin 2t \cos t = 2 \sin t \cos^2 t = \sin t(1 + \cos 2t)$. Similarly $\frac{dy}{dt} = \frac{3}{2}\cos t \cos 2t$. Then $(\frac{dx}{dt})^2 + (\frac{dy}{dt})^2 = (\frac{3}{2}\cos 2t)^2$. So length $= \int_0^{\pi/4} \frac{3}{2}\cos 2t dt = \mathbf{\frac{3}{4}}$. This is the only arc length I have ever personally discovered; the problem was meant to have an asterisk.

18 $\frac{dx}{dt} = -\sin t$ and $\frac{dy}{dt} = 3\cos t$ so length $= \int_0^{2\pi} \sqrt{\sin^2 t + 9\cos^2 t}\, dt =$ perimeter of ellipse. This integral has no closed form. Match it with a table of "elliptic integrals" by writing it as $4\int_0^{\pi/2} \sqrt{9 - 8\sin^2 t}dt = 12\int_0^{\pi/2} \sqrt{1 - \frac{8}{9}\sin^2 t}dt$. The table with $k^2 = \frac{8}{9}$ gives 1.14 for this integral or 12 (1.14) = 13.68 for the perimeter. Numerical integration is the expected route to this answer.

24 The curve $x = y^{3/2}$ is the **mirror image** of $y = x^{3/2}$ in Problem 1: same length $\frac{13^{3/2} - 4^{3/2}}{27}$ (also Problem 2).

28 (a) Length integral $= \int_0^\pi \sqrt{4\cos^2 t \sin^2 t + 4\cos^2 t \sin^2 t}\, dt = \int_0^\pi 2\sqrt{2}|\cos t \sin t|dt = \mathbf{2\sqrt{2}}$. (Notice that $\cos t$ is *negative* beyond $t = \frac{\pi}{2}$: split into $\int_0^{\pi/2} + \int_{\pi/2}^{\pi}$. (b) All points have $x + y = \cos^2 t + \sin^2 t = 1$. (c) The path from (1,0) reaches (0,1) when $t = \frac{\pi}{2}$ and returns to (1,0) at $t = \pi$. Two trips of length $\sqrt{2}$ give $2\sqrt{2}$.

8.3 Area of a Surface of Revolution (page 327)

The area of a surface of revolution is $A = \int 2\pi(\text{radius})\, ds$. Notice ds. The radius is y (or x) if the revolution is around the x (or y) axis. Use the form of ds which is most convenient:

$$ds = \sqrt{1 + (\frac{dy}{dx})^2}\, dx \quad \text{or} \quad \sqrt{(\frac{dx}{dy})^2 + 1}\, dy \quad \text{or} \quad \sqrt{(\frac{dy}{dt})^2 + (\frac{dx}{dt})^2}\, dt.$$

Pick the limits of integration which correspond to your choice of dx, dy, or dt.

1. The mirrored surface behind a searchlight bulb can be modelled by revolving the parabola $y = \frac{1}{10}x^2$ around the y axis. Find the area of the surface from $x = 0$ to $x = 20$.

   - The radius around the y axis is x, and $\frac{dy}{dx} = \frac{x}{5}$. The area formula is $A = \int 2\pi x\, ds = 2\pi \int_0^{20} x\sqrt{1 + \frac{x^2}{25}}\,dx$. Integrate by the substitution $u = 1 + \frac{x^2}{25}$ with $du = \frac{2}{25}x\, dx$ and $x\, dx = \frac{25}{2}\, du$:

$$A = 2\pi(\frac{25}{2}) \int_{u_1}^{u_2} u^{1/2}du = 25\pi(\frac{2}{3}u^{3/2})]_{u_1}^{u_2} = \frac{50}{3}\pi(1+\frac{x^2}{25})^{3/2}]_0^{20} = \frac{50}{3}\pi(17^{3/2}-1) \approx 3618 \text{ square inches.}$$

2. The curve $y = x^{2/3}$ from $(-8, 4)$ to $(8,4)$ is revolved around the x axis. Find the surface area of this hourglass.

   - This problem is trickier than it seems because $dy/dx = \frac{2}{3}x^{-1/3}$ is undefined at (0,0). We prefer to avoid that rough point. Since the curve is symmetrical, we can restrict x to $0 \le x \le 8$ and *double the answer*. Then $x = y^{3/2}$, $dx/dy = \frac{3}{2}y^{1/2}$ and area $= 2\int_0^4 2\pi y\sqrt{\frac{9}{4}y + 1}\, dy$. Substitute $u = \frac{9}{4}y + 1$ and $y = \frac{4}{9}(u - 1)$. Then u goes from 1 to 10:

$$A = 4\pi \int_1^{10} \frac{4}{9}(u-1)\sqrt{u}(\frac{4}{9}du) = \frac{64}{81}\pi \int_1^{10} (u^{3/2} - u^{1/2})du = \frac{64}{81}\pi(\frac{2}{5}u^{5/2} - \frac{2}{3}u^{3/2})]_1^{10} \approx 262.3.$$

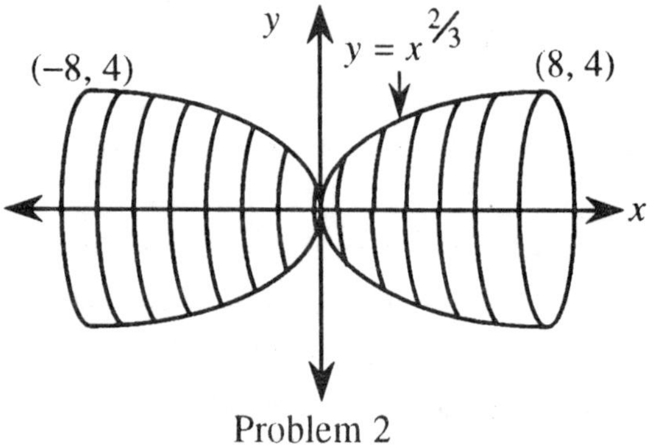

Problem 2

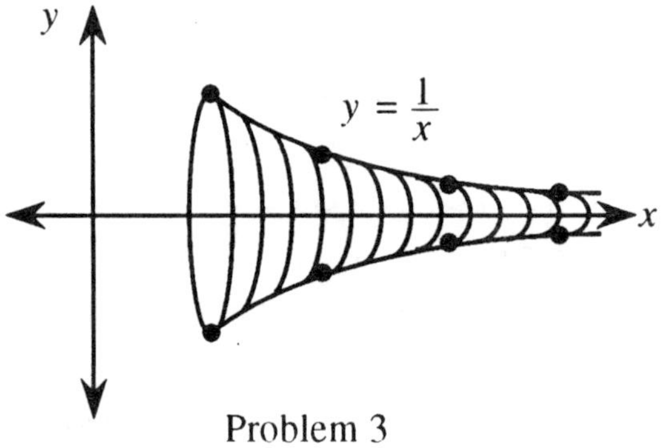

Problem 3

3. (An infinitely long horn with finite volume but infinite area) Show that when $y = 1/x$, $1 \le x \le \infty$ is revolved around the x axis the surface area is infinite but the volume is finite (Problem 8.3.21).

   - Surface area $= \int_1^\infty 2\pi y\sqrt{1 + (dy/dx)^2}dx = \int_1^\infty 2\pi\frac{1}{x}\sqrt{1 + \frac{1}{x^4}}dx = 2\pi \int_1^\infty \frac{\sqrt{x^4+1}}{x^3}dx.$

   This improper integral diverges by comparison with $\int_1^\infty \frac{1}{x}dx = \ln\infty = \infty$. (Note that $\sqrt{x^4+1} > \sqrt{x^4}$.) Therefore the surface area is infinite.

The disk method gives *volume* $= \pi \int_1^\infty y^2 dx = \pi \int_1^\infty \frac{1}{x^2} dx = \pi(-\frac{1}{x})]_1^\infty = \pi$. Therefore π cubic units of paint would fill this infinite bugle, but all the paint in the world won't cover it.

4. The circle with radius 2 centered at (0,6) is given parametrically by $x = 2\cos t$ and $y = 6+2\sin t, 0 \le t \le 2\pi$. Revolving it around the x axis forms a torus (a donut). Find the surface area.

   First a comment. The ordinary xy equation for the circle is $x^2 + (y-6)^2 = 4$ or $y = 6 \pm \sqrt{4-x^2}$. In this form the circle has two parts. The first part $y = 6 + \sqrt{4-x^2}$ is the top half of the circle – rotation gives the outer surface of the donut. The inner surface corresponds to $y = 6 - \sqrt{4-x^2}$. The *parametric form* (using t) takes care of both parts together. As t goes from 0 to 2π, the point $x = 2\cos t, y = 6 + 2\sin t$ goes around the full circle. The surface area is

$$\int_0^{2\pi} 2\pi y \, ds = 2\pi \int_0^{2\pi} (6 + 2\sin t)\sqrt{4\sin^2 t + 4\cos^2 t} \, dt = 2\pi \int_0^{2\pi} (6 + 2\sin t)(2)dt = 48\pi^2.$$

Read-throughs and selected even-numbered solutions :

A surface of revolution comes from revolving a **curve** around **an axis (a line)**. This section computes the **surface area.** When the curve is a short straight piece (length Δs), the surface is a **cone.** Its area is $\Delta S = \mathbf{2\pi r \Delta s}$. In that formula (Problem 13) r is the radius of **the circle traveled by the middle point.** The line from (0,0) to (1,1) has length $\mathbf{\Delta s} = \sqrt{\mathbf{2}}$, and revolving it produces area $\pi\sqrt{\mathbf{2}}$.

When the curve $y = f(x)$ revolves around the x axis, the area of the surface of revolution is the integral $\int \mathbf{2\pi f(x)} \sqrt{\mathbf{1 + (df/dx)^2}}\mathbf{dx}$. For $y = x^2$ the integral to compute is $\int \mathbf{2\pi x^2 \sqrt{1+4x^2}dx}$. When $y = x^2$ is revolved around the y axis, the area is $S = \int \mathbf{2\pi x\sqrt{1+(df/dx)^2}dx}$. For the curve given by $x = 2t, y = t^2$, change ds to $\sqrt{\mathbf{4+4t^2}}\mathbf{dt}$.

2 Area $= \int_0^1 2\pi x^3\sqrt{1+(3x^2)^2}\, dx = [\frac{\pi}{27}(1+9x^4)^{3/2}]_0^1 = \mathbf{\frac{\pi}{27}(10^{3/2}-1)}$

6 Area $= \int_0^1 2\pi \cosh x\sqrt{1+\sinh^2 x}\, dx = \int_0^1 2\pi\cosh^2 x dx = \int_0^1 \frac{\pi}{2}(e^{2x}+2+e^{-2x})dx = [\frac{\pi}{2}(\frac{e^{2x}}{2}+2x+\frac{e^{-2x}}{-2})]_0^1 = \frac{\pi}{2}(\frac{e^2}{2}+2+\frac{e^{-2}}{-2}-1) = \mathbf{\frac{\pi}{2}(\frac{e^2-e^{-2}}{2}+1)}$.

10 Area $= \int_0^1 2\pi x\sqrt{1+\frac{1}{9}x^{-4/3}}dx$. This is unexpectedly difficult (rotation around the x axis is easier). Substitute $u = 3x^{2/3}$ and $du = 2x^{-1/3}dx$ and $x = (\frac{u}{3})^{3/2}$: Area $= \int_0^3 2\pi(\frac{u}{3})^{3/2}\sqrt{1+\frac{1}{u^2}}\frac{du}{2}(\frac{u}{3})^{1/2} = \int_0^3 \frac{\pi}{9}u\sqrt{u^2+1}du = [\frac{\pi}{27}(u^2+1)^{3/2}]_0^3 = \mathbf{\frac{\pi}{27}(10^{3/2}-1)}$. An equally good substitution is $u = x^{4/3}+\frac{1}{9}$.

14 (a) $dS = 2\pi\sqrt{1-x^2}\sqrt{1+\frac{x^2}{1-x^2}}dx = \mathbf{2\pi dx}$. (b) The area between $x = a$ and $x = a+h$ is $\mathbf{2\pi h}$. All slices of thickness h have this area, whether the slice goes near the center or near the outside. (c) $\frac{1}{4}$ of the Earth's area is above latitude **30°** where the height is $R\sin 30° = \frac{R}{2}$. The slice from the Equator up to 30° has the same area (and so do two more slices below the Equator).

20 Area $= \int_{1/2}^1 2\pi x\sqrt{1+\frac{1}{x^4}}dx = \int_{1/2}^1 2\pi\frac{\sqrt{x^4+1}}{x^4}x^3dx$. Substitute $u = \sqrt{x^4+1}$ and $du = 2x^3dx/u$ to find $\int_{\sqrt{17}/4}^{\sqrt{2}} \frac{\pi u^2 du}{u^2-1} = [\pi u - \frac{\pi}{2}\ln\frac{u+1}{u-1}]_{\sqrt{17}/4}^{\sqrt{2}} = \pi(\sqrt{2}-\frac{\sqrt{17}}{4}-\frac{1}{2}\ln\frac{\sqrt{2}+1}{\sqrt{2}-1}+\frac{1}{2}\ln\frac{\sqrt{17}+4}{\sqrt{17}-4}) \approx 5.0$.

8.4 Probability and Calculus (page 334)

1. The probability of a textbook page having no errors is $\frac{4}{5}$. The probability of one error is $\frac{4}{25}$ and the probability of n errors is $\frac{4}{5^{n+1}}$. What is the chance that the page will have (a) at least one error (b) fewer than 3 errors (c) more than 5 errors?

This is a discrete probability problem. The probabilities $\frac{4}{5}, \frac{4}{25}, \frac{4}{125}, \cdots$ add to one.

- (a) Since the probability of no errors is $\frac{4}{5}$, the probability of 1 or more is $1 - \frac{4}{5} = \frac{1}{5}$.
- (b) Add the probabilities of zero, one, and two errors: $\frac{4}{5} + \frac{4}{25} + \frac{4}{125} = \frac{124}{125} = 99.2\%$.
- (c) There are two ways to consider more than five errors. The first is to add the probabilities of 0, 1, 2, 3, 4, and 5 errors. The sum is .999936 and subtraction leaves $1 - .999936 = .000064$. The second way is to add the infinite series $P(6) + P(7) + P(8) + \cdots = \frac{4}{5^7} + \frac{4}{5^8} + \frac{4}{5^9} + \cdots$. Use the formula for a geometric series:

$$\text{sum} = \frac{a}{1-r} = \frac{\text{first term}}{1-(\text{ratio})} = \frac{4/5^7}{1-\frac{1}{5}} = 0.000064.$$

2. If $p(x) = \frac{1}{\pi(1+x^2)}$ for $-\infty < x < \infty$, find the approximate probability that

(a) $-1 \le x \le 1$ (b) $x \ge 10$ (c) $x \ge 0$.

- This is continuous probability. Instead of adding p_n, we integrate $p(x)$:

(a) $\frac{1}{\pi}\int_{-1}^{1} \frac{dx}{1+x^2}$ (b) $\frac{1}{\pi}\int_{10}^{\infty} \frac{dx}{1+x^2}$ (c) $\frac{1}{\pi}\int_{0}^{\infty} \frac{1}{1+x^2}dx$.

The integral of $\frac{1}{1+x^2}$ is $\arctan x$. Remember $\tan\frac{\pi}{4} = 1$ and $\tan(-\frac{\pi}{4}) = -1$:

(a) $\frac{1}{\pi}\arctan x]_{-1}^{1} = \frac{1}{2}$ (b) $\frac{1}{\pi}\arctan x]_{10}^{\infty} \approx 0.0317$ (c) $\frac{1}{\pi}\int_0^{\infty} \frac{dx}{1+x^2} = \frac{1}{2}$.

Part (c) is common sense. Since $\frac{1}{1+x^2}$ is symmetrical (or *even*), the part $x \ge 0$ gives half the area.

3. If $p(x) = \frac{k}{x^4}$ is the probability density for $x \ge 1$, find the constant k.

- The key is to remember that the total probability must be one:

$$\int_1^{\infty} \frac{k}{x^4}dx = 1 \text{ yields } [-\frac{k}{3}x^{-3}]_1^{\infty} = \frac{k}{3} = 1. \text{ Therefore } k = 3.$$

4. Find the *mean* (the expected value = average x) for $p(x) = \frac{3}{x^4}$.

- This is a continuous probability, so use the formula $\mu = \int xp(x)dx$:

$$\text{The mean is } \mu = \int_1^{\infty} \frac{3x}{x^4}dx = -\frac{3}{2}x^{-2}]_1^{\infty} = \frac{3}{2}.$$

5. Lotsa Pasta Restaurant has 50 tables. On an average night, 5 parties are no-shows. So they accept 55 reservations. What is the probability that they will have to turn away a party with a reservation?

- This is like Example 3 in the text, which uses the Poisson model. X is the number of no-shows and the average is $\lambda = 5$. The number of no-shows is n with probability $p_n = \frac{5^n}{n!}e^{-5}$. A party is turned away if $n < 5$. Compute $p_0 + p_1 + p_2 + p_3 + p_4 = e^{-5} + 5e^{-5} + \frac{25}{2}e^{-5} + \frac{125}{6}e^{-5} + \frac{625}{24}e^{-5} \approx 0.434$. In actual fact the overflow parties are sent to the bar for a time $t \to \infty$.

6. Lotsa Pasta takes no reservations for lunch. The average waiting time for a table is 20 minutes. What are your chances of being seated in 15 minutes?

- This is like Example 4. Model the waiting time by $p(x) = \frac{1}{20}e^{-x/20}$ for $0 \le x \le \infty$. The mean waiting time is 20. The chances of being seated in 15 minutes are about 50-50:

$$P(0 \le x \le 15) = \int_0^{15} \frac{1}{20}e^{-x/20}dx = -e^{-x/20}\Big]_0^{15} \approx 0.528.$$

7. A random sample of 1000 baseballs in the factory found 62 to be defective. Estimate with 95% confidence the true percentage of defective baseballs.

- Since a baseball is either defective or not, this is like a yes-no poll (Example 7). The 95% margin of error is never more than $\frac{1}{\sqrt{N}}$, in this case $\frac{1}{\sqrt{1000}} \approx 0.03$. The "poll" of baseballs found 6.2% defective. We are 95% sure that the true percentage is between $6.2 - .03 = 6.17\%$ and $6.2 + .03 = 6.23\%$.

  Amazingly, the error formula $\frac{1}{\sqrt{N}}$ does not depend on the size of the total population. A poll of 400 people has a margin of error no more than $\frac{1}{\sqrt{400}} = \frac{1}{20} = 5\%$, whether the 400 people are from a city of 100,000 or a country of 250 million. The catch is that the sample must be truly random – very hard to do when dealing with people.

8. Grades on the English placement exam at Absorbine Junior College are normally distributed. The mean is $\mu = 72$ and the standard deviation is $\sigma = 10$. The top 15% are placed in Advanced English. What test grade is the cutoff?

   The distribution of grades looks like Figure 8.12a in the text, except that $\mu = 72$ and $\sigma = 10$. Figure 8.12b gives the area under the curve from ∞ to x. From the graph we see that 84% of the scores are at or below $\mu + \sigma = 82$. Therefore 16% are above, close to the desired 15%. The cutoff is a little over 82. (Statisticians have detailed tables for the area $F(x)$.)

Read-throughs and selected even-numbered solutions:

Discrete probability uses counting, **continuous** probability uses calculus. The function $p(x)$ is the probability **density.** The chance that a random variable falls between a and b is $\int_{\mathbf{a}}^{\mathbf{b}} \mathbf{p(x)dx}$. The total probability is $\int_{-\infty}^{\infty} p(x)dx = \mathbf{1}$. In the discrete case $\sum p_n = \mathbf{1}$. The mean (or expected value) is $\mu = \int \mathbf{xp(x)dx}$ in the continuous case and $\mu = \sum np_n$ in the **discrete case.**

The Poisson distribution with mean λ has $p_n = \lambda^{\mathbf{n}}\mathbf{e}^{-\lambda}/\mathbf{n!}$. The sum $\sum p_n = 1$ comes from the **exponential** series. The exponential distribution has $p(x) = e^{-x}$ or $2e^{-2x}$ or $\mathbf{ae^{-ax}}$. The standard Gaussian (or **normal**) distribution has $\sqrt{2\pi}p(x) = e^{-x^2/2}$. Its graph is the well-known **bell-shaped** curve. The chance that the variable falls below x is $F(x) = \int_{-\infty}^{\mathbf{x}} \mathbf{p(x)dx}$. F is the **cumulative** density function. The difference $F(x + dx) - F(x)$ is about $\mathbf{p(x)dx}$, which is the chance that X is between x and $x + dx$.

The *variance,* which measures the spread around μ, is $\sigma^2 = \int \mathbf{(x - \mu)^2 p(x)dx}$ in the continuous case and $\sigma^2 = \sum \mathbf{(n - \mu)^2 p_n}$ in the discrete case. Its square root σ is the **standard deviation.** The normal distribution has $p(x) = \mathbf{e^{-(x-\mu)^2/2\sigma^2}}/\sqrt{2\pi}\sigma$. If $\overline{X}$ is the **average** of N samples from any population with mean μ and variance σ^2, the Law of Averages says that $\overline{X}$ will approach **the mean** μ. The Central Limit Theorem says that the distribution for $\overline{X}$ approaches **a normal distribution.** Its mean is μ and its variance is σ^2/N.

In a yes-no poll when the voters are 50-50, the mean for one voter is $\mu = 0(\frac{1}{2}) + 1(\frac{1}{2}) = \mathbf{\frac{1}{2}}$. The variance is $(0-\mu)^2 p_0 + (1-\mu)^2 p_1 = \mathbf{\frac{1}{4}}$. For a poll with $N = 100, \bar{\sigma}$ is $\sigma/\sqrt{\mathrm{N}} = \mathbf{\frac{1}{20}}$. There is a 95% chance that $\overline{X}$ (the fraction saying yes) will be between $\mu - 2\bar{\sigma} = \mathbf{\frac{1}{2}} - \mathbf{\frac{1}{10}}$ and $\mu + 2\bar{\sigma} = \mathbf{\frac{1}{2}} + \mathbf{\frac{1}{10}}$.

2 The probability of an odd $X = 1, 3, 5, \cdots$ is $\frac{1}{2} + \frac{1}{8} + \frac{1}{32} + \cdots = \frac{\frac{1}{2}}{1-\frac{1}{4}} = \mathbf{\frac{2}{3}}$. The probabilities $p_n = (\frac{1}{3})^n$ **do not add to 1.** They add to $\frac{1}{3} + \frac{1}{9} + \cdots = \frac{1}{2}$ so the adjusted $p_n = 2(\frac{1}{3})^n$ add to 1.

12 $\mu = \int_0^\infty xe^{-x}dx = uv - \int v\,du = -xe^{-x}]_0^\infty + \int_0^\infty e^{-x}dx = \mathbf{1}$.

20 (a) Heads and tails are still equally likely. (b) The coin is still fair so the expected fraction of heads during the second N tosses is $\frac{1}{2}$ and the expected fraction overall is $\mathbf{\frac{1}{2}}(\alpha + \mathbf{\frac{1}{2}})$; which is the average.

28 $\mu = (p_1 + p_2 + p_3 + \cdots) + (p_2 + p_3 + p_4 + \cdots) + (p_3 + p_4 + \cdots) + \cdots = (1) + (\frac{1}{2}) + (\frac{1}{4}) + \cdots = \mathbf{2}$.

32 $2000 \pm 2\sigma$ gives **1700 to 2300** as the 95% confidence interval.

34 The average has mean $\bar{\mu} = 30$ and deviation $\bar{\sigma} = \frac{8}{\sqrt{N}} = 1$. An actual average of $\frac{2000}{64} = 31.25$ is $1.25\,\bar{\sigma}$ above the mean. The probability of exceeding $1.25\,\bar{\sigma}$ is about **.1** from Figure 8.12b. With $N = 256$ we still have $\frac{8000}{256} = 31.25$ but now $\bar{\sigma} = \frac{8}{\sqrt{256}} = \frac{1}{2}$. To go $2.5\,\bar{\sigma}$ above the mean has **probability** $< \mathbf{.01}$.

8.5 Masses and Moments (page 340)

1. Masses of 6, 4, and 3 are placed at $x = 4$, $x = 5$, and $x = 8$. Find the *moment* and the *center of mass.*

   - The total mass is $M = 6 + 4 + 3 = 13$. The moment is $m_1x_1 + m_2x_2 + m_3x_3 = 6 \cdot 4 + 4 \cdot 5 + 3 \cdot 8 = 68$. The center of mass is moment divided by total mass, or $\bar{x} = \frac{68}{13} \approx 5.2$.

2. Continuous density along the axis from $x = 0$ to $x = 4$ is given by $\rho = \sqrt{x}$. Find the moment M_y and the center of mass.

   - Total mass is $M = \int \rho dx = \int_0^4 \sqrt{x}\,dx = \frac{16}{3}$. Total moment is

$$M_y = \int_0^4 x\rho(x)dx = \int_0^4 x^{3/2}dx = \frac{2}{5}x^{5/2}]_0^4 = \frac{64}{5}.$$

   The center of mass $\bar{x}$ is $\frac{M_y}{M} = \frac{64/5}{16/3} = \frac{12}{5} = 2.4$.

3. Four unit masses are placed at $(x, y) = (-1,0)$, $(1,0)$, $(0,1)$ and $(0,2)$. Find M_x, M_y, and $(\bar{x}, \bar{y})$.

   - The total mass is $M = 4$. The moment around the y axis is zero because the placement is symmetric. (Check: $1 \cdot 1 + 1(-1) + 1 \cdot 0 + 1 \cdot 0 = 0$.) The moment around the x axis is the sum of the y coordinates (not x!!) times the masses (all 1): $1 \cdot 0 + 1 \cdot 0 + 1 \cdot 1 + 1 \cdot 2 = 3$. Then the center of mass is $(\frac{0}{4}, \frac{3}{4}) = (\bar{x}, \bar{y})$.

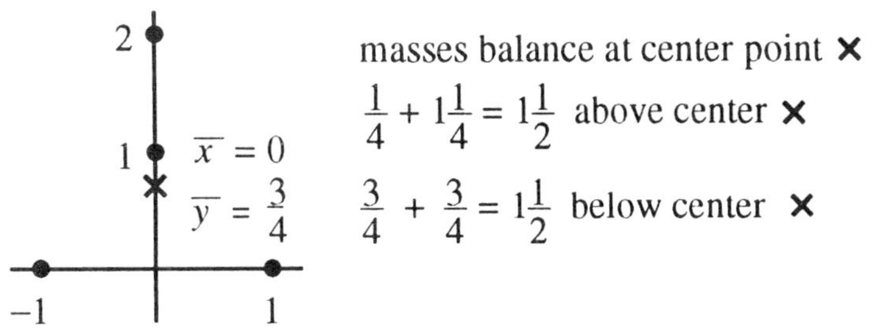

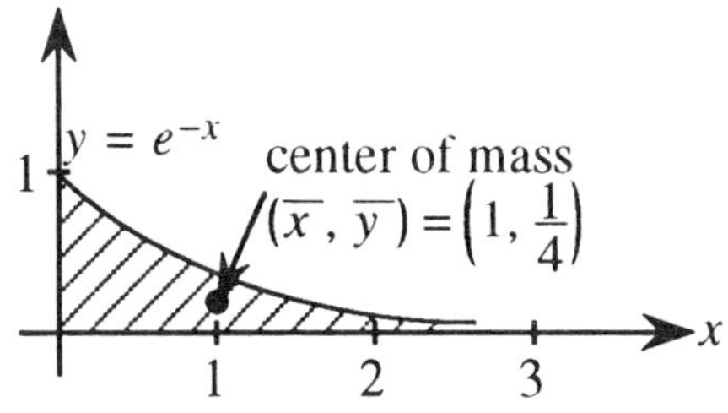

4. Find the area M and the centroid $(\bar{x}, \bar{y})$ inside the curve $y = e^{-x}$ for $0 \le x < \infty$.

- The area M is the value of the improper integral $\int_0^\infty e^{-x}dx = -e^{-x}]_0^\infty = 1$. For M_y we integrate by parts:

$$M_y = \int_0^\infty xy\ dx = \int_0^\infty xe^{-x}dx = [-xe^{-x} - e^{-x}]_0^\infty = 1.$$

Note: The first term $-xe^{-x}$ approaches zero as $x \to \infty$. We can use l'Hôpital's rule from Section 2.6: $\lim \frac{-x}{e^x} = \lim \frac{-1}{e^x} = 0$. The moment M_x around the x axis is an integral *with respect to* y:

$$M_x = \int_0^1 yx\ dy = \int_0^1 y(-\ln y)dy = [-\frac{1}{2}y^2 \ln y + \frac{1}{4}y^2]_0^1 = \frac{1}{4}.$$

We found $x = -\ln y$ from $y = e^{-x}$. The integration was by parts. At $y = 0$ l'Hôpital's rule evaluates $\lim y^2 \ln y = \lim \frac{\ln y}{y^{-2}} = \lim \frac{\frac{1}{y}}{\frac{-2}{y^3}} = 0$. The centroid is $(\bar{x}, \bar{y}) = (\frac{M_y}{M}, \frac{M_x}{M}) = (1, \frac{1}{4})$.

5. Find the surface area of the cone formed when the line $y = 3x$ from $x = 0$ to $x = 2$ is revolved about the y axis. Use the theorem of Pappus in Problem 8.5.28.

- The Pappus formula $A = 2\pi\bar{y}L$ changes to $A = 2\pi\bar{x}L$, since the rotation is about the y axis and not the x axis. The line goes from (0,0) to (2,6) so its length is $L = \sqrt{2^2 + 6^2} = \sqrt{40}$. Then $\bar{x} = 1$ and the surface area is $A = 2\pi\sqrt{40} = 4\pi\sqrt{10}$.

Read-throughs and selected even-numbered solutions :

If masses m_n are at distances x_n, the total mass is $M = \mathbf{\sum m_n}$. The total moment around $x = 0$ is $M_y = \mathbf{\sum m_n x_n}$. The center of mass is at $\bar{x} = \mathbf{M_y/M}$. In the continuous case, the mass distribution is given by the **density** $\rho(x)$. The total mass is $M = \int \boldsymbol{\rho(x)dx}$ and the center of mass is at $\bar{x} = \int \mathbf{x\rho(x)dx/M}$. With $\rho = x$, the integrals from 0 to L give $M = \mathbf{L^2/2}$ and $\int x\rho(x)dx = \mathbf{L^3/3}$ and $\bar{x} = \mathbf{2L/3}$. The total moment is the same as if the whole mass M is placed at $\bar{\mathbf{x}}$.

In a plane with masses m_n at the points (x_n, y_n), the moment around the y axis is $\mathbf{\sum m_n x_n}$. The center of mass has $\bar{x} = \mathbf{\sum m_n x_n / \sum m_n}$ and $\bar{y} = \mathbf{\sum m_n y_n / \sum m_n}$. For a plate with density $\rho = 1$, the mass M equals the **area**. If the plate is divided into vertical strips of height $y(x)$, then $M = \int y(x)dx$ and $M_y = \int \mathbf{xy(x)}dx$.

For a square plate $0 \le x, y \le L$, the mass is $M = \mathbf{L^2}$ and the moment around the y axis is $M_y = \mathbf{L^3/2}$. The center of mass is at $(\bar{x}, \bar{y}) = (\mathbf{L/2}, \mathbf{L/2})$. This point is the **centroid,** where the plate balances.

A mass m at a distance x from the axis has moment of inertia $I = \mathbf{mx^2}$. A rod with $\rho = 1$ from $x = a$ to $x = b$ has $I_y = \mathbf{b^3/3 - a^3/3}$. For a plate with $\rho = 1$ and strips of height $y(x)$, this becomes $I_y = \int \mathbf{x^2 y(x) dx}$. The torque T is **force** times **distance.**

4 $M = \int_0^L x^2 dx = \frac{\mathbf{L^3}}{\mathbf{3}}$; $M_y = \int_0^L x^3 dx = \frac{\mathbf{L^4}}{\mathbf{4}}$; $\bar{x} = \frac{L^4/4}{L^3/3} = \frac{\mathbf{3L}}{\mathbf{4}}$.

10 $M = 3(\frac{1}{2}\mathbf{ab})$; $M_y = \int_0^a 3xb(1 - \frac{x}{a})dx = [\frac{3x^2 b}{2} - \frac{x^3 b}{a}]_0^a = \frac{\mathbf{a^2 b}}{\mathbf{2}}$ and by symmetry $M_x = \frac{\mathbf{b^2 a}}{\mathbf{2}}$; $\bar{x} = \frac{a^2 b/2}{3ab/2} = \frac{\mathbf{a}}{\mathbf{3}}$ and $\bar{y} = \frac{\mathbf{b}}{\mathbf{3}}$. Note that the centroid of the triangle is at $(\frac{a}{3}, \frac{b}{3}.)$

14 Area $M = \int_0^1 (x - x^2)dx = \frac{\mathbf{1}}{\mathbf{6}}$; $M_y = \int_0^1 x(x - x^2)dx = \frac{1}{12}$ and $\bar{x} = \frac{1/12}{1/6} = \frac{\mathbf{1}}{\mathbf{2}}$ (also by symmetry); $M_x = \int_0^1 y(\sqrt{y} - y)dy = \frac{1}{15}$ and $\bar{y} = \frac{1/15}{1/6} = \frac{\mathbf{2}}{\mathbf{5}}$.

16 Area $M = \frac{1}{2}(\pi(2)^2 - \pi(0)^2) = \frac{\mathbf{3\pi}}{\mathbf{2}}$; $M_y = 0$ and $\bar{x} = \mathbf{0}$ by symmetry; M_x for halfcircle of radius 2 minus M_x for halfcircle of radius 1 = (by Example 4) $\frac{2}{3}(2^3 - 1^3) = \frac{14}{3}$ and $\bar{y} = \frac{14/3}{3\pi/2} = \frac{\mathbf{28}}{\mathbf{9\pi}}$.

18 $I_y = \int_{-a/2}^{a/2} x^2$ (strip height) $dx = \int_{-a/2}^{a/2} x^2 a dx = \frac{a^4}{12}$.

32 Torque $= F - 2F + 3F - 4F \cdots + 9F - 10F = \mathbf{-5F}$.

36 $J = \frac{I}{mr^2}$ is smaller for a solid ball than a solid cylinder because the ball has its mass **nearer the center.**

38 Get most of the mass close to the center but keep the radius large.

42 (a) **False** (a solid ball goes faster than a hollow ball) (b) **False** (if the density is varied, the center of mass moves) (c) **False** (you reduce I_x but you increase I_y : the y direction is upward) (d) **False** (imagine the jumper as an arc of a circle going just over the bar: the center of mass of the arc stays below the bar).

8.6 Force, Work, and Energy (page 346)

1. (This is 8.6.8) The great pyramid outside Cairo (height 500 feet) has a square base 800 feet by 800 feet. If the rock weighs $w = 100$ lb/ft^3, how much work did it take to lift it all?

- Imagine the pyramid divided into horizontal square slabs of side s and thickness Δh. The weight of this slab is $ws^2\Delta h$. This slab is carried up to height h. The work done on the slab is therefore $hws^2\Delta h$.

  To get s in terms of h, use a straight line between $s = 800$ when $h = 0$ and $s = 0$ when $h = 500$:

$$s = 800 - \frac{800}{500}h = \frac{8}{5}(500 - h).$$

  Add up the work on all the slabs and let the thickness Δh go to zero (so integrate):

$$W = \int_0^{500} wh(\frac{8}{5}(500 - h))^2 dh = \frac{64}{25}w \int_0^{500} (250{,}000h - 1000h^2 + h^3)dh = 1.3 \times 10^{12} \text{ ft-lbs.}$$

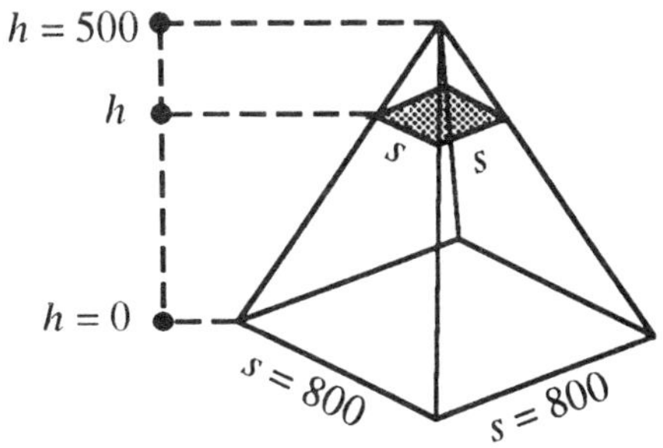

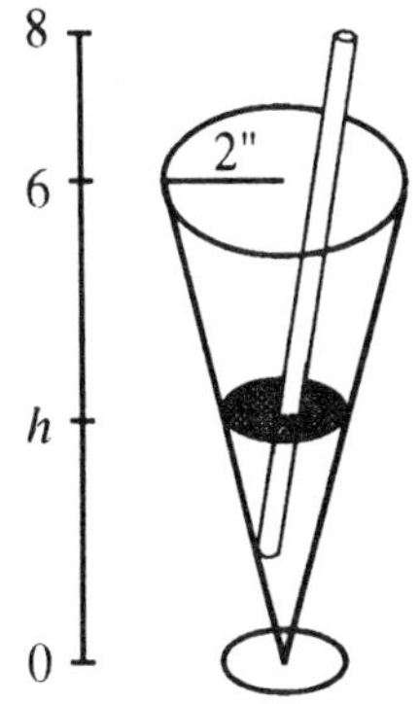

2. A conical glass is filled with water ($6''$ high, $2''$ radius at the top). How much work is done sipping the water out through a straw? The sipper is $2''$ above the glass. Water weighs $w = .036$ lb/in^3.

- As in the previous problem, calculate the work on a slice of water at height h with thickness Δh. The slice volume is $\pi r^2 \Delta h$ and its weight is $w\pi r^2 \Delta h$. The slice must be lifted $(8 - h)$ inches to get sipped, so the work is $(8 - h)w\pi r^2 \Delta h$. (You may object that water must go down to get into the straw and then be lifted from there. But the net effect of water going down to the bottom and back to its starting level is zero work.) From the shape of the glass, $r = \frac{1}{3}h$. The work integral is

$$w\pi \int_0^6 (8-h)(\frac{h}{3})^2 dh = \frac{w\pi}{9}(\frac{8}{3}h^3 - \frac{1}{4}h^4)]_0^6 = 3.2 \text{ inch-pounds.}$$

3. A 10-inch spring requires a force of 25 pounds to stretch it $\frac{1}{2}$ inch. Find (a) the spring constant k in pounds per foot (b) the work done in stretching the spring (c) the work needed to stretch it an additional half-inch.

- (a) Hooke's law $F = kx$ gives $k = \frac{F}{x} = \frac{25}{0.5} = 50$ lbs per foot.
- (b) Work $= \int_0^{1/2} kx\, dx = \frac{1}{2}50x^2]_0^{0.5} = 6.25$ ft-lbs (Not ft/lbs.)
- (c) Work $= \int_{1/2}^1 kx\, dx = \frac{1}{2}k(1)^2 - \frac{1}{2}k(\frac{1}{2})^2 = 18.75$ ft-lbs. The definite integral is the change in potential energy $\frac{1}{2}kx^2$ between $x = \frac{1}{2}$ and $x = 1$.

4. A rectangular water tank is 100 ft long, 30 ft wide, and 20 ft deep. What is the force on the bottom and on each side of a full tank? Water weighs $w = 62.4$ lb/ft^3.

- The force on the bottom is $whA = (62.4)(20)(3000) \approx 3.7 \cdot 10^6$ lbs. On the small ends, which are $20' \times 30'$ rectangles, the force is $\int_0^{20} w \cdot 30 \cdot h\, dh$. The limits of integration are 0 to 20 because the water depth h varies from 0 to 20. The value of the integral is $[15wh^2]_0^{20} \approx 3.7 \times 10^4$ lbs. The larger sides are $100' \times 20'$ rectangles, so the force on them is $\int_0^{20} 100wh\, dh \approx 1.2 \times 10^6$ lbs.

Read-throughs and selected even-numbered solutions :

Work equals **force** times **distance.** For a spring the force $F = \mathbf{kx}$ is proportional to the extension x (this is **Hooke's** law). With this variable force, the work in stretching from 0 to x is $W = \int \mathbf{kx\, dx} = \mathbf{\frac{1}{2}kx^2}$. This equals the increase in the **potential** energy V. Thus W is a **definite** integral and V is the corresponding **indefinite** integral, which includes an arbitrary **constant**. The derivative dV/dx equals **the force**. The force of gravity is $F = \mathbf{GMm/x^2}$ and the potential is $V = \mathbf{-GMm/x}$.

In falling, V is converted to **kinetic** energy $K = \mathbf{\frac{1}{2}mv^2}$. The total energy $K + V$ is **constant** (this is the law of **conservation of energy** when there is no external force).

Pressure is force per unit **area**. Water of density w in a pool of depth h and area A exerts a downward force $F = \mathbf{whA}$ on the base. The pressure is $p = \mathbf{wh}$. On the sides the **pressure** is still wh at depth h, so the total force is $\int whl\,dh$, where l is **the side length at depth h**. In a cubic pool of side s, the force on the base is $F = \mathbf{ws^3}$, the length around the sides is $l = \mathbf{4\pi s}$, and the total force on the four sides is $F = \mathbf{2\pi ws^3}$. The work to pump the water out of the pool is $W = \int whA\,dh = \mathbf{\frac{1}{2}ws^4}$.

2 (a) Spring constant $k = \frac{75 \text{ pounds}}{3 \text{ inches}} = $ **25 pounds per inch**

(b) Work $W = \int_0^3 kx\,dx = 25(\frac{9}{2}) = \mathbf{\frac{225}{2}}$ **inch-pounds** or $\frac{225}{24}$ foot-pounds (integral starts at no stretch)

(c) Work $W = \int_3^6 kx\,dx = 25(\frac{36-9}{2}) = \mathbf{\frac{675}{2}}$ **inch-pounds.**

10 The change in $V = -\frac{GmM}{x}$ is $\Delta V = GmM(\frac{1}{R-10} - \frac{1}{R+10}) = GmM\frac{20}{R^2-10^2} = \frac{20GmM}{R^2}\frac{R^2}{R^2-10^2}$. The first factor is the distance (20 feet) times the force (30 pounds). The second factor is the correction (practically 1.)

12 If the rocket starts at R and reaches x, its potential energy increases by $GMm(\frac{1}{R} - \frac{1}{x})$. This equals $\frac{1}{2}mv^2$ (gain in potential = loss in kinetic energy) so $\mathbf{\frac{1}{R} - \frac{1}{x} = \frac{v^2}{2GM}}$ and $\mathbf{x = (\frac{1}{R} - \frac{v^2}{2GM})^{-1}}$. If the rocket reaches $x = \infty$ then $\frac{1}{R} = \frac{v^2}{2GM}$ or $v = \sqrt{\frac{2GM}{R}} = 25{,}000$ mph.

14 A horizontal slice with radius 1 foot, height h feet, and density ρ lbs/ft^3 has potential energy $\pi(1)^2 h\rho dh$. Integrate from $h = 0$ to $h = 4$: $\int_0^4 \pi\rho h dh = \mathbf{8\pi\rho}$.

20 Work to empty a cone-shaped tank: $W = \int wAhdh = \int_0^H w\pi r^2 \frac{h^3}{H^2} dh = \mathbf{w\pi r^2 \frac{H^2}{4}}$. For a cylinder (Problem 17) $W = \frac{1}{2}wAH^2 = w\pi r^2 \frac{H^2}{2}$. So the work for a cone is half of the work for a cylinder, even though the volume is only one third. (The cone-shaped tank has more water concentrated near the bottom.)

8 Chapter Review Problems

Review Problems

R1 How do you find the area between the graphs of $y_1(x)$ and $y_2(x)$?

R2 Write the definite integrals for the volumes when an arch of $y = x \sin x$ is revolved about the x axis and the y axis.

R3 Write the definite integrals for the volumes when the region above that arch and below $y = 2$ is revolved about the x axis and y axis.

R4 Write the three formulas for ds. How is ds used in finding the length of a curve and the surface area of a solid of revolution?

R5 What is the difference between discrete and continuous probability? Compare the discrete and continuous formulas for the mean and for the sum of all probabilities.

R6 What happens to the cumulative density function in Figure 8.12 if we let the x axis extend toward $-\infty$ and $+\infty$?

R7 How do the "moments" M_x and M_y differ from the "*moment of inertia*"?

R8 How do you find the work done by a constant force F? By a variable force $F(x)$?

Drill Problems

Find the areas of the regions bounded by the curves in **D1** to **D7**. Answers are not 100% guaranteed.

D1 $y = 4 - x^2, y = x - 2$ Ans $\frac{125}{6}$

D2 $y = x^3, y = x$ Ans $\frac{1}{2}$

D3 $x = \sqrt{y}, x = \sqrt[3]{y}$ Ans $\frac{1}{12}$

D4 $y = x^2, y = \frac{1}{x}, x = 2$ Ans $\frac{7}{13} + \ln 2$

D5 $y = \frac{1}{x^2+4x+5}, x = 0, y = 0$ Ans $\frac{\pi}{2} - \tan^{-1} 2$

D6 $y = \tan^{-1} x, y = \sec^{-1} x, y = 0$ Ans $\ln 2$

D7 $xy = 8, y = 3x - 10, x = 1$ Ans $7.5 + 16 \ln 2$

D8 The region between the parabola $y = x^2$ and the line $y = 4x$ is revolved about the x axis. Find the volume. Ans 8π

D9 The region bounded by $y = x^{-1/2}, x = \frac{1}{4}, x = 1$, and $y = 0$ is revolved about the x axis. Find the volume.
Ans $2\pi \ln 2$

D10 The region between $y = e^{x/2}, x = 0$, and $y = e$ is revolved about the x axis. Find $\pi(e^2 + 1)$ as volume.

D11 The region bounded by $y = x \ln x, x = 0, x = \pi$, and the x axis is revolved about the y axis. Find the volume.
Ans $2\pi^2$

D12 The region bounded by $y = \sin x^2, y = 0$, and $x = \sqrt{\pi/2}$ is revolved about the y axis. Find the volume.
Ans π

D13 The right half of the ellipse $\frac{x^2}{4} + \frac{y^2}{9} = 1$ is revolved about the y axis. Find the volume. Ans 16π

D14 A chopped-off cone is 15′ high. The upper radius is 3′ while the lower radius is 10′. Find its volume by horizontal slices.
Ans 695π

D15 Find the length of $y = \frac{x^4}{8} + \frac{1}{4x^2}$ from $x = 1$ to $x = 2$. Ans $\frac{17}{12}$

D16 Find the length of $y = 2x^{2/3}$ from $x = 1$ to $x = 8$. Ans $\frac{17}{12}$

D17 Set up an integral for the length of $y = \ln x$ from $x = 1$ to $x = e$.

D18 Find the length of the curve $e^x = \sin y$ for $\frac{\pi}{6} \le y \le \frac{\pi}{2}$. Ans $\ln(2 + \sqrt{3})$

D19 Find the length of the curve $x = \frac{1}{2}\ln(1 + t^2),\ y = \tan^{-1} t$ from $t = 0$ to $t = \pi$. Ans $\sqrt{2}(e^\pi - 1)$

D20 Find the length of the curve $x = \frac{1}{2}t^2,\ y = \frac{1}{15}(10t + 25)^{3/2}$ from $t = 0$ to $t = 4$. Ans 28

D21 Find the length of the curve $x = t^{3/2},\ y = 4t,\ 0 \le t \le 4$. Ans $\frac{488}{27}$

Find the areas of the surfaces formed by revolving the curves **D22** to **D26**.

D22 $y = x^3$ around the x axis, $0 \le x \le 2$ Ans $\frac{\pi}{27}(145^{3/2} - 1)$

D23 $x = 2\sqrt{y}$ around the y axis, $1 \le y \le 4$ Ans $\frac{8\pi}{3}(5^{3/2} - 2^{3/2})$

D24 $y = \sin x^2$ around the x axis, $0 \le x \le \frac{\sqrt{\pi}}{2}$ Ans $\frac{\pi}{2}(2 - \sqrt{2})$

D25 $y = \frac{x^3}{6} + \frac{1}{2x}$ around the y axis, $1 \le x \le 2$ Ans $\pi(3.75 + \ln 2)$

D26 $x = \ln\sqrt{y} - \frac{1}{4}y^2$ around the y axis, $0 \le y \le 3$ Ans $\frac{32\pi}{3}$

D27 The line at a post office has the distribution $p(x) = 0.4e^{-0.4x}$, where x is the waiting time in minutes for a random customer. (a) What is the mean waiting time? (b) What percentage of customers wait longer than six minutes? Ans (a) 4 min (b) $e^{-2.4} \approx 9.1\%$

D28 The weights of oranges vary normally with a mean of 5 ounces and a standard deviation of 9.8 ounces. 95% of the oranges lie between what weights? Ans 3.4 oz and 6.6 oz

D29 Show that $M_x = \frac{1}{10}, M_y = \frac{1}{4}$, and $(\bar{x}, \bar{y}) = (\frac{3}{4}, \frac{3}{10})$ for the region between $y = x^2, y = 0$, and $x = 1$.

D30 Find the moments of inertia I_x and I_y for the triangle with corners (0,0), (2,0) and (0,3).

D31 A force of 50 pounds stretches a spring 4 inches. Find the work required to stretch it an additional 4 inches. Ans 25 ft - lbs.

D32 A 10′ hanging rope weighs 2 lbs. How much work to wind up the rope? Ans 10 ft - lbs.

Calculator Problems

C1 Find the length of the graph $y = x^3$ from (0,0) to (2,8). Ans 8.3

C2 Compute the surface area when the arc $x^2 + y^2 = 25$ from (3,4) to (4,3) is revolved about the x axis.

C3 Find the distance around the ellipse $x^2 + 4y^2 = 1$.

CHAPTER 9 POLAR COORDINATES AND COMPLEX NUMBERS

9.1 Polar Coordinates (page 350)

Circles around the origin are so important that they have their own coordinate system – *polar coordinates.* The center at the origin is sometimes called the "pole." A circle has an equation like $r = 3$. Each point on that circle has two coordinates, say $r = 3$ and $\theta = \frac{\pi}{2}$. This angle locates the point 90° around from the x axis, so it is on the y axis at distance 3.

The connection to x and y is by the equations $x = r\cos\theta$ and $y = r\sin\theta$. Substituting $r = 3$ and $\theta = \frac{\pi}{2}$ as in our example, the point has $x = 3\cos\frac{\pi}{2} = 0$ and $y = 3\sin\frac{\pi}{2} = 3$. The polar coordinates are $(r,\theta) = (3, \frac{\pi}{2})$ and the rectangular coordinates are $(x, y) = (0, 3)$.

1. Find polar coordinates for these points – first with $r \geq 0$ and $0 \leq \theta < 2\pi$, then three other pairs (r,θ) that give the same point:

   (a) $(x,y) = (\sqrt{3}, 1)$ (b) $(x,y) = (-1,1)$ (c) $(x,y) = (-3,-4)$

   - (a) $r^2 = x^2 + y^2 = 4$ yields $r = 2$ and $\frac{y}{x} = \frac{1}{\sqrt{3}} = \tan\theta$ leads to $\theta = \frac{\pi}{6}$. The polar coordinates are $(2, \frac{\pi}{6})$. Other representations of the same point are $(2, \frac{\pi}{6} + 2\pi)$ and $(2, \frac{\pi}{6} - 2\pi)$. Allowing $r < 0$ we have $(-2, -\frac{5\pi}{6})$ and $(-2, \frac{7\pi}{6})$. There are an infinite number of possibilities.
   - (b) $r^2 = x^2 + y^2$ yields $r = \sqrt{2}$ and $\frac{y}{x} = \frac{1}{-1} = \tan\theta$. Normally the arctan function gives $\tan^{-1}(-1) = -\frac{\pi}{4}$. But that is a fourth quadrant angle, while the point $(-1,1)$ is in the second quadrant. The choice $\theta = \frac{3\pi}{4}$ gives the "standard" polar coordinates $(\sqrt{2}, \frac{3\pi}{4})$. Other representations are $(\sqrt{2}, \frac{11\pi}{4})$ and $(\sqrt{2}, -\frac{5\pi}{4})$. Allowing negative r we have $(-\sqrt{2}, -\frac{\pi}{4})$ and $(-\sqrt{2}, \frac{7\pi}{4})$.
   - (c) The point $(-3,-4)$ is in the third quadrant with $r = \sqrt{9+16} = 5$. Choose $\theta = \pi + \tan^{-1}(\frac{-4}{-3}) \approx \pi + 0.9 \approx 4$ radians. Other representations of this point are $(5, 2\pi + 4)$ and $(5, 4\pi + 4)$, and $(-5, 0.9)$.

2. Convert $(r,\theta) = (6, -\frac{\pi}{2})$ to rectangular coordinates by $x = r\cos\theta$ and $y = r\sin\theta$.

   - The x coordinate is $6\cos(-\frac{\pi}{2}) = 0$. The y coordinate is $6\sin(-\frac{\pi}{2}) = -6$.

3. The Law of Cosines in trigonometry states that $c^2 = a^2 + b^2 - 2ab\cos C$. Here a, b and c are the side lengths of the triangle and C is the angle opposite side c. Use the Law of Cosines to find the distance between the points with polar coordinates (r,θ) and (R,φ).

   Does it ever happen that c^2 is larger than $a^2 + b^2$?

   - In the figure, the desired distance is labeled d. The other sides of the triangle have lengths R and r. The angle opposite d is $(\varphi - \theta)$. The Law of Cosines gives $d = \sqrt{R^2 + r^2 - 2Rr\cos(\varphi - \theta)}$.

   Yes, c^2 is larger than $a^2 + b^2$ when the angle $C = \varphi - \theta$ is larger than 90°. Its cosine is negative. The next problem is an example. When the angle C is *acute* (smaller than 90°) then the term $-2ab\cos C$ reduces c^2 below $a^2 + b^2$.

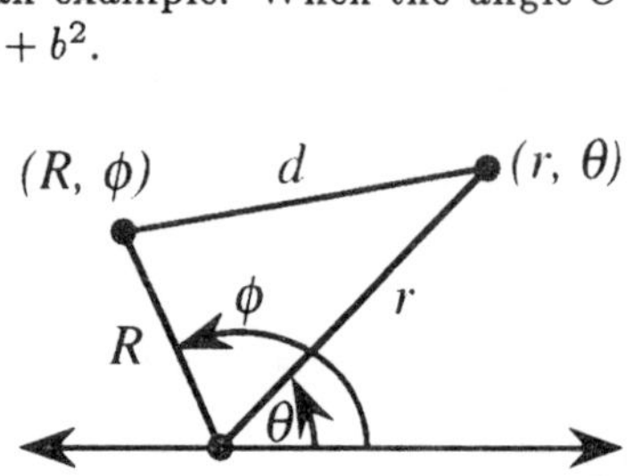

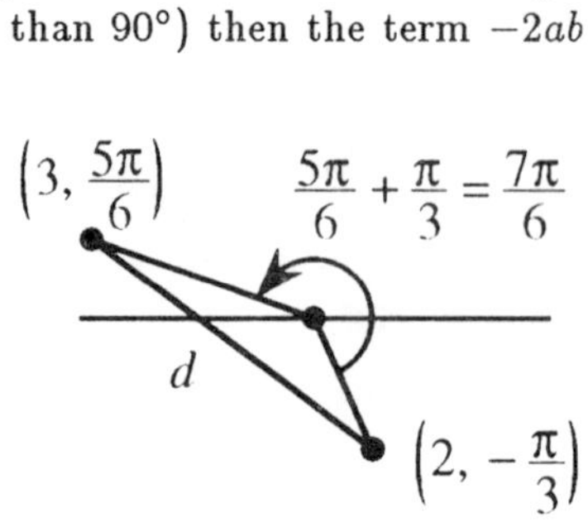

3′. Use the formula in Problem 3 to find the distance between the polar points $(3, \frac{5\pi}{6})$ and $(2, -\frac{\pi}{3})$.

- $d = \sqrt{3^2 + 2^2 - 2\cdot 3\cdot 2\cos(\frac{5\pi}{6} - (-\frac{\pi}{3}))} = \sqrt{13 - 12\cos\frac{7\pi}{6}} = \sqrt{13 + 6\sqrt{3}} \approx 4.8.$

4. Sketch the regions that are described in polar coordinates by
   (a) $r \geq 0$ and $\frac{\pi}{3} < \theta < \frac{2\pi}{3}$ (b) $1 \leq r \leq 2$ (c) $0 \leq \theta < \frac{\pi}{3}$ and $0 \leq r < 3$.

- The three regions are drawn. For (a), the dotted lines mean that $\theta = \frac{\pi}{3}$ and $\theta = \frac{2\pi}{3}$ are not included. If $r < 0$ were also allowed, there would be a symmetric region below the axis – a shaded X instead of a shaded V.

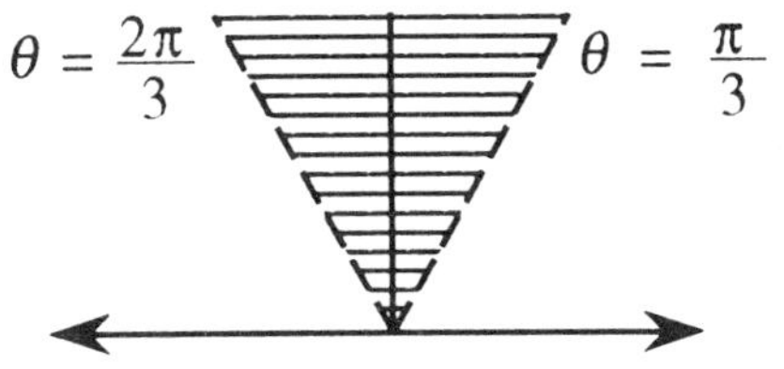

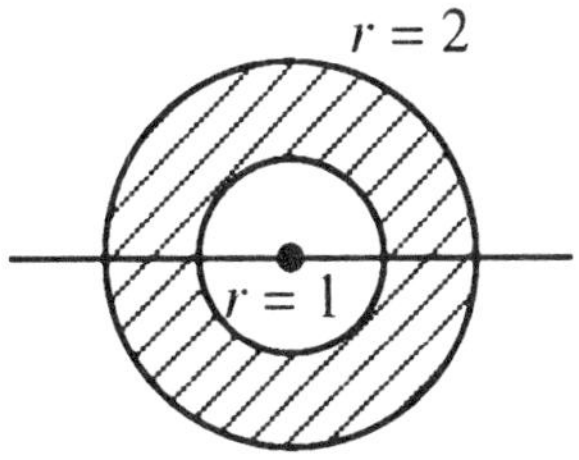

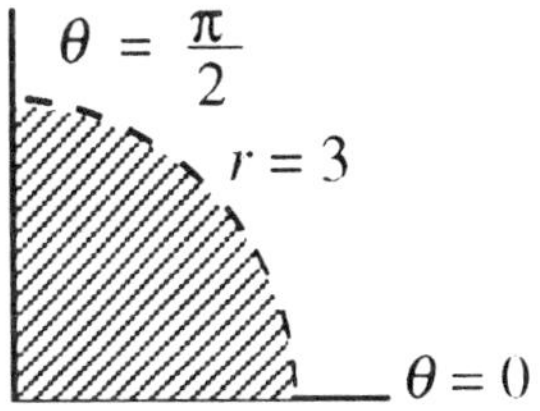

5. Write the polar equation for the circle centered at $(x, y) = (1, 1)$ with radius $\sqrt{2}$.

- The rectangular equation is $(x-1)^2 + (y-1)^2 = 2$ or $x^2 - 2x + y^2 - 2y = 0$. Replace x with $r\cos\theta$ and y with $r\sin\theta$. Always replace $x^2 + y^2$ with r^2. The equation becomes $r^2 = 2r\cos\theta + 2r\sin\theta$. Divide by r to get $r = 2\,(\cos\theta + \sin\theta)$.

Note that $r = 0$ when $\theta = -\frac{\pi}{4}$. The circle goes through the origin.

6. Write the polar equations for these lines: (a) $x = 3$ (b) $y = -1$ (c) $x + 2y = 5$.

- (a) $x = 3$ becomes $r\cos\theta = 3$ or $r = 3\sec\theta$. Remember: $r = 3\cos\theta$ is a circle.
- (b) $y = -1$ becomes $r\sin\theta = -1$ or $r = -\csc\theta$. But $r = -\sin\theta$ is a circle.
- (c) $x + 2y = 5$ becomes $r\cos\theta + 2r\sin\theta = 5$. Again $r = \cos\theta + 2\sin\theta$ is a circle.

Read-throughs and selected even-numbered solutions :

Polar coordinates r and θ correspond to $x = \mathbf{r\cos}\,\theta$ and $y = \mathbf{r\sin}\,\theta$. The points with $r > 0$ and $\theta = \pi$ are located **on the negative x axis**. The points with $r = 1$ and $0 \leq \theta \leq \pi$ are located on **a semicircle**. Reversing the sign of θ moves the point (x, y) to $(\mathbf{x}, -\mathbf{y})$.

Given x and y, the polar distance is $r = \sqrt{\mathbf{x^2 + y^2}}$. The tangent of θ is $\mathbf{y/x}$. The point (6,8) has $r = \mathbf{10}$ and $\theta = \mathbf{tan^{-1}\frac{8}{6}}$. Another point with the same θ is $(\mathbf{3}, \mathbf{4})$. Another point with the same r is $(\mathbf{10{,}0})$. Another point with the same r and $\tan\theta$ is $(-\mathbf{6}, -\mathbf{8})$.

The polar equation $r = \cos\theta$ produces a shifted **circle**. The top point is at $\theta = \pi/\mathbf{4}$, which gives $r = \sqrt{\mathbf{2}}/\mathbf{2}$. When θ goes from 0 to 2π, we go **two** times around the graph. Rewriting as $r^2 = r\cos\theta$ leads to the xy equation $\mathbf{x^2 + y^2} = \mathbf{x}$. Substituting $r = \cos\theta$ into $x = r\cos\theta$ yields $x = \mathbf{cos^2}\theta$ and similarly $y = \mathbf{cos}\,\theta\ \mathbf{sin}\,\theta$. In this form x and y are functions of the **parameter** θ.

10 $r = 3\pi, \theta = 3\pi$ has rectangular coordinates $\mathbf{x = -3\pi, y = 0}$

16 (a) $(-1, \frac{\pi}{2})$ is the same point as $(\mathbf{1}, \mathbf{\frac{3\pi}{2}})$ or $(-1, \frac{5\pi}{2})$ or $\cdots$ (b) $(-1, \frac{3\pi}{4})$ is the same point as $(\mathbf{1}, \mathbf{\frac{7\pi}{4}})$ or $(-1, -\frac{\pi}{4})$ or $\cdots$ (c) $(1, -\frac{\pi}{2})$ is the same point as $(-1, \frac{\pi}{2})$ or $(\mathbf{1}, \mathbf{\frac{3\pi}{2}})$ or $\cdots$ (d) $r = 0, \theta = 0$ is the same point as $\mathbf{r} = \mathbf{0}$, $\theta =$ **any angle.**

18 (a) **False** ($r = 1, \theta = \frac{\pi}{4}$ is a different point from $r = -1, \theta = -\frac{\pi}{4}$) (b) **False** (for fixed r we can add any multiple of 2π to θ) (c) **True** ($r \sin\theta = 1$ is the horizontal line $y = 1$).

22 Take the line from (0,0) to (r_1, θ_1) as the base (its length is r_1). The height of the third point (r_2, θ_2), measured perpendicular to this base, is r_2 times $\sin(\theta_2 - \theta_1)$.

26 From $x = \cos^2\theta$ and $y = \sin\theta\cos\theta$, square and add to find $\mathbf{x^2 + y^2} = \cos^2\theta(\cos^2\theta + \sin^2\theta) = \cos^2\theta = \mathbf{x}$.

28 Multiply $r = a\cos\theta + b\sin\theta$ by r to find $x^2 + y^2 = ax + by$. Complete squares in $x^2 - ax = (x - \frac{a}{2})^2 - (\frac{a}{2})^2$ and similarly in $y^2 - by$ to find $(x - \frac{a}{2})^2 + (y - \frac{b}{2})^2 = (\frac{a}{2})^2 + (\frac{b}{2})^2$. This is a circle centered at $(\frac{a}{2}, \frac{b}{2})$ with radius $r = \sqrt{(\frac{a}{2})^2 + (\frac{b}{2})^2} = \mathbf{\frac{1}{2}\sqrt{a^2 + b^2}}$.

9.2 Polar Equations and Graphs (page 355)

The polar equation $r = F(\theta)$ is like $y = f(x)$. For each angle θ the equation tells us the distance r (which is now allowed to be negative). By connecting those points we get a polar curve. Examples are $r = 1$ and $r\cos = \theta$ (circles) and $r = 1 + \cos\theta$ (cardioid) and $r = 1/(1 + e\cos\theta)$ (parabola, hyperbola, or ellipse, depending on e). These have nice-looking polar equations – because the origin is a special point for those curves.

Note $y = \sin x$ would be a disaster in polar coordinates. Literally it becomes $r\sin\theta = \sin(r\cos\theta)$. This mixes r and θ together. It is comparable to $x^3 + xy^2 = 1$, which mixes x and y. (For mixed equations we need implicit differentiation.) Equations in this section are not mixed, they are $r = F(\theta)$ and sometimes $r^2 = F(\theta)$.

Part of drawing the picture is recognizing the symmetry. One symmetry is "through the pole." If r changes to $-r$, the equation $r^2 = F(\theta)$ stays the same – this curve has *polar symmetry.* But $r = \tan\theta$ also has polar symmetry, because $\tan\theta = \tan(\theta + \pi)$. If we go around by 180°, or π radians, we get the same result as changing r to $-r$.

The three basic symmetries are across the x axis, across the y axis, and through the pole. Each symmetry has two main tests. (This is not clear in some texts I consulted.) Since one test could be passed without the other, I think you need to try both tests:

- x axis symmetry: θ to $-\theta$ (test 1) or θ to $\pi - \theta$ and r to $-r$ (test 2)
- y axis symmetry: θ to $\pi - \theta$ (test 1) or θ to $-\theta$ and r to $-r$ (test 2)
- polar symmetry: θ to $\pi + \theta$ (test 1) or r to $-r$ (test 2).

1. Sketch the polar curve $r^2 = 4\sin\theta$ after a check for symmetry.

   - When r is replaced by $-r$, the equation $(-r)^2 = 4\sin\theta$ is the same. This means polar symmetry (through the origin). If θ is replaced by $(\pi - \theta)$, the equation $r^2 = 4\sin(\pi - \theta) = 4\sin\theta$ is still the same. There is symmetry about the y axis. *Any two symmetries* (out of three) *imply the third.* This graph must be symmetric across the x axis. (θ to $-\theta$ doesn't show it, because $\sin\theta$ changes. But r to $-r$ and θ to $\pi - \theta$ leaves $r^2 = 4\sin\theta$ the same.) We can plot the curve in the first quadrant and reflect it to get the complete graph. Here is a table of values for the first quadrant and a sketch of the curve. The two closed parts (not circles) meet at $r = 0$.

θ	0	$\frac{\pi}{6}$	$\frac{\pi}{4}$	$\frac{\pi}{3}$	$\frac{\pi}{2}$
$4\sin\theta$	0	2	$2\sqrt{2}$	$2\sqrt{3}$	4
$r = \sqrt{4\sin\theta}$	0	$\sqrt{2} \approx 1.4$	$\sqrt{2\sqrt{2}} \approx 1.7$	$\sqrt{2\sqrt{3}} \approx 1.9$	2

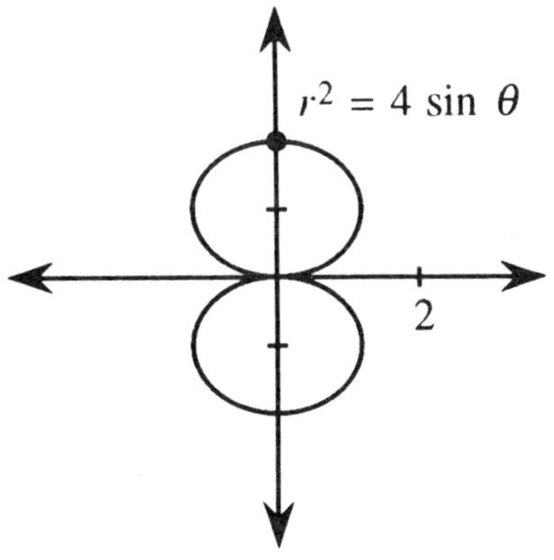

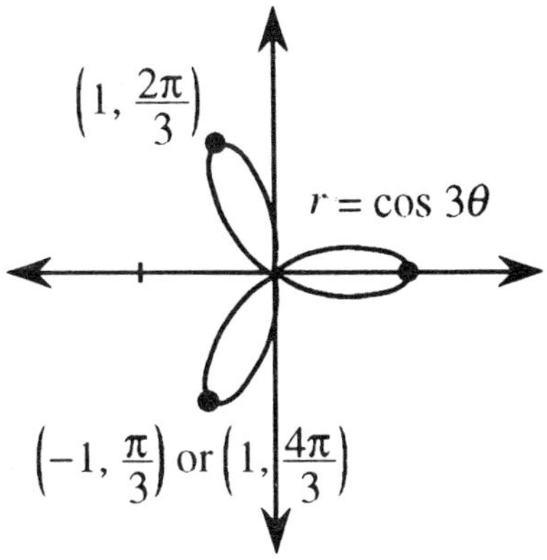

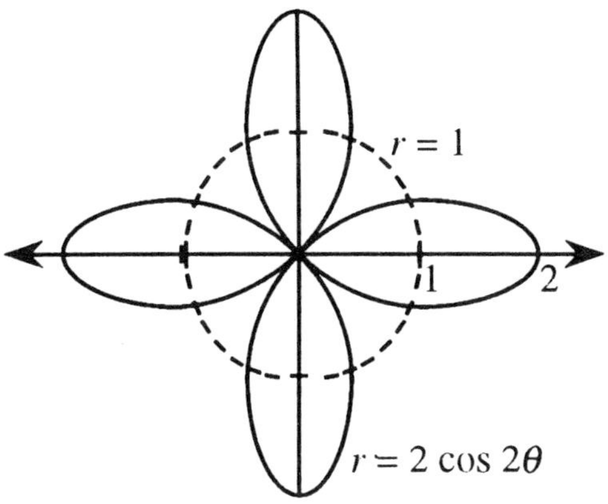

2. (This is Problem 9.2.9) Check $r = \cos 3\theta$ for symmetry and sketch its graph.

   - The cosine is even, $\cos(-3\theta) = \cos 3\theta$, so this curve is symmetric across the x axis (where θ goes to $-\theta$). The other symmetry tests fail. For θ up to $\frac{\pi}{2}$ we get a loop and a half in the figure. Reflection across the x axis yields the rest. The curve has three petals.

θ	0	$\frac{\pi}{12}$	$\frac{\pi}{6}$	$\frac{\pi}{4}$	$\frac{\pi}{3}$	$\frac{5\pi}{12}$	$\frac{\pi}{2}$
$r\cos 3\theta$	1	$\frac{\sqrt{2}}{2}$	0	$-\frac{\sqrt{2}}{2}$	-1	$-\frac{\sqrt{2}}{2}$	0

3. Find the eight points where the four petals of $r = 2\cos 2\theta$ cross the circle $r = 1$.

   - Setting $2\cos 2\theta = 1$ leads to four crossing points $(1, \frac{\pi}{6})$, $(1, \frac{7\pi}{6})$, $(1, -\frac{\pi}{6})$, and $(1, -\frac{7\pi}{6})$. The sketch shows *four other crossing points*: $(1, \frac{\pi}{3})$, $(1, \frac{2\pi}{3})$, $(1, \frac{4\pi}{3})$ and $(1, \frac{5\pi}{3})$. These coordinates *do not* satisfy $r = 2\cos 2\theta$. But $r < 0$ yields other names $(-1, \frac{4\pi}{3})$, $(-1, \frac{5\pi}{3})$, $(-1, \frac{\pi}{3})$ and $(-1, \frac{2\pi}{3})$ for these points, that do satisfy the equation.

In general, you need a sketch to find all intersections.

4. Identify these five curves:

   (a) $r = 5\csc\theta$ (b) $r = 6\sin\theta + 4\cos\theta$ (c) $r = \frac{9}{1+6\cos\theta}$ (d) $r = \frac{4}{2+\cos\theta}$ (e) $r = \frac{1}{3-3\sin\theta}$.

   - (a) $r = \frac{5}{\sin\theta}$ is $r\sin\theta = 5$. This is the *horizontal line* $y = 5$.
   - Multiply equation (b) by r to get $r^2 = 6r\sin\theta + 4r\cos\theta$, or $x^2 + y^2 = 6y + 4x$. Complete squares to $(x-2)^2 + (y-3)^2 = 2^2 + 3^2 = 13$. This is a *circle* centered at (2,3) with radius $\sqrt{13}$.
   - (c) The pattern for conic sections (ellipse, parabola, and hyperbola) is $r = \frac{A}{1+e\cos\theta}$. Our equation has $A = 9$ and $e = 6$. The graph is a *hyperbola* with one focus at (0,0). The directrix is the line $x = \frac{9}{6} = \frac{3}{2}$.
   - (d) $r = \frac{4}{2+\cos\theta}$ doesn't exactly fit $\frac{A}{1+e\cos\theta}$ because of the 2 in the denominator. Factor it out: $\frac{2}{1+\frac{1}{2}\cos\theta}$ is an *ellipse* with $e = \frac{1}{2}$.
   - (e) $r = \frac{1}{3-3\sin\theta}$ is actually a *parabola*. To recognize the standard form, remember that $-\sin\theta = \cos(\frac{\pi}{2}+\theta)$. So $r = \frac{\frac{1}{3}}{1+\cos(\frac{\pi}{2}+\theta)}$. Since θ is replaced by $(\frac{\pi}{2}+\theta)$, the standard parabola has been rotated.

5. Find the length of the major axis (the distance between vertices) of the hyperbola $r = \frac{A}{1+e\cos\theta}$.

   - Figure 9.5c in the text shows the vertices on the x axis: $\theta = 0$ gives $r = \frac{A}{1+e}$ and $\theta = \pi$ gives $r = \frac{A}{1-e}$. (The hyperbola has $A > 0$ and $e > 1$.) Notice that $(\frac{A}{1-e}, \pi)$ is on the *right* of the origin because $r = \frac{A}{1-e}$ is negative. The distance between the vertices is $\frac{A}{e-1} - \frac{A}{e+1} = \frac{2A}{e^2-1}$.

   Compare with exercise 9.2.35 for the ellipse. The distance between its vertices is $2a = \frac{2A}{1-e^2}$. The distance between vertices of a parabola ($e = 1$) is $\frac{2A}{0}$ = infinty! One vertex of the parabla is out at infinity.

Read-throughs and selected even-numbered solutions :

The circle of radius 3 around the origin has polar equation $\mathbf{r = 3}$. The 45° line has polar equation $\boldsymbol{\theta = \pi/4}$. Those graphs meet at an angle of **90°**. Multiplying $r = 4\cos\theta$ by r yields the xy equation $\mathbf{x^2 + y^2 = 4x}$. Its graph is a **circle** with center at **(2,0)**. The graph of $r = 4/\cos\theta$ is the line $x = \mathbf{4}$. The equation $r^2 = \cos 2\theta$ is not changed when $\theta \to -\theta$ (symmetric across **the x axis**) and when $\theta \to \pi + \theta$ (or $r \to \mathbf{-r}$). The graph of $r = 1 + \cos\theta$ is a **cardioid**.

The graph of $r = A/(\mathbf{1} + \mathbf{e}\cos\theta)$ is a conic section with one focus at $(\mathbf{0}, \mathbf{0})$. It is an ellipse if $\mathbf{e < 1}$ and a hyperbola if $\mathbf{e > 1}$. The equation $r = 1/(1+\cos\theta)$ leads to $r + x = 1$ which gives a **parabola**. Then r = distance from origin equals $1 - x$ = distance from **directrix** $\mathbf{y = 1}$. The equations $r = 3(1-x)$ and $r = \frac{1}{3}(1-x)$ represent a **hyperbola** and an **ellipse**. Including a shift and rotation, conics are determined by **five** numbers.

6 $r = \frac{1}{1+2\cos\theta}$ is the hyperbola of Example 7 and Figure 9.5c: $r + 2r\cos\theta = 1$ is $r = 1 - 2x$ or $x^2 + y^2 = 1 - 4x + 4x^2$. The figure should show $r = -1$ and $\theta = \pi$ on the right branch.

14 $r = 1 - 2\sin 3\theta$ has **y axis symmetry**: change θ to $\pi - \theta$, then $\sin 3(\pi - \theta) = \sin(\pi - 3\theta) = \sin 3\theta$.

22 If $\cos\theta = \frac{r^2}{4}$ and $\cos\theta = 1 - r$ then $\frac{r^2}{4} = 1 - r$ and $r^2 + 4r - 4 = 0$. This gives $r = -2 - \sqrt{8}$ and $\mathbf{r = -2 + \sqrt{8}}$. The first r is negative and cannot equal $1 - \cos\theta$. The second gives $\cos\theta = 1 - r = 3 - \sqrt{8}$ and $\theta \approx$ **80° or** $\theta \approx$ **−80°**. The curves also meet at **the origin** $\mathbf{r = 0}$ and at the point $\mathbf{r = -2}, \boldsymbol{\theta = 0}$ which is also $\mathbf{r = +2}, \boldsymbol{\theta = \pi}$.

26 The other 101 petals in $r = \cos 101\theta$ are **duplicates of the first 101**. For example $\theta = \pi$ gives $r = \cos 101\pi = -1$ which is also $\theta = 0, r = +1$. (Note that $\cos 100\pi = +1$ gives a new point.)

28 (a) Yes, x and y symmetry imply r symmetry. Reflections across the x axis and then the y axis take (x, y) to $(x, -y)$ to $(-x, -y)$ which is reflection through the origin. (b) The point $r = -1, \theta = \frac{3\pi}{2}$ satisfies the equation $r = \cos 2\theta$ and it is the same point as $r = 1, \theta = \frac{\pi}{2}$.

32 (a) $\theta = \frac{\pi}{2}$ gives $r = 1$; this is $x = 0, y = 1$ (b) The graph crosses the x axis at $\theta = 0$ and π where $x = \frac{1}{1+e}$ and $x = \frac{-1}{1-e}$. The center of the graph is halfway between at $x = \frac{1}{2}(\frac{1}{1+e} - \frac{1}{1-e}) = \frac{-e}{1-e^2}$. The second focus is twice as far from the origin at $\mathbf{\frac{-2e}{1-e^2}}$. (Check: $e = 0$ gives center of circle, $e = 1$ gives second focus of parabola at infinity.)

9.3 Slope, Length, and Area for Polar Curves (page 359)

This section does *calculus in polar coordinates*. All the calculations for $y = f(x)$ – its slope $\frac{dy}{dx}$ and area

$\int y\,dx$ and arc length $\int \sqrt{1+(\frac{dy}{dx})^2}\,dx$ – can also be done for polar curves $r = F(\theta)$. But the formulas are a little more complicated! The slope is not $\frac{dF}{d\theta}$ and the area is not $\int F(\theta)d\theta$. These problems give practice with the polar formulas for slope, area, arc length, and surface area of revolution.

1. (This is 9.3.5) Draw the 4-petaled flower $r = \cos 2\theta$ and find the area inside. The petals are along the axes.

   - We compute the area of one petal and multiply by 4. The right-hand petal lies between the lines $\theta = -\frac{\pi}{4}$ and $\theta = \frac{\pi}{4}$. Those are the limits of integration:

$$\text{Area} = 4\int_{-\pi/4}^{\pi/4} \frac{1}{2}(\cos 2\theta)^2 d\theta = \int_{-\pi/4}^{\pi/4} (1+\cos 4\theta)d\theta = \frac{\pi}{2}.$$

2. Find the area inside $r = 2(1+\cos\theta)$ and outside $r = 2(1-\cos\theta)$. Sketch those cardioids.

   - In the figure, half the required area is shaded. *Take advantage of symmetries!* A typical line through the origin is also sketched. Imagine this line sweeping from $\theta = 0$ to $\theta = \frac{\pi}{2}$ – the whole shaded area is covered. The outer radius is $2(1+\cos\theta)$, the inner radius is $2(1-\cos\theta)$. The shaded area is

$$\int_0^{\pi/2} \frac{1}{2}[4(1+\cos\theta)^2 - 4(1-\cos\theta)^2]d\theta = 8\int_0^{\pi/2} \cos\theta\, d\theta = 8. \qquad \text{Total area 16.}$$

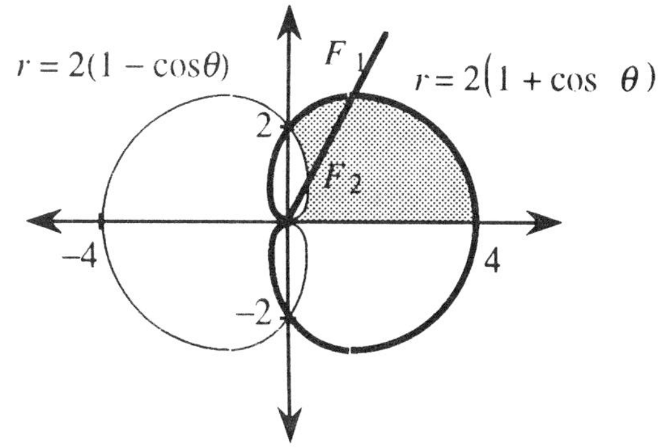

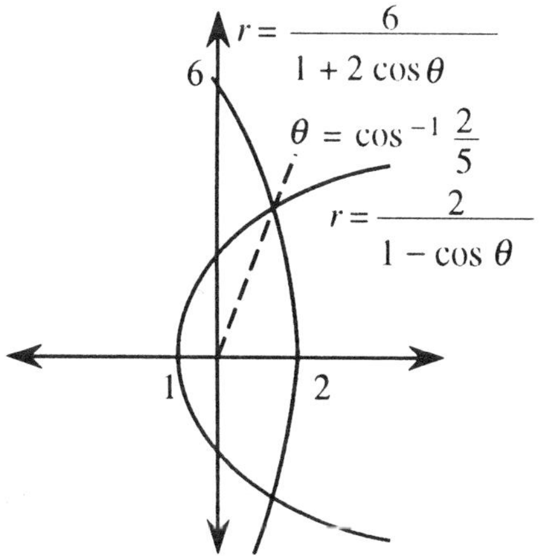

3. Set up the area integral(s) between the parabola $r = \frac{2}{1-\cos\theta}$ and the hyperbola $r = \frac{6}{1+2\cos\theta}$.

   - The curves are shown in the sketch. We need to find where they cross. Solving $\frac{6}{1+2\cos\theta} = \frac{2}{1-\cos\theta}$ yields $6(1-\cos\theta) = 2(1+2\cos\theta)$ or $\cos\theta = \frac{2}{5} = .4$. At that angle $r = \frac{6}{1+2(\frac{2}{5})} = \frac{6}{1.8}$.

   Imagine a ray sweeping around the origin from $\theta = 0$ to $\theta = \pi$. From $\theta = 0$ to $\theta = \cos^{-1}.4$, the ray crosses the *hyperbola*. Then it crosses the *parabola*. That is why the area must be computed in two parts. Using symmetry we find only the top half:

$$\text{Half-area} = \int_0^{\cos^{-1}.4} \frac{1}{2}\left(\frac{6}{1+2\cos\theta}\right)^2 d\theta \quad + \quad \int_{\cos^{-1}.4}^{\pi} \frac{1}{2}\left(\frac{2}{1-\cos\theta}\right)^2 d\theta.$$

   Simpson's rule gives the total area (top half doubled) as approximately 12.1.

Problems* 4 *and* 5 *are about lengths of curves.

4. Find the distance around the cardioid $r = 1 + \cos\theta$.

   - Length in polar coordinates is $ds = \sqrt{(\frac{dr}{d\theta})^2 + r^2}\, d\theta$. For the cardioid this square root is

$$\sqrt{(-\sin\theta)^2 + (1+\cos\theta)^2} = \sqrt{\sin^2\theta + \cos^2\theta + 1 + 2\cos\theta} = \sqrt{2 + 2\cos\theta}.$$

Half the curve is traced as θ goes from 0 to π. The total length is $\int ds = 2\int_0^\pi \sqrt{2+2\cos\theta}\, d\theta$. Evaluating this integral uses the trick $1 + \cos\theta = 2\cos^2\frac{\theta}{2}$. Thus the cardioid length is

$$2\int_0^\pi \sqrt{4\cos^2\frac{\theta}{2}}\, d\theta = 4\int_0^\pi \cos\frac{\theta}{2}\, d\theta = 8\sin\frac{\theta}{2}\Big]_0^\pi = 8.$$

5. Find the length of the spiral $r = e^{\theta/2}$ as θ goes from 0 to 2π.

   - For this curve $ds = \sqrt{(\frac{dr}{d\theta})^2 + r^2}\, d\theta$ is equal to $\sqrt{\frac{1}{4}e^\theta + e^\theta}\, d\theta = \sqrt{\frac{5}{4}e^\theta}\, d\theta = \frac{\sqrt{5}}{2}e^{\theta/2}d\theta$:

$$\text{Length} = \int_0^{2\pi} \frac{\sqrt{5}}{2}e^{\theta/2}d\theta = \sqrt{5}\, e^{\theta/2}\Big]_0^{2\pi} = \sqrt{5}(e^\pi - 1) \approx 49.5.$$

Problems* 6 *and* 7 *ask for the areas of surfaces of revolution.

6. Find the surface area when the spiral $r = e^{\theta/2}$ between $\theta = 0$ and $\theta = \pi$ is revolved about the horizontal axis.

   - From Section 8.3 we know that the area is $\int 2\pi y\, ds$. For this curve the previous problem found $ds = \frac{\sqrt{5}}{2}e^{\theta/2}d\theta$. The factor y in the area integral is $r\sin\theta = e^{\theta/2}\sin\theta$. The area is

$$\begin{aligned}\textstyle\int_0^\pi 2\pi(e^{\theta/2}\sin\theta)\frac{\sqrt{5}}{2}e^{\theta/2}d\theta &= \textstyle\sqrt{5}\pi\int_0^\pi e^\theta \sin\theta d\theta \\ &= \textstyle\frac{\sqrt{5}\pi}{2}e^\theta(\sin\theta - \cos\theta)\Big]_0^\pi = \frac{\sqrt{5}\pi}{2}(e^\pi + 1) \approx 84.8.\end{aligned}$$

7. Find the surface area when the curve $r^2 = 4\sin\theta$ is revolved around the y axis.

   - The curve is drawn in Section 9.2 of this guide (Problem 1).
   - If we revolve the piece from $\theta = 0$ to $\theta = \pi/2$, and double that area, we get the total surface area. In the integral $\int_{\theta=0}^{\pi/2} 2\pi x\, ds$ we replace x by $r\cos\theta = 2\sqrt{\sin\theta}\cos\theta$. Also $ds = \sqrt{(\frac{dr}{d\theta})^2 + r^2}\, d\theta = \sqrt{\frac{\cos^2\theta}{\sin\theta} + 4\sin\theta}\, d\theta$. The integral for surface area is not too easy:

$$4\pi\int_0^{\pi/2} 2\sqrt{\sin\theta}\cos\theta\sqrt{\frac{\cos^2\theta}{\sin\theta} + 4\sin\theta}\, d\theta = 8\pi\int_0^{\pi/2} \cos\theta\sqrt{\cos^2\theta + 4\sin^2\theta}\, d\theta$$

$$= 8\pi\int_0^{\pi/2} \cos\theta\sqrt{1 + 3\sin^2\theta}\, d\theta = 8\pi\int_0^1 \sqrt{1+3u^2}\, du \text{ (where } u = \sin\theta).$$

A table of integrals gives $8\pi\sqrt{3}(\frac{u}{2}\sqrt{\frac{1}{3}+u^2} + \frac{1}{6}\ln(u + \sqrt{\frac{1}{3}+u^2})]_0^1 = 8\pi\sqrt{3}(\frac{1}{\sqrt{3}} + \frac{1}{6}\ln(2+\sqrt{3})) \approx 34.1.$

8. Find the slope of the three-petal flower $r = \cos 3\theta$ at the tips of the petals.

- The flower is drawn in Section 9.2. The tips are at $(1,0)$, $(-1, \frac{\pi}{3})$, and $(-1, -\frac{\pi}{3})$. Clearly the tangent line at $(1,0)$ is vertical (infinite slope). For the other two slopes, find $\frac{dy}{dx} = \frac{dy/d\theta}{dx/d\theta}$. From $y = r\sin\theta$ we get $\frac{dy}{d\theta} = r\cos\theta + \sin\theta\frac{dr}{d\theta}$. Similarly $x = r\cos\theta$ gives $\frac{dx}{d\theta} = -r\sin\theta + \cos\theta\frac{dr}{d\theta}$. Substitute $\frac{dr}{d\theta} = -3\sin 3\theta$ for this flower, and set $r = -1$, $\theta = \frac{\pi}{3}$:

$$\frac{dy}{dx} = \frac{r\cos\theta - 3\sin 3\theta\sin\theta}{-r\sin\theta - 3\sin 3\theta\cos\theta} = \frac{(-1)\cos\pi/3 - 3\sin\pi\sin\pi/3}{\sin\pi/3 - 3\sin\pi\cos\pi/3} = \frac{-1/2}{\sqrt{3}/2} = -\frac{1}{\sqrt{3}}.$$

9. If $F(3) = 0$, show that the graph of $r = F(\theta)$ at $r = 0$, $\theta = 3$ has slope $\tan 3$.

- As an example of this idea, look at the graph of $r = \cos 3\theta$ (Section 9.1 of this guide). At $\theta = \pi/6$, $\theta = \pi/2$, and $\theta = -\pi/6$ we find $r = 0$. The rays out from the origin at those three angles are tangent to the graph. In other words the slope of $r = \cos 3\theta$ at $(0, \pi/6)$ is $\tan(\pi/6)$, the slope at $(0, \pi/2)$ is $\tan(\pi/2)$ and the slope at $(0, -\pi/6)$ is $\tan(-\pi/6)$.
- To prove the general statement, write $\frac{dy}{dx} = \frac{r\cos\theta + \sin\theta dr/d\theta}{-r\sin\theta + \cos\theta dr/d\theta}$ as in Problem 8. With $r = F(\theta)$ and $F(3) = 0$, substitute $\theta = 3$, $r = 0$, and $dr/d\theta = F'(3)$. The slope at $\theta = 3$ is $\frac{dy}{dx} = \frac{\sin(3)F'(3)}{\cos(3)F'(3)} = \tan(3)$.

Read-throughs and selected even-numbered solutions :

A circular wedge with angle $\Delta\theta$ is a fraction $\mathbf{\Delta\theta/2\pi}$ of a whole circle. If the radius is r, the wedge area is $\mathbf{\frac{1}{2}r^2\Delta\theta}$. Then the area inside $r = F(\theta)$ is $\int \mathbf{\frac{1}{2}r^2 d\theta} = \int \mathbf{\frac{1}{2}(F(\theta))^2 d\theta}$. The area inside $r = \theta^2$ from 0 to π is $\mathbf{\pi^5/10}$. That spiral meets the circle $r = 1$ at $\theta = \mathbf{1}$. The area inside the circle and outside the spiral is $\mathbf{\frac{1}{2} - \frac{1}{10}}$. A chopped wedge of angle $\Delta\theta$ between r_1 and r_2 has area $\mathbf{\frac{1}{2}r_2^2\Delta\theta - \frac{1}{2}r_1^2\Delta\theta}$.

The curve $r = F(\theta)$ has $x = r\cos\theta = \mathbf{F(\theta)\cos\theta}$ and $y = \mathbf{F(\theta)\sin\theta}$. The slope dy/dx is $dy/d\theta$ divided by $\mathbf{dx/d\theta}$. For length $(ds)^2 = (dx)^2 + (dy)^2 = \mathbf{(dr)^2 + (rd\theta)^2}$. The length of the spiral $r = \theta$ to $\theta = \pi$ is $\int \mathbf{\sqrt{1+\theta^2}d\theta}$. The surface area when $r = \theta$ is revolved around the x axis is $\int 2\pi y\, ds = \int \mathbf{2\pi\theta\sin\theta\sqrt{1+\theta^2}d\theta}$. The volume of that solid is $\int \pi y^2 dx = \int \mathbf{\pi\theta^2\sin^2\theta(\cos\theta - \theta\sin\theta)d\theta}$.

4 The inner loop is where $r < 0$ or $\cos\theta < -\frac{1}{2}$ or $\frac{2\pi}{3} < \theta < \frac{4\pi}{3}$. Its area is $\int \frac{r^2}{2}d\theta = \int \frac{1}{2}(1 + 4\cos\theta + 4\cos^2\theta)d\theta = [\frac{\theta}{2} + 2\sin\theta + \theta + \cos\theta\sin\theta]_{2\pi/3}^{4\pi/3} = \frac{\pi}{3} - 2(\sqrt{3}) + \frac{2\pi}{3} + \frac{1}{2}\sqrt{3} = \pi - \mathbf{\frac{3}{2}\sqrt{3}}$.

16 The spiral $r = e^{-\theta}$ starts at $r = 1$ and returns to the x axis at $r = e^{-2\pi}$. Then it goes inside itself (no new area). So area $= \int_0^{2\pi}\frac{1}{2}e^{-2\theta}d\theta = [-\frac{1}{4}e^{-2\theta}]_0^{2\pi} = \mathbf{\frac{1}{4}(1 - e^{-4\pi})}$.

20 Simplify $\frac{\tan\phi - \tan\theta}{1 + \tan\phi\tan\theta} = \frac{\frac{F+\tan\theta F'}{-\tan\theta F + F'} - \tan\theta}{1 + \frac{F+\tan\theta F'}{-\tan\theta F+F'}\tan\theta} = \frac{F + \tan\theta F' - \tan\theta(-\tan\theta F + F')}{-\tan\theta F + F' + \tan\theta(F + \tan\theta F')} = \frac{(1+\tan^2\theta)F}{(1+\tan^2\theta)F'} = \mathbf{\frac{F}{F'}}$.

22 $r = 1 - \cos\theta$ is the mirror image of Figure 9.4c across the y axis. By Problem 20, $\tan\psi = \frac{F}{F'} = \frac{1-\cos\theta}{\sin\theta}$. This is $\frac{\frac{1}{2}\sin^2\frac{\theta}{2}}{\frac{1}{2}\sin\frac{\theta}{2}\cos\frac{\theta}{2}} = \tan\frac{\theta}{2}$. So $\psi = \frac{\theta}{2}$ (check at $\theta = \pi$ where $\psi = \frac{\pi}{2}$).

24 By Problem 18 $\frac{dy}{dx} = \frac{\cos\theta + \tan\theta(-\sin\theta)}{-\cos\theta\tan\theta - \sin\theta} = \frac{\cos^2\theta - \sin^2\theta}{\cos\theta(-2\sin\theta)} = -\frac{\cos 2\theta}{\sin 2\theta} = -\frac{1}{\sqrt{3}}$ at $\theta = \frac{\pi}{6}$. At that point $x = r\cos\theta = \cos^2\frac{\pi}{6} = (\frac{\sqrt{3}}{2})^2$ and $y = r\sin\theta = \cos\frac{\pi}{6}\sin\frac{\pi}{6} = \frac{1}{2}(\frac{\sqrt{3}}{2})$. The tangent line is $\mathbf{y - \frac{\sqrt{3}}{4} = -\frac{1}{\sqrt{3}}(x - \frac{3}{4})}$.

26 $r = \sec\theta$ has $\frac{dr}{d\theta} = \sec\theta\tan\theta$ and $\frac{ds}{d\theta} = \sqrt{\sec^2\theta + \sec^2\theta\tan^2\theta} = \sqrt{\sec^4\theta} = \sec^2\theta$. Then arc length $= \int_0^{\pi/4}\sec^2\theta\, d\theta = \tan\frac{\pi}{4} = \mathbf{1}$. Note: $r = \sec\theta$ is the line $r\cos\theta = 1$ or $x = 1$ from $y = 0$ up to $y = 1$.

32 $r = 1 + \cos\theta$ has $\frac{ds}{d\theta} = \sqrt{(1 + 2\cos\theta + \cos^2\theta) + \sin^2\theta} = \sqrt{2 + 2\cos\theta}$. Also $y = r\sin\theta = (1 + \cos\theta)\sin\theta$. Surface area $\int 2\pi y\, ds = 2\pi\sqrt{2}\int_0^{\pi}(1+\cos\theta)^{3/2}\sin\theta\, d\theta = [2\pi\sqrt{2}(-\frac{2}{5})(1 + \cos\theta)^{5/2}]_0^{\pi} = \frac{32\pi}{5}$.

40 The parameter θ along the ellipse $x = 4\cos\theta, y = 3\sin\theta$ is *not* the angle from the origin. For example at $\theta = \frac{\pi}{4}$ the point (x, y) is *not* on the 45° line. So the area formula $\int \frac{1}{2}r^2 d\theta$ does not apply. The correct area is 12π.

9.4 Complex Numbers (page 364)

There are two important forms for every complex number: the *rectangular form* $x+iy$ and the *polar form* $re^{i\theta}$. Converting from one to the other is like changing between rectangular and polar coordinates. In one direction use $r = \sqrt{x^2+y^2}$ and $\tan\theta = \frac{y}{x}$. In the other direction (definitely easier) use $x = r\cos\theta$ and $y = r\sin\theta$. Problem 1 goes to polar and Problem 2 goes to rectangular.

1. Convert these complex numbers to polar form: (a) $3+4i$ (b) $-5-12i$ (c) $i\sqrt{3}-1$.

- (a) $r = \sqrt{3^2+4^2} = 5$ and $\theta = \tan^{-1}\frac{4}{3} \approx .93$. Therefore $3+4i \approx 5e^{.93i}$.
- (b) $-5-12i$ lies in the third quadrant of the complex plane, so $\theta = \pi + \arctan^{-1}\frac{-12}{-5} \approx \pi + 1.17 \approx 4.3$. The distance from the origin is $r = \sqrt{(-5)^2+(-12)^2} = 13$. Thus $-5-12i \approx 13e^{4.3i}$.
- (c) $i\sqrt{3}-1$ is not exactly in standard form: rewrite as $-1+i\sqrt{3}$. Then $x = -1$ and $y = \sqrt{3}$ and $r = \sqrt{1+3} = 2$. This complex number is in the second quadrant of the complex plane, since $x < 0$ and $y > 0$. The angle is $\theta = \frac{2\pi}{3}$. Then $-1+i\sqrt{3} = 2e^{2\pi/3}$.

We chose the standard polar form, with $r > 0$ and $0 \le \theta < 2\pi$. Other polar forms are allowed. The answer for (c) could also be $2e^{(2\pi+2\pi/3)i}$ or $2e^{-4\pi i/3}$.

2. Convert these complex numbers to rectangular form: (a) $6e^{i\pi/4}$ (b) $e^{-7\pi/6}$ (c) $3e^{\pi/3}$

- (a) The point $z = 6e^{i\pi/4}$ is 6 units out along the ray $\theta = \pi/4$. Since $x = 6\cos\frac{\pi}{4} = 3\sqrt{2}$ and $y = 6\sin\frac{\pi}{4} = 3\sqrt{2}$, the rectangular form is $3\sqrt{2}+3\sqrt{2}i$.
- (b) We have $r = 1$. The number is $\cos(-\frac{7\pi}{6}) + i\sin(-\frac{7\pi}{6}) = -\frac{\sqrt{3}}{2} + \frac{i}{2}$.
- (c) *There is no i in the exponent!* $3e^{\pi/3}$ is just a plain real number (approximately 8.5). Its rectangular form is $3e^{\pi/3} + 0i$.

3. For each pair of numbers find z_1+z_2 and z_1-z_2 and z_1z_2 and z_1/z_2:

(a) $z_1 = 4-3i$ and $z_2 = 12+5i$ (b) $z_1 = 3e^{i\pi/6}$ and $z_2 = 2e^{i7\pi/4}$.

- (a) Add $z_1+z_2 = 4-3i+12+5i = 16+2i$. Subtract $(4-3i)-(12+5i) = -8-8i$. Multiply:

$$(4-3i)(12+5i) = 48-36i+20i-15i^2 = 63-16i$$

To divide by $12+5i$, multiply top and bottom by its complex conjugate $12-5i$.
Then the bottom is real:

$$\frac{4-3i}{12+5i}\cdot\frac{12-5i}{12-5i} = \frac{33-56i}{12^2+5^2} = \frac{33}{169} - \frac{56}{169}i.$$

- You could choose to multiply in polar form. First convert $4-3i$ to $re^{i\theta}$ with $r = 5$ and $\tan\theta = -\frac{3}{4}$. Also $12+5i$ has $r = 13$ and $\tan\theta = \frac{5}{12}$. Multiply the r's to get $5\cdot 13 = 65$. Add the θ's. This is hard without a calculator that knows $\tan^{-1}(-\frac{3}{4})$ and $\tan^{-1}(\frac{5}{12})$. Our answer is $\theta_1+\theta_2 \approx -.249$.

So multiplication gives $65e^{-.249i}$ which is close to the first answer $63-16i$. Probably a trig identity would give $\tan^{-1}(-\frac{3}{4})+\tan^{-1}(\frac{5}{12})=\tan^{-1}(-\frac{16}{63})$.
For division in polar form, divide r's and subtract angles: $\frac{5}{13}e^{i(\theta_1-\theta_2)}\approx\frac{5}{13}e^{-i}$. This is $\frac{z_1}{z_2}=\frac{5}{13}\cos(-1)+\frac{5}{13}i\sin(-1)\approx .2-.3i\approx\frac{33}{169}-\frac{56}{169}i$.

- (b) Numbers in polar form are not easy to add. Convert to rectangular form:

$$3e^{i\pi/6} \text{ equals } 3\cos\frac{\pi}{6}+3i\sin\frac{\pi}{6}=\frac{3\sqrt{3}}{2}+\frac{3i}{2}. \text{ Also } 2e^{i7\pi/4} \text{ equals } 2\cos\frac{7\pi}{4}+2i\sin\frac{7\pi}{4}=\sqrt{2}-i\sqrt{2}.$$

The sum is $(\frac{3\sqrt{3}}{2}+\sqrt{2})+(\frac{3}{2}-\sqrt{2})i$. The difference is $(\frac{3\sqrt{3}}{2}-\sqrt{2})+(\frac{3}{2}+\sqrt{2})i$.
Multiply and divide in polar form whenever possible. *Multiply* r's *and add* θ's:

$$z_1z_2=(3\cdot 2)e^{i(\frac{\pi}{6}+\frac{7\pi}{4})}=6e^{\frac{23\pi i}{12}} \text{ and } \frac{z_1}{z_2}=\frac{3}{2}e^{i(\frac{\pi}{6}-\frac{7\pi}{4})}=\frac{3}{2}e^{-19\pi i/24}.$$

3. Find $(2-2\sqrt{3}i)^{10}$ in polar and rectangular form.

   - DeMoivre's Theorem is based on the polar form: $2-2\sqrt{3}i=4e^{-i\pi/3}$. The tenth power is $(4e^{-i\pi/3})^{10}=4^{10}e^{-10\pi/3}$. In rectangular form this is

$$4^{10}(\cos\frac{-10\pi}{3}+i\sin\frac{-10\pi}{3})=4^{10}(\cos\frac{2\pi}{3}+i\sin\frac{2\pi}{3})=2^{20}-(\frac{1}{2}+i\frac{\sqrt{3}}{2})=-2^{19}+2^{19}i\sqrt{3}.$$

4. (This is 9.4.3) Plot $z=2e^{i\pi/6}$ and its reciprocal $\frac{1}{z}=\frac{1}{2}e^{-i\pi/6}$ and their squares.

   - The squares are $(2e^{i\pi/6})^2=4e^{i\pi/3}$ and $(\frac{1}{2}e^{-i\pi/6})^2=\frac{1}{4}e^{-i\pi/3}$. The points z, $\frac{1}{z}$, z^2, $\frac{1}{z^2}$ are plotted.

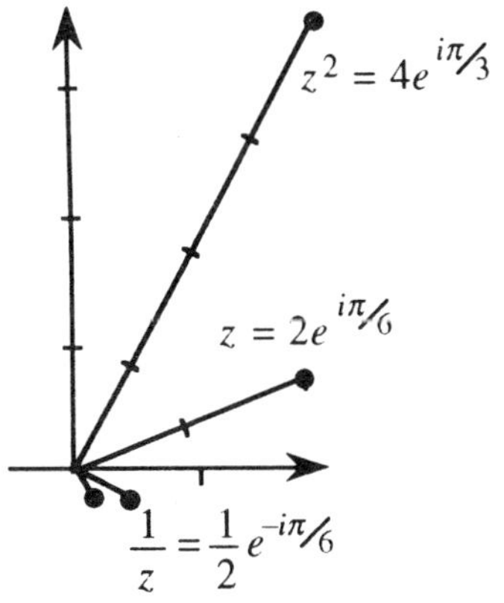

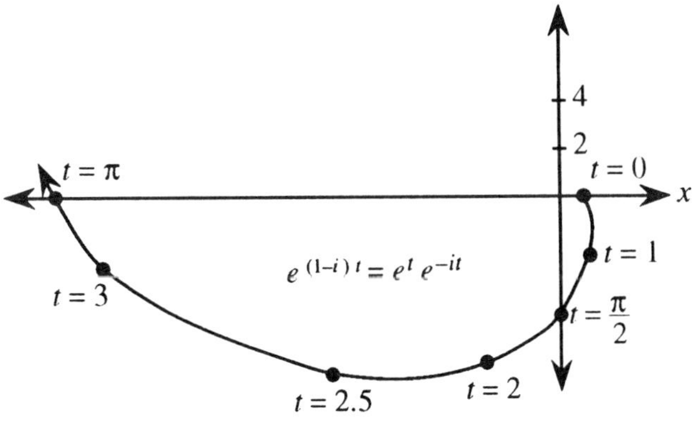

5. (This is 9.4.25) For $c=1-i$, sketch the path of $y=e^{ct}$ as t increases from 0.

   - The moving point e^{ct} is $e^{(1-i)t}=e^te^{-it}=e^t(\cos(-t)+i\sin(-t))$. The table gives x and y:

t	0	.5	1.0	$\pi/2$	2.0	2.5	3.0	π
$x=e^t\ \cos(-t)$	1	1.4	1.5	0	−3.1	−9.8	−19.9	23.1
$y=e^t\ \sin(-t)$	0	−.8	−2.3	−4.8	−6.7	−7.3	−2.8	0

The sketch shows how e^{ct} spirals rapidly outwards from $e^0=1$.

6. For the differential equation $y'' + 4y' + 3y = 0$, find all solutions of the form $y = e^{ct}$.

- The derivatives of $y = e^{ct}$ are $y' = ce^{ct}$ and $y'' = c^2e^{ct}$. The equation asks for $c^2e^{ct} + 4ce^{ct} + 3e^{ct} = 0$. This means that $e^{ct}(c^2 + 4c + 3) = 0$. Factor $c^2 + 4c + 3$ into $(c + 3)(c + 1)$. This is zero for $c = -3$ and $c = -1$. The pure exponential solutions are $y = e^{-3t}$ and $y = e^{-t}$. Any combination like $2e^{-3t} + 7e^{-t}$ also solves the differential equation.

7. Construct two real solutions of $y'' + 2y' + 5y = 0$. Start with solutions of the form $y = e^{ct}$.

- Substitute $y'' = c^2e^{ct}$ and $y' = ce^{ct}$ and $y = e^{ct}$. This leads to $c^2 + 2c + 5 = 0$ or $c = -1 \pm 2i$. The pure (but complex) exponential solutions are $y = e^{(-1+2i)t}$ and $y = e^{(-1-2i)t}$. The first one is $y = e^{-t}(\cos 2t + i \sin 2t)$. The real part is $x = e^{-t}\cos 2t$; the imaginary part is $y = e^{-t}\sin 2t$. (*Note:* The imaginary part is without the i.) Each of these is a real solution, as may be checked by substitution into $y'' + 2y' + 5y = 0$.

The other exponential is $y = e^{(-1-2i)t} = e^{-t}(\cos(-2t) + i\sin(-2t))$. Its real and imaginary parts are the *same real solutions* – except for the minus sign in $\sin(-2t) = -\sin 2t$.

Read-throughs and selected even-numbered solutions :

The complex number $3 + 4i$ has real part **3** and imaginary part **4**. Its absolute value is $r =$ **5** and its complex conjugate is $\mathbf{3 - 4i}$. Its position in the complex plane is at **(3,4)**. Its polar form is $r\cos\theta + ir\sin\theta = \mathbf{re^{i\theta}}$ **(or $\mathbf{5e^{i\theta}}$)**. Its square is $\mathbf{-7 - 14i}$. Its nth power is $\mathbf{r^n}e^{in\theta}$.

The sum of $1 + i$ and $1 - i$ is **2**. The product of $1 + i$ and $1 - i$ is **2**. In polar form this is $\sqrt{2}e^{i\pi/4}$ times $\mathbf{\sqrt{2}e^{-i\pi/4}}$. The quotient $(1+i)/(1-i)$ equals the imaginary number **i**. The number $(1+i)^8$ equals **16**. An eighth root of 1 is $w = \mathbf{(1 + i)/\sqrt{2}}$. The other eighth roots are $\mathbf{w^2, w^3, \cdots, w^7, w^8 = 1}$.

To solve $d^8y/dt^8 = y$, look for a solution of the form $y = \mathbf{e^{ct}}$. Substituting and canceling e^{ct} leads to the equation $\mathbf{c^8 = 1}$. There are **eight** choices for c, one of which is $(-1 + i)/\sqrt{2}$. With that choice $|\, e^{ct} \,| = \mathbf{e^{-t/\sqrt{2}}}$. The real solutions are Re $e^{ct} = \mathbf{e^{-t/\sqrt{2}}\cos\frac{t}{\sqrt{2}}}$ and Im $e^{ct} = \mathbf{e^{-t/\sqrt{2}}\sin\frac{t}{\sqrt{2}}}$.

10 $e^{ix} = i$ yields $\mathbf{x = \frac{\pi}{2}}$ (note that $\frac{i\pi}{2}$ becomes $\ln i$); $e^{ix} = e^{-1}$ yields $\mathbf{x = i}$, second solutions are $\frac{\pi}{2} + 2\pi$ and $i + 2\pi$.

14 The roots of $c^2 - 4c + 5 = 0$ must multiply to give **5**. Check: The roots are $\frac{4\pm\sqrt{16-20}}{2} = 2 \pm i$. Their product is $(2 + i)(2 - i) = 4 - i^2 = 5$.

18 The fourth roots of $re^{i\theta}$ are $r^{1/4}$ times $e^{i\theta/4}, e^{i(\theta+2\pi)/4}, e^{i(\theta+4\pi)/4}, e^{i(\theta+6\pi)/4}$. Multiply $(r^{1/4})^4$ to get r. Add angles to get $(4\theta + 12\pi)/4 = \theta + 3\pi$. The product of the 4 roots is $re^{i(\theta+3\pi)} = -re^{i\theta}$.

28 $\frac{dy}{dt} = iy$ leads to $y = e^{it} = \cos t + i\sin t$. Matching real and imaginary parts of $\frac{d}{dt}(\cos t + i\sin t) = i(\cos t + i\sin t)$ yields $\frac{d}{dt}\cos t = -\sin t$ and $\frac{d}{dt}\sin t = \cos t$.

34 Problem 30 yields $\cos ix = \frac{1}{2}(e^{i(ix)} + e^{-i(ix)}) = \frac{1}{2}(e^{-x} + e^{x}) = \cosh x$; similarly $\sin ix = \frac{1}{2i}(e^{i(ix)} - e^{-i(ix)}) = \frac{i}{2i}(e^{-x} - e^{x}) = i\sinh x$. With $x = 1$ the cosine of i equals $\frac{1}{2}(e^{-1} + e^{1})=$ **3.086**. The cosine of i is larger than 1!

9 Chapter Review Problems

Review Problems

R1 Express the point (r, θ) in rectangular coordinates. Express the point (a, b) in polar coordinates. Express the point (r, θ) with three other pairs of polar coordinates.

R2 As θ goes from 0 to 2π, how often do you cover the graph of $r = \cos\theta$? $r = \cos 2\theta$? $r = \cos 3\theta$?

R3 Give an example of a polar equation for each of the conic sections, including circles.

R4 How do you find the area between two polar curves $r = F(\theta)$ and $r = G(\theta)$ if $0 < F < G$?

R5 Write the polar form for ds. How is this used for surface areas of revolution?

R6 What is the polar formula for slope? Is it $dr/d\theta$ or dy/dx?

R7 Multiply $(a + ib)(c + id)$ and divide $(a + ib)/(c + id)$.

R8 Sketch the eighth roots of 1 in the complex plane. How about the roots of -1?

R9 Starting with $y = e^{ct}$, find two real solutions to $y'' + 25y = 0$.

R10 How do you test the symmetry of a polar graph? Find the symmetries of

(a) $r = 2\cos\theta + 1$ (b) $= 8\sin\theta$ (c) $r = \frac{6}{1-\cos\theta}$ (d) $r = \sin 2\theta$ (e) $r = 1 + 2\sin\theta$

Drill Problems

D1 Show that the area inside $r^2 = \sin 2\theta$ and outside $r = \frac{\sqrt{2}}{2}$ is $\frac{\sqrt{3}}{2} - \frac{\pi}{6}$.

D2 Find the area inside both curves $r = 2 - \cos\theta$ and $r = 3\cos\theta$.

D3 Show that the area enclosed by $r = 2\cos 3\theta$ is π.

D4 Show that the length of $r = 4\sin^3\frac{\theta}{3}$ between $\theta = 0$ and $\theta = \pi$ is $2\pi - \frac{3}{2}\sqrt{3}$.

D5 Confirm that the length of the spiral $r = 3\theta^2$ from $\theta = 0$ to $\theta = \frac{5}{3}$ is $\frac{7}{3}$.

D6 Find the slope of $r = \sin 3\theta$ at $\theta = \frac{\pi}{6}$.

D7 Find the slope of the tangent line to $r = \tan\theta$ at $(1, \frac{\pi}{2})$.

D8 Show that the slope of $r = 1 + \sin\theta$ at $\theta = \frac{\pi}{6}$ is $\frac{2}{\sqrt{3}}$.

D9 The curve $r^2 = \cos 2\theta$ from $(1, -\frac{\pi}{4})$ to $(1, \frac{\pi}{4})$ is revolved around the y axis. Show that the surface area is $2\sqrt{2}\pi$.

D10 Sketch the parabola $r = 4/(1 + \cos\theta)$ to see its focus and vertex.

D11 Find the center of the ellipse whose polar equation is $r = \frac{6}{2 - \cos\theta}$. What is the eccentricity e?

D12 The asymptotes of the hyperbola $r = \frac{6}{1 + 3\cos\theta}$ are the rays where $1 + 3\cos\theta = 0$. Find their slopes.

D13 Find all the sixth roots (two real, four complex) of 64.

D14 Find four roots of the equation $z^4 - 2z^2 + 4 = 0$.

D15 Add, subtract, multiply, and divide $1 + \sqrt{3}\, i$ and $1 - \sqrt{3}\, i$.

D16 Add, subtract, multiply, and divide $e^{i\pi/4}$ and $e^{-i\pi/4}$.

D17 Find all solutions of the form $y = e^{ct}$ for $y'' - y' - 2y = 0$ and $y''' - 2y' - 3y = 0$.

D18 Construct real solutions of $y'' - 4y' + 13y = 0$ from the real and imaginary parts of $y = e^{ct}$.

D19 Use a calculator or an integral to estimate the length of $r = 1 + \sin\theta$ (near 2.5?).

Graph Problems (intended to be drawn by hand)

G1 $r^2 = \sin 2\theta$

G2 $r = 6\sin\theta$

G3 $r = \sin 4\theta$

G4 $r = 5\sec\theta$

G5 $r = e^{\theta/2}$

G6 $r = 2 - 3\cos\theta$

G7 $r = \frac{6}{1 + 2\cos\theta}$

G8 $r = \frac{1}{1 - \sin\theta}$

CHAPTER 10 INFINITE SERIES

10.1 The Geometric Series (page 373)

The advice in the text is: *Learn the geometric series.* This is the most important series and also the simplest. The pure geometric series starts with the constant term 1: $1+x+x^2+\cdots=\frac{1}{1-x}$. With other symbols $\sum_{n=0}^{\infty} r^n=\frac{1}{1-r}$. The ratio between terms is x (or r). Convergence requires $|x|<1$ (or $|r|<1$).

A closely related geometric series starts with any number a, instead of 1. Just multiply everything by a : $a+ax+ax^2+\cdots=\frac{a}{1-x}$. With other symbols $\sum_{n=0}^{\infty} ar^n=\frac{a}{1-r}$. As a particular case, choose $a=x^{N+1}$. The geometric series has its first terms chopped off:

$$\textit{Tail end}:\ x^{N+1}+x^{N+2}+\cdots=\frac{x^{N+1}}{1-x} \quad\text{or}\quad \sum_{N+1}^{\infty} r^n=\frac{r^{N+1}}{1-r}.$$

Now do the opposite. *Keep the first terms and chop off the last terms.* Instead of an infinite series you have a finite series. The sum looks harder at first, but not after you see where it comes from:

$$\textit{Front end}:\ 1+x+\cdots+x^N=\frac{1-x^{N+1}}{1-x} \quad\text{or}\quad \sum_{n=0}^{N} r^n=\frac{1-r^{N+1}}{1-r}.$$

The proof is in one line. *Start with the whole series and subtract the tail end.* That leaves the front end. It is the difference between $\frac{1}{1-x}$ and $\frac{x^{N+1}}{1-x}$ – complete sum minus tail end.

Again you can multiply everything by a. The front end is a finite sum, and the restriction $|x|<1$ or $|r|<1$ is now omitted. We only avoid $x=1$ or $r=1$, because the formula gives $\frac{0}{0}$. (But with $x=1$ the front end is $1+1+\cdots+1$. We don't need a formula to add up 1's.)

Personal note: I just heard a lecture on the mathematics of finance. After two minutes the lecturer wrote down a series – you can guess what it was. The geometric series is the most important there too. It gives the cost of a contract to pay you \$1 every year, *forever.* You don't have to pay an infinite amount $1+1+1+\cdots$ up front, because of interest. Suppose the interest rate per year is i. What you pay now is multiplied by $(1+i)$ every year. By paying $\frac{1}{1+i}$ at present, you can collect \$1 after a year. Also pay $\frac{1}{(1+i)^2}$ to have another \$1 waiting after two years. The "present value" of \$1 to be collected after every year comes out to be $\frac{1}{i}$:

$$\frac{1}{1+i}+\frac{1}{(1+i)^2}+\frac{1}{(1+i)^3}+\cdots=\frac{\frac{1}{1+i}}{1-\frac{1}{1+i}}.\quad \text{Multiply by}\frac{1+i}{1+i}\text{ to find}\frac{1}{i}.$$

Here is a quick way to explain that. If you pay $\frac{1}{i}$ now, it earns $i(\frac{1}{i})=\$1$ interest after a year. So you collect that dollar, and leave the $\frac{1}{i}$. The second year interest is again \$1, which you collect. This goes on forever. We are assuming constant interest rate and no bank charge, both unfortunately false.

What if you only want \$1 for N years? Now the annuity is not "perpetual" – it ends. What you pay now is smaller. Use the formula for a finite sum $a+a^2+\cdots+a^N$ (the front end, starting with $a=\frac{1}{1+i}$). The N-year annuity costs

$$\frac{1}{1+i}+\frac{1}{1+i^2}+\cdots+\frac{1}{(1+i)^N}=\frac{\frac{1}{1+i}-\frac{1}{(1+i)^{N+1}}}{1-\frac{1}{1+i}}=\frac{1}{i}[1-\frac{1}{(1+i)^N}].$$

Again the management professor had a good way to understand this. The contract that pays forever costs $\frac{1}{i}$. But you only hold it for N years. Then you give the tail end back and stop collecting the annual dollar. At that future time the tail end is worth $\frac{1}{i}$, because it still pays \$1 forever. But something worth $\frac{1}{i}$ then is only worth $\frac{1}{i}\frac{1}{(1+i)^N}$ now. That is the present value of what you are giving back, so subtract it from the present value $\frac{1}{i}$ of the infinite contract to get your cost $\frac{1}{i}(1-\frac{1}{(1+i)^N})$.

EXAMPLE. Suppose the interest rate is 100%, so $i = 1$. An investment of $\frac{1}{i} = \$1$ will bring you a dollar each year forever. If you only want that dollar for $N = 10$ years, your present investment can be a little smaller. It is smaller by $\frac{1}{i}\frac{1}{(1+i)^N} = \frac{1}{2^{10}}$, which is not much! This is $\frac{1}{1024}$ dollars or about $\frac{1}{10}$ cent. For a ten-year annuity you pay 99.9 cents, for a perpetual annuity you pay the full 100 cents.

The reason for those strange numbers is the 100% interest rate. A dollar in ten years is only worth $\frac{1}{10}$ cent now. With galloping inflation like that, all calculations look unreasonable. You see why 100% interest or more (which does happen!) can destroy an economy.

NOTE: Annuities are also discussed in Section 6.6. You could redo the calculation based on 10% interest, so $i = 0.1$. How much is a contract for \$1 a year in perpetuity? (That means forever.) How much do you pay now to receive \$1 a year for $N = 10$ years?

In Problems 1 – 5, find the sum of the geometric series from the formula $\sum_{n=0}^{\infty} ar^n = \frac{a}{1-r}$. The ratio r of successive terms must satisfy $|r| < 1$ for the series to converge.

1. $24 - 12 + 6 - 3 + \cdots$ This has $r = -\frac{1}{2}$.
2. $\sqrt{125} + \sqrt{25} + \sqrt{5} + \cdots$ This has $r = \frac{1}{\sqrt{5}}$.
3. $1 + (x-\pi) + (x-\pi)^2 + (x-\pi)^3 + \cdots$ This has $r = x - \pi$.
4. $6 - \frac{12}{x} + \frac{24}{x^2} - \frac{48}{x^3} + \cdots$ This has $r = -\frac{2}{x}$.
5. $5.282828\cdots$ (subtract 5, then add it back) This has $r = \frac{1}{100}$.

- 1. Starting from $a = 24$ with $r = -\frac{1}{2}$ the sum is $\frac{24}{1-(-\frac{1}{2})} = \frac{24}{\frac{3}{2}} = (\frac{2}{3})24 = 16$.
- 2. The common ratio is $\frac{1}{\sqrt{5}}$. The sum is $\frac{\sqrt{125}}{1-\frac{1}{\sqrt{5}}} = \frac{25}{\sqrt{5}-1}$.
- 3. Each term equals the previous term multiplied by $(x-\pi)$, so $r = x - \pi$. The sum is $\frac{1}{1-(x-\pi)} = \frac{1}{\pi+1-x}$. This does not work for all values of x! The formula $\frac{a}{1-r}$ is only correct when $|r| = |x-\pi| < 1$, or $\pi - 1 < x < \pi + 1$. For other values of x the series does not have a sum.
- 4. The common ratio is $r = \frac{-2}{x}$, so the sum is $\frac{6}{1+\frac{2}{x}} = \frac{6x}{x+2}$. This is valid only when $|r| = |\frac{2}{x}| < 1$, or $|x| > 2$. For example, the series diverges for $x = 1.5$ but for $x = -4$ the sum is 12.
- 5. Every repeating decimal involves a geometric series. The number $5.282828\cdots$ can be written $5 + \frac{28}{100} + \frac{28}{10000} + \cdots$. Subtract the 5 because it is not part of the pattern. The series $\frac{28}{100} + \frac{28}{10000} + \cdots$ converges to $\frac{\frac{28}{100}}{1-\frac{1}{100}} = \frac{28}{99}$. Now add back the 5 to get the answer $5\frac{28}{99}$.

6. A ball is dropped from a height of 8 feet and rebounds 6 feet. After each bounce it rebounds $\frac{3}{4}$ of the distance it fell. How far will the ball travel before it comes to rest?

- The ball goes down 8, up 6, down 6, up $\frac{3}{4}(6) = \frac{9}{2}$, down $\frac{9}{2}$, up $\frac{27}{8}, \cdots$. Every distance after the original 8 is repeated. The total is $8 + 2(6) + 2(\frac{9}{2}) + 2(\frac{27}{8}) + \cdots$. Subtract the 8, because it doesn't fit the pattern. The rest of the series is $2[6 + 6(\frac{3}{4}) + 6(\frac{3}{4})^2 + \cdots] = 2[\frac{6}{1-\frac{3}{4}}] = 48$. The total distance is $8 + 48 = 56$ feet.

7. (Draw your own figure above) A robot is programmed to move in this way: First go 1 km east. Turn 90° to go $\frac{1}{2}$ km north. Turn left again to go $\frac{1}{4}$ km west. After each left turn, the distance is half the previous step. If the robot starts at $(0,0)$ where does it end up?

- First look at the east-west coordinate x. This changes every other step following the pattern $1, 1-\frac{1}{4}, 1-\frac{1}{4}+\frac{1}{16}, 1-\frac{1}{4}+\frac{1}{16}-\frac{1}{64}, \cdots$ The limiting x coordinate is $1-\frac{1}{4}+\frac{1}{16}-\cdots = \frac{1}{1+\frac{1}{4}} = \frac{4}{5}$.

  Similarly, the y coordinate ends up at $\frac{1}{2}-\frac{1}{8}+\frac{1}{32}-\cdots = \frac{\frac{1}{2}}{1+\frac{1}{4}} = \frac{2}{5}$. The robot "spirals" inward to the point $(\frac{4}{5}, \frac{2}{5})$.

8. (This is 10.1.6) Multiply $(1+x+x^2+\cdots)$ times $(1-x+x^2-\cdots)$. Compare with equation (14).

- The first terms are $1(1-x+x^2-\cdots)+x(1-x+x^2-\cdots)$. Those combine to give 1.
  The next terms are $x^2(1-x+x^2-\cdots)+x^3(1-x+x^2-\cdots)$. Those combine to give x^2.
  By grouping the multiplication two steps at a time, the final answer becomes $1+x^2+x^4+\cdots$.
  This answer equals $\frac{1}{1-x^2}$. It is the geometric series in equation (14) with $r=x^2$.
  Check: The original series add to $\frac{1}{1+x}$ and $\frac{1}{1-x}$. Multiplication gives $\frac{1}{1-x^2}$.

9. Find two infinite series that add up to $\ln 7$.

- There are three candidates: (10 a), (10 b), and (13) on pages 370 – 371. Formula (10 b) cannot be used directly for $\ln(1+x) = \ln 7$, because we would need $x=6$. *These formulas do not apply for* $|x|>1$. One way out is to use (10a) with $x=\frac{6}{7}$. This would give $-\ln(1-\frac{6}{7}) = -\ln(\frac{1}{7}) = \ln 7$. So one answer is $(\frac{6}{7})+\frac{1}{2}(\frac{6}{7})^2+\frac{1}{3}(\frac{6}{7})^3+\cdots = 1.946$.
  A faster method is to use the series (13) for $\ln(\frac{1+x}{1-x})$. Set $\frac{1+x}{1-x}=7$ to get $x=\frac{3}{4}$. Substitute this x in the series: $\ln 7 = 2[(\frac{3}{4})+\frac{1}{3}(\frac{3}{4})^3+\frac{1}{5}(\frac{3}{4})^5+\cdots] = 1.946$.
  For any positive number a, we solve $\frac{1+x}{1-x}=a$ to find $x=\frac{a-1}{a+1}$. Substitute this x into the series (13) to get $\ln(\frac{1+x}{1-x}) = \ln a$. The number x will always between -1 and 1.

10. Find a series for $\frac{1}{(1-x)^3}$ that converges for $|x|<1$.

- Here are two ways. Equation (5) on page 369 gives a series for $\frac{1}{(1-x)^2}$. Multiply by the series $1+x+x^2+\cdots = \frac{1}{1-x}$ to obtain the series for $\frac{1}{(1-x)^3}$. A better way is to take the *derivative* of the series for $\frac{1}{(1-x)^2}$ to get $\frac{2}{(1-x)^3}$:
  The derivative of $1+2x+3x^2+4x^3+\cdots+(n+1)x^n+\cdots$ is $2+6x+12x^2+\cdots+n(n+1)x^{n-1}+\cdots$.
  Dividing by 2 gives $\frac{1}{(1-x)^3} = 1+3x+6x^2+\cdots+\frac{n(n+1)}{2}x^{n-1}+\cdots$.

10. What is the average number of times you would have to roll a pair of dice to get a seven?

- Six out of the 36 combinations on the two dice add to seven: $1+6, 2+5, 3+4, \cdots, 6+1$. On the first roll, your chances of getting a seven are $p_1 = \frac{1}{6}$. The probability of *not* getting a seven on the first roll but getting one on the second roll is $p_2 = (\frac{5}{6})(\frac{1}{6})$. The chance of two failures followed by success is $p_3 = (\frac{5}{6})^2(\frac{1}{6})$.
- The mean number of rolls is $\sum np_n = 1(\frac{1}{6}) + 2(\frac{5}{6})(\frac{1}{6}) + 3(\frac{5}{6})^2(\frac{1}{6}) + \cdots$. This infinite series fits the pattern of equation (5) on page 369 with $x = \frac{5}{6}$:
$\frac{1}{6}(1 + 2(\frac{5}{6}) + 3(\frac{5}{6})^2 + 4(\frac{5}{6})^3 + \cdots) = \frac{1}{6}(\frac{1}{(1-\frac{5}{6})^2}) = \frac{1}{6}(6^2) =$ average of 6 rolls.

11. Use a series to compute $\int_0^{1/2} \tan^{-1} x\, dx$, correct to 4 places.

- Equation (18) on page 371 tells us that $\tan^{-1} x = x - \frac{1}{3}x^3 + 15x^5 - \cdots$ provided $|x| < 1$. Integrate from 0 to $\frac{1}{2}$:

$$\int_0^{1/2} \tan^{-1} x\, dx = [\tfrac{1}{2}x^2 - \tfrac{1}{3\cdot 4}x^3 + \tfrac{1}{5\cdot 6}x^6 - \cdots]_0^{1/2}$$
$$\approx \tfrac{1}{2}(\tfrac{1}{2})^2 - \tfrac{1}{3\cdot 4}(\tfrac{1}{2})^4 + \tfrac{1}{5\cdot 6}(\tfrac{1}{2})^6 - \tfrac{1}{7\cdot 8}(\tfrac{1}{2})^8 + \tfrac{1}{9\cdot 10}(\tfrac{1}{2})^{10} \approx .1203$$

Read-throughs and selected even-numbered solutions :

The geometric series $1 + x + x^2 + \cdots$ adds up to $\mathbf{1/(1-x)}$. It converges provided $|x| < \mathbf{1}$. The sum of n terms is $\mathbf{(1-x^n)/(1-x)}$. The derivatives of the series match the derivatives of $1/(1-x)$ at the point $x = \mathbf{0}$, where the nth derivative is $\mathbf{n!}$ The decimal 1.111... is the geometric series at $x = \mathbf{.1}$ and equals the fraction $\mathbf{10/9}$. The decimal .666... multiplies this by $\mathbf{.6}$. The decimal .999... is the same as $\mathbf{1}$.

The derivative of the geometric series is $\mathbf{1/(1-x)^2} = \mathbf{1 + 2x + 3x^2} + \cdots$. This also comes from squaring the **geometric** series. By choosing $x = .01$, the decimal 1.02030405 is close to $\mathbf{(100/99)^2}$. The differential equation $dy/dx = y^2$ is solved by the geometric series, going term by term starting from $y(0) = \mathbf{1}$.

The integral of the geometric series is $\mathbf{-\ln(1-x)} = \mathbf{x + x^2/2} + \cdots$. At $x = 1$ this becomes the **harmonic** series, which diverges. At $x = \mathbf{\frac{1}{2}}$ we find $\ln 2 = \mathbf{\frac{1}{2} + (\frac{1}{2})^2/2 + (\frac{1}{2})^3/3} + \cdots$. The change from x to $-x$ produces the series $1/(1+x) = \mathbf{1 - x + x^2 - x^3} + \cdots$ and $\ln(1+x) = \mathbf{x - x^2/2 + x^3/3} \cdots$.

In the geometric series, changing to x^2 or $-x^2$ gives $1/(1-x^2) = \mathbf{1 + x^2 + x^4} + \cdots$ and $1/(1+x^2) = \mathbf{1 - x^2 + x^4} - \cdots$. Integrating the last one yields $x - \frac{1}{3}x^3 + \frac{1}{5}x^5 \cdots = \mathbf{\tan^{-1}x}$. The angle whose tangent is $x = 1$ is $\tan^{-1} 1 = \mathbf{1 - \frac{1}{3} + \frac{1}{5}} - \cdots$. Then substituting $x = 1$ gives the series $\pi = \mathbf{4(1 - \frac{1}{3} + \frac{1}{5}} - \cdots)$.

4 $1 + (1-x) + (1-x)^2 + \cdots = \frac{1}{1-(1-x)} = \frac{1}{x}$; integration gives $\ln x = x - \frac{(1-x)^2}{2} - \frac{(1-x)^3}{3} - \cdots + C$ and at $x = 1$ we find $C = -1$. Therefore $\mathbf{\ln x} = -[\mathbf{(1-x)} + \mathbf{\frac{(1-x)^2}{2}} + \mathbf{\frac{(1-x)^3}{3}} + \cdots]$. At $x = 0$ this is $-\infty = -\infty$.

18 $\frac{1}{2}x - \frac{1}{4}x^2 + \frac{1}{8}x^3 - \cdots = \frac{x}{2} - (\frac{x}{2})^2 + (\frac{x}{2})^3 - \cdots = \frac{\frac{x}{2}}{1+\frac{x}{2}} = \mathbf{\frac{x}{2+x}}$.

20 $x - 2x^2 + 3x^3 - \cdots = x(1 - 2x + 3x^2 - \cdots) =$ [change x to $-x$ in equation (5)] $= \mathbf{\frac{x}{(1+x)^2}}$.

26 $\int(1 + x^2 + x^4 + \cdots)dx = x + \frac{x^3}{3} + \frac{x^5}{5} + \cdots = \mathbf{\frac{1}{2} \ln \frac{1+x}{1-x}}$ by equation (13). This is $\int \frac{1}{1-x^2}dx$ which is also $\mathbf{\tanh^{-1}x}$ (+ C).

36 $y'(0) = [y(0)]^2 = \mathbf{b^2}$; $y''(0) = 2y(0)y'(0) = \mathbf{2b^3}$; $y'''(0) =$ (from second derivative of $y' = y^2$) $= 2y(0)y''(0) + 2y'(0)y'(0) = \mathbf{6b^4}$. Then $y(x) = b + b^2x + 2b^3(\frac{x^2}{2}) + 6b^4(\frac{x^3}{6}) + \cdots = b(1 + bx + b^2x^2 + b^3x^3 + \cdots)$ which is $\mathbf{y(x)} = \mathbf{\frac{b}{1-bx}}$.

40 Note: The equations referred to should be (10) and (13). Choose $x = \frac{2}{3}$ in (10a): $\frac{2}{3} + \frac{1}{2}(\frac{2}{3})^2 + \cdots = -\ln\frac{1}{3}$.

In equation (13) choose $x = \frac{1}{2}$ so that $\frac{1+x}{1-x} = 3$. Then $2(\frac{1}{2} + \frac{1}{3}(\frac{1}{2})^3 + \frac{1}{5}(\frac{1}{2})^5 + \cdots) = \ln 3$. This converges faster because of the factor $(\frac{1}{2})^2$ between successive terms, compared to $\frac{2}{3}$ in the first series. **The series are equal** because $-\ln\frac{1}{3} = \ln 3$.

42 Equation (18) gives $\tan^{-1}\frac{1}{10} \approx \frac{1}{10} - \frac{1}{300} + \frac{1}{50000} \approx .10000 - .00333 + .00002 \approx \mathbf{.09669}$ (which is .0967 to four decimal places).

46 Equation (20) is $\pi = 4(\tan^{-1}\frac{1}{2} + \tan^{-1}\frac{1}{3}) \approx 4(\frac{1}{2} - \frac{1}{3}(\frac{1}{2})^3 + \frac{1}{5}(\frac{1}{2})^5 - \frac{1}{7}(\frac{1}{2})^7 + \frac{1}{9}(\frac{1}{2})^9 - \frac{1}{11}(\frac{1}{2})^{11}$ $+\frac{1}{3} - \frac{1}{3}(\frac{1}{3})^3 + \frac{1}{5}(\frac{1}{3})^5 - \frac{1}{7}(\frac{1}{3})^7) = 2 - .16667 + .02500 - .00446 + .00087 - .00018 + 1.33333 - .04938 + .00329$ $-.00026 = 3.1415^+ = \mathbf{3.142}$. Note: $\frac{1}{13}(\frac{1}{2})^{13}$ and $\frac{1}{9}(\frac{1}{3})^9$ will increase the total toward 3.1416.

10.2 Convergence Tests: Positive Series (page 380)

The goal of this section is to be able to decide whether a series Σa_n converges or diverges. That goal is not completely reached. It never will be! There is no single test to apply to the numbers a_n which is necessary and sufficient for convergence. But there is a series of six tests, which often produces a definite answer. This section shows by example how they work:

10A	Shrinking Test	10C	Integral Test	10E	Root Test
10B	Comparison Test	10D	Ratio Test	10F	Limit Comparison Test

We suggest applying the tests more or less in that order. But this depends on the series. When the terms a_n are complicated (we assume they are positive), look first for a simpler series to compare with. For the most basic series – the geometric series Σx^n and the p-series $\Sigma\frac{1}{n^p}$ - the convergence conditions $|x| < 1$ and $p > 1$ are known exactly. *Remember to use those series in comparisons.*

Shrinking Test : The terms of the series must approach zero. If not, the series diverges. You know right away $\sum\frac{n^2+1}{n^2}$ diverges, because $\frac{n^2+1}{n^2} \to 1$. The series $\sum\frac{n+1}{n^2}$ passes this shrinking test but it fails a later test and diverges.

1. Which of these positive series diverge because the terms fail to approach zero?

   (a) $\sum(\frac{4}{3})^n$ (b) $\sum\frac{\sqrt{n^2+3}}{n-\pi}$ (c) $\sum\frac{n!}{10^n}$ (d) $\sum\frac{e^n}{n^5}$ (e) $\sum\frac{\ln(n+1)}{\ln(n^3)}$

   - *All of these series fail the shrinking test* 10A *and diverge.* In (a) the terms $(\frac{4}{3})^n$ approach $+\infty$. The terms in (b) behave like $\frac{\sqrt{n^2}}{n} = 1$, so their limit is one. In (c), each term $\frac{n!}{10^n}$ multiplies the previous term by $\frac{n}{10}$. As soon as n is larger than 10, the terms start growing. (In fact they approach infinity.) For (d), l'Hôpital's Rule shows that $\lim_{n\to\infty}\frac{e^n}{n^5}$ is infinite. Finally $\ln(n^3) = 3\ln n$ so all terms in $\frac{\ln(n+1)}{\ln(n^3)}$ are greater than $\frac{1}{3}$.

Comparison Test : For a geometric series $\sum ar^n$ or a p-series $\sum\frac{a}{n^p}$, you can decide convergence immediately: $|r| < 1$ and $p > 1$ are complete tests, necessary and sufficient. Many other series look and behave like geometric or p-series. A smaller series than a convergent series also converges. A larger series than a divergent series also diverges.

2. Apply a comparison test to decide the convergence or divergence of

   (a) $\sum_4^\infty\frac{3}{n-3}$ (b) $\sum_1^\infty\frac{1}{n2^2}$ (c) $\sum\frac{1}{(n^4+n)^{1/3}}$ (d) $\sum_1^\infty\frac{n}{e^n}$ (e) $\sum\frac{\sqrt{n+6}}{n(n+1)(n+2)}$ (f) $\sum\frac{\tan^{-1}n}{n^2}$.

   - (a)$\sum_4^\infty\frac{3}{n-3} = 3(1 + \frac{1}{2} + \frac{1}{3} + \frac{1}{4} + \cdots)$. This is a divergent p-series ($p = 1$, the harmonic series).

- (b) Each term of $\sum_1^\infty \frac{1}{n2^n} = \frac{1}{2} + \frac{1}{2}\cdot\frac{1}{4} + \frac{1}{3}\cdot\frac{1}{8} + \frac{1}{4}\cdot\frac{1}{16} + \cdots$ is smaller than the corresponding term of $\sum_1^\infty \frac{1}{2^n} = \frac{1}{2} + \frac{1}{4} + \frac{1}{8} + \frac{1}{16} + \cdots$. The latter is a convergent geometric series. Therefore the series $\sum_1^\infty \frac{1}{n2^n}$ also converges.
- (c) Each term $\frac{1}{(n^4+n)^{1/3}}$ is smaller than $\frac{1}{n^{4/3}}$. Comparison with p-series $\sum \frac{1}{n^{4/3}}$ with $p = \frac{4}{3}$ gives convergence.
  Other series like $\sum \frac{1}{(n^4-2n)^{1/3}}$ or $\sum \frac{1}{(n^4-n^3-1)^{1/3}}$ converge by the same comparison. (These are slightly *larger* than $\sum \frac{1}{n^{4/3}}$ but as $n \to \infty$ the behavior is the same. The "limit comparison test" applies.)
- (d) $\sum \frac{1}{e^n}$ is a convergent geometric series, because $r = \frac{1}{e}$ is less than 1. Our series is $\sum \frac{n}{e^n}$. When n gets large, e^n completely dominates n and the series remains convergent. To be more precise, notice that $n < 2^n$. Then $\frac{n}{e^n} < \frac{2^n}{e^n} = (\frac{2}{e})^n$. Since $\frac{2}{e} < 1$, we know that $\sum(\frac{2}{e})^n$ is a convergent geometric series. By comparison $\sum \frac{n}{e^n}$ also converges.
- (e) The terms $\frac{\sqrt{n+6}}{n(n+1)(n+2)}$ behave like $\frac{\sqrt{n}}{n^3} = \frac{1}{n^{5/2}}$ when n gets large. Since $\sum \frac{1}{n^{5/2}}$ converges, so does the original series. Use the limit comparison test or show directly that $\sum \frac{3}{n^{5/2}}$ is larger. Increase $\sqrt{n+6}$ to $3\sqrt{n}$. Decrease the denominator $n(n+1)(n+2)$ to n^3 :
  $\frac{\sqrt{n+6}}{(n)(n+1)(n+2)} < \frac{3\sqrt{n}}{n^3} = \frac{3}{n^{5/2}}$. The larger series $\sum \frac{3}{n^{5/2}}$ converges. Therefore (e) converges.
- (f) The angle $\tan^{-1} n$ approaches $\frac{\pi}{2}$ from below. Therefore $\sum \frac{\tan^{-1} n}{n^2}$ is less than $\frac{\pi}{2}\sum \frac{1}{n^2}$. This is a convergent p-series (with $p = 2$.)

10C. Replace n by x and use the ***Integral Test*** 10C – provided the resulting function of x is decreasing to zero, and you can find its integral. The series $\sum_1^\infty f(n)$ converges or diverges together with the integral $\int_1^\infty f(x)dx$.

3. Determine the convergence of (a) $\sum_2^\infty \frac{1}{n(\ln n)^2}$ (b) $\sum_1^\infty \frac{n}{e^n}$ (c) $\sum \frac{6n}{(n^2+8)^{2/3}}$.

- (a) $\int_2^\infty \frac{dx}{x(\ln x)^2} = \int_{\ln 2}^\infty \frac{du}{u^2} = [-\frac{1}{u}]_{\ln 2}^\infty = \frac{1}{\ln 2}$. The integral converges so the series converges.
- (b) $\int_1^\infty \frac{x\,dx}{e^x} = \int_1^\infty xe^{-x}dx = [-xe^{-x} - e^{-x}]_1^\infty = \frac{2}{e}$. The series converges because the integral converges. Compare with problem 2(d) above. *There we had to find a comparison series. Here the integral test gives a comparison automatically.*
- (c) $\int_1^\infty \frac{6x\,dx}{(x^2+8)^{2/3}} = 3\int_9^\infty u^{-2/3}du = [9u^{1/3}]_9^\infty = \infty$. The series diverges because the integral diverges. It could have been compared to the divergent p-series $\sum \frac{1}{n^{1/3}}$.

10D. ***Ratio test*** Compute the limit L of $\frac{a_{n+1}}{a_n}$, the ratio of the $(n+1)$st term to the nth term. If $L < 1$ then the series converges. If $L > 1$ then the series diverges. If $L = 1$, the ratio test doesn't decide; look for a more careful comparison test. The ratio test is generally successful when the terms include a factorial. (The ratio of $(n+1)!$ to $n!$ is $n+1$).

4. Use the ratio test to determine the convergence of these five series:

(a) $\sum \frac{2^n}{(n+1)!}$ (b) $\sum n(\frac{2}{3})^n$ (c) $\sum \frac{n^3}{(\ln 2)^n}$ (d) $\sum \frac{(n+2)!5!}{n!5^n}$ (e) $\sum_1^\infty \frac{n^n}{n!}$.

- (a) The ratio $\frac{a_{n+1}}{a_n}$ is $\frac{2^{n+1}}{(n+2)!} \div \frac{2^n}{(n+1)!} = \frac{2^{n+1}(n+1)!}{(n+2)!2^n} = \frac{2}{(n+2)}$. This ratio $\frac{2}{n+2}$ goes to zero. Since the limit is $L = 0$, the series converges.
- (b) The ratio is $(n+1)(\frac{2}{3})^{n+1}$ divided by $(n)(\frac{2}{3})^n$. This is $(\frac{n+1}{n})\cdot\frac{2}{3} \to \frac{2}{3}$. Since $L = \frac{2}{3}$, the series converges.

- (c) $\frac{(n+1)^3}{(\ln 2)^{n+1}} \div \frac{n^3}{(\ln 2)^n} = (\frac{n+1}{n})^3 \cdot \frac{1}{\ln 2}$. The limit is $L = \frac{1}{\ln 2} > 1$, so this series diverges. (In fact the terms $\frac{n^3}{(\ln 2)^n}$ approach infinity. They fail the very first test, that $a_n \to 0$.)
- (d) Look at the terms $\frac{(n+2)!5!}{n!5^n}$. The first part $\frac{(n+2)!}{n!}$ is $(n+2)(n+1)$. This is about n^2. But 5^n grows much faster. We expect convergence – now prove it by the ratio test:
$\frac{a_{n+1}}{a_n} = \frac{(n+3)!5!}{(n+1)!5^{n+1}} \cdot \frac{n!5^n}{(n+2)!5!} = \frac{(n+3)}{5(n+1)} \to L = \frac{1}{5}$. Convergence.
- (e) The ratio test for $\frac{n^n}{n!}$ leads to $L = e$ (divergence because n^n grows faster than $n!$).
$\frac{(n+1)^{n+1}}{(n+1)!} \div \frac{n^n}{n!} = \frac{(n+1)^{n+1}}{n^n} \cdot \frac{n!}{(n+1)!} = (\frac{n+1}{n})^n \cdot \frac{n+1}{n+1} = (1+\frac{1}{n})^n \to e$. And $Le > 1$.

10E. The ***Root Test*** is useful when the terms of the series include an expression $(\ \ )^n$. The test looks at the limit of $(a_n)^{1/n}$. If the limit L is less than 1, the series converges. If $L > 1$ the series diverges. As with the ratio test, the root test gives no answer if $L = 1$ – and this often happens.

5. Test these series for convergence: (a) $\sum \frac{1}{(\ln n)^n}$ (b) $\sum \frac{2^{n+1}}{n^n}$ (c) $\sum_1^\infty \frac{2^n}{n^{10}}$

- (a) The nth root of $a_n = \frac{1}{(\ln n)^n}$ is $\frac{1}{\ln n}$. Since $\frac{1}{\ln n} \to L = 0$, the series converges.
- (b) The nth root of $\frac{2^{n+1}}{n^n}$ is $\frac{2^n\sqrt{2}}{n} \to L = 0$. (Certainly $\sqrt[n]{2}$ approaches 1.) The series converges.
- (c) The nth root of $\frac{2^n}{n^{10}}$ is $\frac{2}{n^{10/n}}$. Its limit is $L = 2$ and the series diverges. (We used the fact that $n^{1/n} \to 1$. Its tenth power $n^{10/n}$ goes to $1^{10} = 1$.) This series also fails test 10A –its terms don't go to zero. That is always true if $L > 1$!

10F. The ***Limit Comparison Test*** shows that $\sum \frac{\ln n}{n^{1.2}}$ converges. Compare $a_n = \frac{\ln n}{n^{1.2}}$ with $c_n = \frac{1}{n^{1.1}}$. Note that $\sum c_n$ is convergent (p-series with $p = 1.1$). Since $\frac{a_n}{c_n} = \frac{\ln n}{n^{.1}} \to 0$, the original series $\frac{\ln n}{n^{1.2}}$ converges.

In general we look at $L = \lim \frac{a_n}{c_n}$. Look for a comparison series $\sum c_n$ so that L is a finite number. *This test does not require $L < 1$.* If $\sum c_n$ converges then $\sum a_n$ converges. If $\sum c_n$ diverges and $L > 0$, the test gives the opposite answer: $\sum a_n$ also diverges.

6. Apply the limit comparison test to (a) $\sum \frac{n}{n^n+2}$ (divergent) and (b) $\sum \frac{n+7}{n^3-2}$ (convergent).
The ratio test and root test gives no decision. They find $L = 1$. ***Comparison tests and integral tests are more precise than ratio and root tests.***

- (a) The terms $a_n = \frac{n}{n^2+2}$ behave like $c_n = \frac{1}{n}$ when n is large. The ratio is $\frac{a_n}{c_n} = \frac{n}{n^2+2} \div \frac{1}{n} = \frac{n^2}{n^2+2} \to L = 1$. The comparison series $\sum \frac{1}{n}$ diverges (p-series with $p = 1$). Then $\sum a_n$ diverges.
- (b) The terms $a_n = \frac{n+7}{n^3-2}$ behave like $c_n = \frac{1}{n^2}$ when n is large. The ratio is $\frac{a_n}{c_n} = \frac{n+7}{n^3-2} \div \frac{1}{n^2} = \frac{n^3+7n^2}{n^3-2} \to L = 1$. The comparison series $\sum \frac{1}{n^2}$ converges (p-series with $p = 2$). Then $\sum a_n$ converges.

Read-throughs and selected even-numbered solutions :

The convergence of $a_1 + a_2 + \cdots$ is decided by the partial sums $s_n = \mathbf{a_1} + \cdots + \mathbf{a_n}$. If the s_n approach s, then $\sum a_n = \mathbf{s}$. For the **geometric** series $1 + x + \cdots$ the partial sums are $s_n = \mathbf{(1-x^n)/(1-x)}$. In that case $s_n \to 1/(1-x)$ if and only if $\mathbf{|x| < 1}$. In all cases the limit $s_n \to s$ requires that $a_n \to \mathbf{0}$. But the harmonic series $a_n = 1/n$ shows that we can have $a_n \to \mathbf{0}$ and still the series **diverges.**

The comparison test says that if $0 \le a_n \le b_n$ then $\sum \mathbf{a_n}$ **converges if** $\sum \mathbf{b_n}$ **converges.** In case a decreasing $y(x)$ agrees with a_n at $x = n$, we can apply the **integral** test. The sum $\sum a_n$ converges if and only if

$\int_1^\infty$ **y(x)dx converges.** By this test the p-series $\sum 1/n^p$ converges if and only if p is **greater than 1.** For the harmonic series $(p = 1)$, $s_n = 1 + \cdots + 1/n$ is near the integral $f(n) = \ln n$.

The **ratio** test applies when $a_{n+1}/a_n \to L$. There is convergence if $|L| < 1$, divergence if $|L| > 1$, and no decision if $L = 1$ **or** -1. The same is true for the **root** test, when $(a_n)^{1/n} \to L$. For a geometric-p-series combination $a_n = x^n/n^p$, the ratio a_{n+1}/a_n equals $x(n+1)^p/n^p$. Its limit is $L = x$ so there is convergence if $|x| < 1$. For the exponential $e^x = \sum x^n/n!$ the limiting ratio a_{n+1}/a_n or $x/(n+1)$ is $L = 0$. This series always **converges** because $n!$ grows faster than any x^n or n^p.

There is no sharp line between **convergence** and **divergence.** But if $\sum b_n$ converges and a_n/b_n approaches L, it follows from the **limit comparison** test that $\sum a_n$ also converges.

2 The series $\frac{a}{10} + \frac{b}{100} + \frac{c}{1000} + \cdots$ converges because it is below the comparison series $\frac{9}{10} + \frac{9}{100} + \frac{9}{1000} \cdots = .999 \cdots = 1$.

12 $\sum \frac{1}{\sqrt{n^2+10}}$ **diverges** by comparison with the smaller series $\sum \frac{1}{n+10}$ which diverges. Check that $\sqrt{n^2+10}$ is less than $n + 10$.

24 The terms are $(\frac{n-1}{n})^n = (1 - \frac{1}{n})^n \to e^{-1}$ so the ratio approaches $L = \frac{e^{-1}}{e^{-1}} = 1$. (Divergent series because its terms don't approach zero.)

38 $\sum \frac{\ln n}{n}$ **diverges** by comparison with $\int_1^\infty \frac{\ln x}{x} dx = [\frac{1}{2}(\ln x)^2]_1^\infty = \infty$. ($\frac{\ln x}{x}$ decreases for $x > e$ and these later terms decide divergence. Another comparison is with $\sum \frac{1}{n}$.)

46 The first term is $\frac{1}{2 \ln 2}$. After that $\frac{1}{n \ln n} < \int_{n-1}^n \frac{dx}{x \ln x}$. The sum from 3 to n is below $\int_2^n \frac{dx}{x \ln x} = \ln(\ln n) - \ln(\ln 2)$. By page 377 the computer has not reached, $n = 3.2 \cdot 10^{19}$ in a million years. So the sum has not reached $\frac{1}{2 \ln 2} + \ln(\ln 3.2 \cdot 10^{19}) - \ln(\ln 2) < 5$.

58 $\sum \frac{1}{n^{\ln n}}$ converges by comparison with $\sum \frac{2}{n^2}$ (note $\ln n > 2$ beyond the 8th term). (Also $\sum \frac{1}{(\ln n)^{\ln n}}$ converges!)

60 $\ln 10^n = n \ln 10$ so the sum is $\frac{1}{\ln 10} \sum \frac{1}{n} = \infty$ (harmonic series diverges)

62 $n^{-1/n}$ approaches 1 so the series cannot converge

10.3 Convergence Tests: All Series (page 384)

We turn to series whose terms may be positive or negative. If you take absolute values to make all terms positive, you have the "absolute series." When the absolute series converges, the original series also converges. In this case the original series is said to ***converge absolutely.*** Some series converge but do not converge absolutely. These are ***conditionally convergent.*** They depend on some plus-minus cancellation for their partial sums to approach a limit. It is easiest to prove convergence when their signs are alternating and they pass the *alternating series test: The terms decrease in absolute value to zero.*

Such a series is at least conditionally convergent. It may or may not be absolutely convergent.

Remember to think about the limit of the terms: Unless $a_n \to 0$, the series cannot possibly converge. The shrinking test comes first! In problems 1-6, decide if the alternating series converges: absolutely or conditionally?

1. $\sum(-1)^n(\frac{n^2-1}{n^2+1})$ 2. $\sum(-1)^n \frac{n^2-1}{n^3+1}$ 3. $\sum(-1)^n(\frac{n^2-1}{\sqrt{n^7}+1})$.

- Series 1 fails the shrinking test because $\frac{n^2-1}{n^2+1}$ does not approach zero.
- For series 2, the absolute series $\sum \frac{n^2-1}{n^3+1}$ diverges by comparison with $\sum \frac{1}{n}$. *The alternating series 2 is conditionally convergent.* Its terms decrease in absolute value, approaching zero. The decrease is not fast enough for the absolute series to converge. But the alternating series test is passed.
- Series 3 is absolutely convergent. Compare the terms with $\frac{n^2}{n^{7/2}} = \frac{1}{n^{3/2}}$. And $\sum \frac{1}{n^{3/2}}$ converges.

4. $\sum_{n=2}^{\infty}(-1)^n \frac{1}{\ln n}$ 5. $\sum_{n=2}^{\infty}(-1)^n \frac{1}{n \ln n}$ 6. $\sum_2^{\infty} \frac{\sin n}{n \ln n}$.

- The absolute values $\frac{1}{\ln n}$ are decreasing to zero. By the alternating series test, series 4 converges. However, since $\sum \frac{1}{\ln n} > \sum \frac{1}{n} = \infty$, series 4 is not absolutely convergent.
- The integral test shows that $\sum \frac{1}{n \ln n}$ is divergent, because $\int \frac{dx}{x \ln x} = \int \frac{du}{u} = \ln u = \ln(\ln x) \to \infty$. But series 5 is conditionally convergent. The alternating terms approach zero.
- Series 6 has both positive and negative terms. *But it is not strictly alternating.* Is this one that you and your instructor cannot decide?

  Important fact when the alternating series test is passed: If you use the first n terms to estimate the sum, the error can be no larger than the $(n+1)$st term. By "error" we mean $|s_n - s| = |$ partial sum minus infinite sum$|$. Figure 10.3 on page 383 explains why this is true. Suppose you use $a_1 - a_2 + a_3 - a_4$ to estimate the sum s. *The error is smaller than* a_5 (because adding a_5 throws us beyond the final sum s).

  Another way to express it: *s is between the partial sums* s_n *and* s_{n+1}. Use this in problems 7-9.

7. Section 10.1 derived $\ln(1+x) = x - \frac{1}{2}x^2 + \frac{1}{3}x^3 - \cdots$. The quadratic approximation is $\ln(1+x) \approx x - \frac{1}{2}x^2$. How close is this for $\ln 1.2$, when x is 0.2?

   - The number $\ln 1.2$ is $\ln(1+0.2) = (0.2) - \frac{1}{2}(0.2)^2 + \frac{1}{3}(0.2)^3 - \cdots$. This is a convergent alternating series. Its terms decrease to zero. The quadratic approximation $(0.2) - \frac{1}{2}(0.2)^2 = 0.18$ uses the first two terms. The error is no larger than the next term and has the same positive sign: $0 <$ error $< \frac{1}{3}(0.2)^3 < 0.0027$.

     In other words $0.18 < \ln 1.2 < 0.1827$. The third term 0.0027 throws us beyond $\ln 1.2 = 0.1823\cdots$.

8. Section 10.4 will show that $\frac{1}{e} = 1 - \frac{1}{1} + \frac{1}{2!} - \frac{1}{3!} + \frac{1}{4!} - \frac{1}{5!} + \cdots$. Compute $\frac{1}{e}$ to 4 places.

   - The question is, how many terms do we need? The error is to be less than $0.00005 = 5 \times 10^{-5}$. Since $\frac{1}{7!} \approx 2.0 \times 10^{-4}$ and $\frac{1}{8!} \approx 2.5 \times 10^{-5}$, we do *not* need $\frac{1}{8!}$. Then $\frac{1}{e} \approx \frac{1}{2!} - \frac{1}{3!} + \frac{1}{4!} \cdots - \frac{1}{7!} \approx 0.36786$.

9. How many terms of $\sum(-1)^{n+1}\frac{1}{n^3}$ give the sum to five decimal place accuracy?

   - Five-place accuracy means error $< 0.000005 = 5 \times 10^{-6}$. If the series stops at $\pm\frac{1}{n^3}$, the error is less than the next term $\frac{1}{(n+1)^3}$. Set $\frac{1}{(n+1)^3} < 5 \times 10^{-6}$ to find $(n+1)^3 > 2 \times 10^5$ and $n+1 > 58.8$. Using $n = 58$ terms gives five-place accuracy.

Read-throughs and selected even-numbered solutions :

The series $\sum a_n$ is absolutely convergent if the series $\sum |\mathbf{a_n}|$ is convergent. Then the original series $\sum a_n$ is also **convergent.** But the series $\sum a_n$ can converge without converging absolutely. That is called **conditional** convergence, and the series $\mathbf{1 - \frac{1}{2} + \frac{1}{3}} - \cdots$ is an example.

For alternating series, the sign of each a_{n+1} is **opposite** to the sign of a_n. With the extra conditions that $|\mathbf{a_{n+1}}| \le |\mathbf{a_n}|$ and $\mathbf{a_n} \to \mathbf{0}$, the series converges (at least conditionally). The partial sums $s_1, s_3, \cdots$ are **decreasing** and the partial sums $s_2, s_4, \cdots$ are **increasing.** The difference between s_n and s_{n-1} is $\mathbf{a_n}$. Therefore the two series converge to the same number s. An alternating series that converges absolutely [conditionally] (not at all) is $\sum(-\mathbf{1})^{\mathbf{n+1}}/\mathbf{n^2}[\sum(-\mathbf{1})^{\mathbf{n+1}}/\mathbf{n}](\sum(-\mathbf{1})^{\mathbf{n+1}})$. With absolute [conditional] convergence a reordering cannot [can] change the sum.

10 $\sum(-1)^{n+1}2^{1/n}$ **diverges** (terms don't approach zero)

16 The terms alternate in sign but do not decrease to zero. The positive terms $\frac{2}{3}, \frac{4}{7}, \frac{6}{11}, \cdots$ approach $\frac{1}{2}$ and so does the sequence $\frac{3}{5}, \frac{5}{9}, \frac{7}{13}, \cdots$

20 The difference between s and s_{100} is less than $\mathbf{\frac{1}{101^2}}$, the next term in the series (because after that term comes $-\frac{1}{102^2}$ and the sums stay between s_{100} and s_{101}).

24 The series $a_1+a_2-a_3+a_4+a_5-a_6+\cdots$ is sure to converge (*conditionally*) if $0 \le a_{3n+3} < a_{3n+1}+a_{3n+2} < a_{3n}$ for every n. Then it passes the alternating series test when each pair of positive terms is combined. (The series could converge without passing this particular test.)

28 Shorter answer than expected: $1+\frac{1}{3}+\frac{1}{5}-\frac{1}{2}-\frac{1}{4}-\frac{1}{6}$ comes from rearranging $1-\frac{1}{2}+\frac{1}{3}-\frac{1}{4}+\frac{1}{5}-\frac{1}{6}$. Continue this way, **six terms at a time.** The partial sums $s_6, s_{12}, \cdots$ are not changed and still approach $\ln 2$. The partial sums in between also approach $\ln 2$ because the six terms in each group approach zero.

36 If $\sum a_n$ is conditionally but not absolutely convergent, take positive terms until the sum exceeds 10. Then take one negative term. Then positive terms until the sum exceeds 20. Then one negative term, and so on. The partial sums approach $+\infty$ (because the single negative terms go to zero, otherwise no conditional convergence in the first place).

40 For $s = -1$ choose all minus signs: $-\frac{1}{2}-\frac{1}{4}-\cdots = -1$. For $s = 0$ choose one plus sign and then all minus: $\frac{1}{2}-\frac{1}{4}-\frac{1}{8}-\cdots = 0$. For $s = \frac{1}{3}$ choose alternating signs: $\frac{1}{2}-\frac{1}{4}+\frac{1}{8}-\cdots = \frac{\frac{1}{2}}{1+\frac{1}{2}} = \frac{1}{3}$.

10.4 The Taylor Series for e^x, sin x and cos x (page 390)

In Problems 1 – 4, begin with these three known series (valid for all x):

$$\begin{aligned} e^x &= 1+\frac{x}{1!}+\frac{x^2}{2!}+\cdots+\frac{x^n}{n!}+\cdots \\ \sin x &= x-\frac{x^3}{3!}+\frac{x^5}{5!}-\cdots+(-1)^n\frac{x^{2n+1}}{(2n+1)!}+\cdots \\ \cos x &= 1-\frac{x^2}{2!}+\frac{x^4}{4!}-\cdots+(-1)^n\frac{x^{2n}}{(2n)!}+\cdots \end{aligned}$$

Write down the corresponding power series for these four functions:

1. e^{-x} 2. $\sin 2x$ 3. $\cos^2 x$ 4. $x \sin x^2$

- Substitute $-x$ for x in the series for e^x to get

$$e^{-x} = 1 - \frac{x}{1!} + \frac{x^2}{2!} - \frac{x^3}{3!} + \cdots + \frac{(-1)^n x^n}{n!} + \cdots$$

- Substitute $2x$ for x in the series for $\sin x$ to get

$$\sin 2x = 2x - \frac{2^3x^3}{3!} + \frac{2^5x^5}{5!} - \cdots + \frac{(-1)^n 2^{2n+1}x^{2n+1}}{(2n+1)!} + \cdots$$

- You don't want to square the series for $\cos x$ if you can avoid it. Here is how to avoid it. Use $\cos 2x = 2\cos^2 x - 1$ or $\cos^2 x = \frac{1+\cos 2x}{2}$:

$$\cos^2 x = \tfrac{1}{2} + \tfrac{1}{2}\cos 2x = \tfrac{1}{2} + \tfrac{1}{2}\left(1 - \tfrac{(2x)^2}{2!} + \tfrac{(2x)^4}{4!} - \cdots + \tfrac{(-1)^n(2x)^{2n}}{(2n)!}\right) + \cdots$$

Take $x = \frac{\pi}{4}$ as a check. We expect $\cos^2(\frac{\pi}{4}) = (\frac{\sqrt{2}}{2})^2 = \frac{1}{2}$. The first four terms give 0.5000.

- To obtain $x\sin x^2$, substitute x^2 for x in the series for $\sin x$, then multiply each term by x:

$$x\sin x^2 = x\left(x^2 - \tfrac{(x^2)^3}{3!} + \tfrac{(x^2)^5}{5!} - \cdots\right) = x^3 - \tfrac{x^7}{3!} + \tfrac{x^{11}}{5!} - \cdots.$$

5. Find the Taylor series for $f(x) = \frac{1}{1-x}$ about the basepoint $x = 0$. Compare with Section 10.1.

Formula 10K on page 387 gives the pattern for Taylor series. We need the value of $f(x)$ and all its derivatives at $x = 0$, so the series can be made to agree with the function. A table helps to organize this:

n	nth derivative	set $x = 0$ for $f^{(n)}(0)$	nth term	$= \frac{f^n(0)}{n!}x^n$
0	$\frac{1}{1-x}$	$1 = 0!$	$\frac{0!}{0!}x^0$	$= 1$
1	$\frac{1}{(1-x)^2}$	$1 = 1!$	$\frac{1!}{1!}x$	$= x$
2	$\frac{2}{(1-x)^3}$	$2 = 2!$	$\frac{2!x^2}{2!}$	$= x^2$
3	$\frac{2\cdot 3}{(1-x)^4}$	$2\cdot 3 = 3!$	$\frac{3!x^3}{3!}$	$= x^3$
4	$\frac{2\cdot 3\cdot 4}{(1-x)^5}$	$2\cdot 3\cdot 4 = 4!$	$\frac{4!x^4}{4!}$	$= x^4$

The nth derivative is $f^n(0) = n!$ so the nth term of the Taylor series is x^n. This means that $f(x) = \frac{1}{1-x} = 1 + x + x^2 + x^3 + \cdots$. That is the geometric series from Section 10.1.

6. Find the Taylor series for $y = \ln x$ about the point $x = 5$. You need to know all the derivatives of $f(x) = \ln x$ evaluated at $x = 5$:

n	$f^{(n)}(x)$	$f^{(n)}(5)$	nth term $= \frac{f^{(n)}(5)}{n!}(x-5)^n$
0	$\ln x$	$\ln 5$	$\ln 5$
1	x^{-1}	$\frac{1}{5}$	$\frac{1}{5}(x-5)$
2	$-x^{-2}$	$\frac{-1}{5^2}$	$\frac{-1}{2\cdot 5^2}(x-5)^2$
3	$2x^{-3}$	$\frac{2}{5^3}$	$\frac{2!}{3!5^3}(x-5)^3$

The pattern is $\ln x = \ln\ 5 + \frac{1}{5}(x-5) - \frac{1}{2\cdot 5^2}(x-5)^2 + \frac{1}{3\cdot 5^3}(x-5)^3 + \cdots$

7. (This is 10.4.40) Find the first two nonzero terms of the Taylor series for $f(x) = \ln(\cos x)$ around $x = 0$.

- The derivatives of $f(x) = \ln\cos x$ are $f'(x) = -\tan x$, $f''(x) = -\sec^2 x$, $f'''(x) = -2\sec^2 x\tan x$, $f^4(x) = -2\sec^4 x - 4\sec^2 x\tan^2 x$. Their values at $x = 0$ are $0, 0, -1, 0, -2$. (We go out to the fourth power of x in order to get two nonzero terms.) So the series for $\ln\cos x$ begins $0 + 0 + \frac{(-1)}{2!}x^2 + 0 + \frac{(-2)x^4}{4!} = -\frac{x^2}{2} - \frac{x^4}{12}$.
  If you have a graphing calculator, graph $y = \ln\cos x$ together with $y = -\frac{x^2}{2} - \frac{x^4}{12}$ on the domain $-\frac{\pi}{2} < x < \frac{\pi}{2}$. You will see how closely the polynomial approximates the function.

8. (This is 10.4.13) Find the infinite series (powers of x) that solves $y'' = 2y' - y$ starting from $y = 0$ and $y' = 1$ at $x = 0$.

- The equation $y'' = 2y' - y$ will let us figure out the nth derivative of $y = f(x)$ at $x = 0$:

n	$f^{(n)}$	$f^{(n)}$ at $x=0$	nth term $= \frac{f^{(n)}(0)}{n!}x^n$
0	y	0(given)	0
1	y'	1(given)	x
2	$y'' = 2y' - y$	$2(1) - 0 = 2$	$\frac{2}{2!}x^2 = x^2$
3	$y''' = 2y'' - y'$	$2(2) - 1 = 3$	$\frac{3}{3!}x^3 = \frac{1}{2!}x^3$
4	$y'''' = 2y''' - y''$	$2(3) - 2 = 4$	$\frac{4}{4!}x^4 = \frac{1}{3!}x^4$

The nth derivative of $y(x)$ at $x = 0$ is n. The nth term is $\frac{n}{n!}x^n = \frac{1}{(n-1)!}x^n$. Then

$$y(x) = 0 + x + x^2 + \frac{1}{2!}x^3 + \frac{1}{3!}x^4 + \cdots + \frac{1}{(n-1)!}x^n + \frac{1}{n!}x^{n+1} + \cdots.$$

This is the series for e^x multiplied by x. Therefore $y = xe^x$ solves the differential equation. You can check by substituting $y' = xe^x + e^x$ and $y'' = xe^x + 2e^x$ into the equation.

9. (This is part of 10.4.47) Find the real and imaginary parts and the 99th power of $e^{-i\pi/6}$.

- Use Euler's formula $e^{i\theta} = \cos\theta + i\sin\theta$ to find the real and imaginary parts:

$$e^{i\pi/6} = \cos(\tfrac{\pi}{6}) + i\sin(-\tfrac{\pi}{6}) = \tfrac{\sqrt{3}}{2} - \tfrac{i}{2}.\text{The 99th power is } e^{-99i\pi/6} = \cos(-\tfrac{99\pi}{6}) + i\sin(-\tfrac{99\pi}{6}).$$

But $\frac{99\pi}{6} = \frac{33\pi}{2} = 16\pi + \frac{\pi}{2}$. We can ignore 16π (*why?*). The answer is $\cos(-\frac{\pi}{2}) + i\sin(-\frac{\pi}{2}) = -i$.

Read-throughs and selected even-numbered solutions :

The **Taylor** series is chosen to match $f(x)$ and all its **derivatives** at the basepoint. Around $x = 0$ the series begins with $f(0) + \mathbf{f'(0)}x + \frac{\mathbf{1}}{\mathbf{2}}\mathbf{f''(0)}x^2$. The coefficient of x^n is $\mathbf{f^{(n)}(0)/n!}$. For $f(x) = e^x$ this series is $\sum \mathbf{x^n/n!}$. For $f(x) = \cos x$ the series is $\mathbf{1 - x^2/2! + x^4/4!} - \cdots$. For $f(x) = \sin x$ the series is $\mathbf{x - x^3/3!} + \cdots$. If the signs were positive in those series, the functions would be $\cosh x$ and **sinh x**. Addition gives $\cosh x + \sinh x = \mathbf{e^x}$.

In the Taylor series for $f(x)$ around $x = a$, the coefficient of $(x-a)^n$ is $b_n = \mathbf{f^{(n)}(a)/n!}$. Then $b_n(x-a)^n$ has the same **derivatives** as f at the basepoint. In the example $f(x) = x^2$, the Taylor coefficients are $b_0 = \mathbf{a^2}$, $b_1 = \mathbf{2a}$, $b_2 = \mathbf{1}$. The series $b_0 + b_1(x-a) + b_2(x-a)^2$ agrees with the original $\mathbf{x^2}$. The series for e^x around $x = a$ has $b_n = \mathbf{e^a/n!}$. Then the Taylor series reproduces the identity $e^x = \mathbf{(e^a)(e^{x-a})}$.

We define e^x, $\sin x$, $\cos x$, and also $e^{i\theta}$ by their series. The derivative $d/dx(1 + x + \frac{1}{2}x^2 + \cdots) = 1 + x + \cdots$ translates to $\mathbf{d/dx(e^x) = e^x}$. The derivative of $1 - \frac{1}{2}x^2 + \cdots$ is $\mathbf{-x + x^3/3!} - \cdots$. Using $i^2 = -1$ the series

$1 + i\theta + \frac{1}{2}(i\theta)^2 + \cdots$ splits into $e^{i\theta} = \mathbf{cos}\ \theta + \mathbf{i\ sin}\ \theta$. Its square gives $e^{2i\theta} = \mathbf{cos\ 2}\theta + \mathbf{i\ sin\ 2}\theta$. Its reciprocal is $e^{-i\theta} = \mathbf{cos}\ \theta - \mathbf{i\ sin}\ \theta$. Multiplying by r gives $re^{i\theta} = \mathbf{r\ cos}\ \theta + i\mathbf{r\ sin}\ \theta$, which connects the polar and rectangular forms of a **complex** number. The logarithm of $e^{i\theta}$ is $i\theta$.

14 At $x = 0$ the equation gives $y'' = y = 1$, $y''' = y' = 0$ and $y'''' = y'' = y = 1$ (even derivatives equal 1, odd derivatives equal 0). The Taylor series is $y(x) = 1 + \frac{x^2}{2!} + \frac{x^4}{4!} + \cdots = \frac{1}{2}(e^x + e^{-x}) = \cosh x$.

18 At $x = 2\pi$ the cosine and its derivatives are $1, 0, -1, 0, 1, \cdots$. The Taylor series is $\cos x = 1 - \frac{(x-2\pi)^2}{2!} + \frac{(x-2\pi)^4}{4!} - \cdots$. At $x = 0$ the function $\cos(x - 2\pi)$ and its derivatives again equal $1, 0, -1, 0, 1, \cdots$. Now the Taylor series is $\cos(x - 2\pi) = 1 - \frac{x^2}{2!} + \frac{x^4}{4!} - \cdots$.

22 At $x = 1$ the function x^4 and its derivatives equal 1,4,12,24,0,0,$\cdots$. The Taylor series has five nonzero terms: $x^4 = 1 + 4(x-1) + \frac{12}{2}(x-1)^2 + \frac{24}{6}(x-1)^3 + \frac{24}{24}(x-1)^4$.

26 $\cos\sqrt{x} = 1 - \frac{(\sqrt{x})^2}{2!} + \frac{(\sqrt{x})^4}{4!} - \cdots = 1 - \frac{x}{2} + \frac{x^2}{24} - \cdots$. (Note that $\sin\sqrt{x}$ would not succeed; the terms $\sqrt{x}, (\sqrt{x})^3, \cdots$ are not acceptable in a Taylor series. The function has no derivative at $x = 0$.)

28 $\frac{\sin x}{x} = \frac{x - \frac{x^3}{6} + \frac{x^5}{120} - \cdots}{x} = 1 - \frac{x^2}{6} + \frac{x^4}{120} - \cdots$.

46 $(e^{i\theta})^2 = e^{2i\theta}$ equals $\cos\ 2\theta + i \sin\ 2\theta$, so **neither** of the proposed answers is correct.

50 $(2e^{i\pi/3})^2 = 4e^{2\pi i/3}$ and also $(1 + \sqrt{3}i)(1 + \sqrt{3}i) = 1 + 2\sqrt{3}i - 3 = -2 + 2\sqrt{3}i$; $(4e^{i\pi/4})^2 = 16e^{i\pi/2}$ and also $(2\sqrt{2} + i2\sqrt{2})(2\sqrt{2} + i2\sqrt{2}) = 8 + 16i - 8 = 16i$.

10.5 Power Series (page 395)

1. This problem compares the convergence of a power series when x is close to and far from the basepoint. How many terms are needed to estimate $\sin(\frac{1}{2})$ and $\sin 10$ to 6-place accuracy?

   - With $x = 0$ as basepoint, the series is $\sin x = x - \frac{x^3}{3!} + \frac{x^5}{5!} - \frac{x^7}{7!} + \cdots$. This is a convergent alternating series when x is positive. *The first unused term is larger than the error* (provided the terms decrease as in the alternating series test). We want the error to be less than 5×10^{-7}.
   For $\sin\frac{1}{2} = \frac{1}{2} - \frac{1}{3!}(\frac{1}{2})^3 + \frac{1}{5!}(\frac{1}{2})^5 - \frac{1}{7!}(\frac{1}{2})^7$, the error is less than $\frac{1}{9!}(\frac{1}{2})^9 \approx 5 \times 10^{-9}$. These four terms give the desired accuracy. However, for $\sin(10) = 10 - \frac{10^3}{3!} + \frac{10^5}{5!} - \cdots$ we are forced to go out to $\frac{-10^{37}}{37!}$. It requires nineteen nonzero terms to guarantee an error below $\frac{10^{39}}{39!} \approx 4 \times 10^{-8}$.
   To estimate $\sin 10$, it would be much more efficient to move the basepoint to 3π (because 3π is close to 10). The series has powers of $(x - 3\pi)$. The constant term is $\sin 3\pi = 0$. Then $\sin 10 =$???

2. Find the convergence interval of x's for the series $(x+2) + \frac{1}{\sqrt{2}}(x+2)^2 + \frac{1}{\sqrt{3}}(x+2)^3 + \cdots$.

   - This is a power series with basepoint $a = -2$ (it has powers of $x + 2$ which is $x - a$). ***The interval of convergence is always is centered at the basepoint.*** Perform the ratio test on the *absolute series*, treating x as a constant:

   $$\left|\frac{(x+2)^{n+1}}{\sqrt{n+1}} \Big/ \frac{(x+2)^n}{\sqrt{n}}\right| = \frac{\sqrt{n}}{\sqrt{n+1}}|x+2| \text{ approaches } L = |x+2| \text{ as } n \to \infty.$$

   When the limit $L = |x+2|$ is below 1, we have absolute convergence. This interval is from $x = -3$ to $x = -1$. The radius of convergence is $r = 1$, the half-length of the interval. The endpoints $x = -3$ (conditional convergence) and $x = -1$ (no convergence) need separate discussion.

3. Find the interval of x's for which $\sum_{n=1}^{\infty} \frac{(-1)^n x^n}{n3^n}$ converges.

- The absolute value ratios

$$\left|\frac{x^{n+1}}{(n+1)3^{n+1}}\Big/\frac{x^n}{n\,3^n}\right| = \frac{n\,3^n}{(n+1)3^{n+1}}\left|\frac{x^{n+1}}{x^n}\right| \text{ approach } L = \frac{|x|}{3}.$$

Absolute convergence occurs when $\frac{|x|}{3} < 1$, or $|x| < 3$. The series diverges when $|x| > 3$. At the right hand endpoint $x = 3$, the series is $\sum_{n=1}^{\infty} \frac{(-1)^n\,3^n}{n\,3^n} = \sum_1^{\infty} \frac{(-1)^n}{n}$. This is the alternating harmonic series, which converges conditionally to ln 2. (See pages 370-371.) At the left endpoint $x = -3$, the series is $\sum_1^{\infty} \frac{(-1)^n(-3)^n}{n\,3^n} = \sum_1^{\infty} \frac{1}{n}$, the harmonic series that diverges.
In summary, $\sum_1^{\infty} \frac{(-1)^n\,x^n}{n\,3^n}$ converges if $-3 < x \le 3$. The convergence radius is $r = 3$.

4. The series for $y = \cos x$ around the basepoint $x = \pi$ is $-1 + 0 + \frac{1}{2}(x-\pi)^2 + 0 - \frac{1}{4!}(x-\pi)^4 + R_4$. Estimate the remainder R_4 when $x = 3$.

- Equation (2) on page 392 gives $R_4(x) = \frac{f^{(5)}(c)(x-\pi)^5}{5!}$. The unknown point c is between x and π. For the function $y = \cos x$, the fifth derivative is $f^{(5)}(x) = -\sin x$. We don't know $\sin c$, but we know it lies between -1 and 1. Therefore $|R_4(x)| \le \frac{1}{5!}|x-\pi|^5$. For $x = 3$ this remainder $|R_4(3)|$ is less than $\frac{1}{5!}|3-\pi|^5 < 5 \times 10^{-7}$.

5. Use the binomial series on page 394 to estimate $(242)^{1/5}$ to six-place accuracy.

- With $p = \frac{1}{5}$, the binomial series gives $(1+x)^{1/5}$. But we can't use formula (6)! The series for $(1+x)^p$ has the requirement $|x| < 1$, and for $242^{1/5} = (1+241)^{1/5}$ we would have $x = 241$. The series would quickly diverge – terms would get bigger. We need to start from a known answer near $242^{1/5}$ (much nearer than $1^{1/5}$), and then use the series to make a small correction.
The nearest convenient value is $(243)^{1/5}$ which is exactly 3. The fifth power is $3 \cdot 3 \cdot 3 \cdot 3 \cdot 3 = 243$. So move the problem to $(242)^{1/5} = (243)^{1/5}(\frac{242}{243})^{1/5} = 3(1 - \frac{1}{243})^{1/5}$. Now we have $x = \frac{-1}{243}$ and the binomial series is good:

$$3(1+x)^{1/5} = 3(1 + \tfrac{1}{5}x + \tfrac{1}{5}(\tfrac{-4}{5})\tfrac{x^2}{2} + \cdots) = 3 + \tfrac{3}{5}(\tfrac{-1}{243}) - \tfrac{12}{50}(\tfrac{-1}{243})^2 + \cdots = 2.99753.$$

6. Integrate the binomial series for $\frac{1}{\sqrt{1-x^2}}$ to find the power series for $\sin^{-1} x$.
Since $\frac{1}{\sqrt{1-x^2}} = (1-x^2)^{-1/2}$, use the binomial series (6) with $p = -\frac{1}{2}$. And also replace x in that series by $-x^2$. The result is

$$\begin{aligned}(1-x^2)^{-1/2} &= 1 + (-\tfrac{1}{2})(-x^2) + \frac{(-\frac{1}{2})(-\frac{3}{2})}{2}(-x^2)^2 + \frac{(-\frac{1}{2})(-\frac{3}{2})(-\frac{5}{2})}{3!}(-x^2)^3 + \cdots \\ &= 1 + \frac{x^2}{2} + \frac{3}{8}x^4 + \frac{5}{16}x^6 + \cdots\end{aligned}$$

Integrating each term gives $\sin^{-1} x = \int_0^x \frac{1}{\sqrt{1-t^2}}dt = x + \frac{x^3}{6} + \frac{3}{40}x^5 + \frac{5}{7\cdot 16}x^7 + \cdots$.

7. (This is 10.5.46.) Compute $\int_0^1 e^{-x^2}\,dx$ to three decimal places by integrating the power series.

- Start with $e^x = 1 + \frac{x}{1} + \frac{x^2}{2!} + \frac{x^3}{3!} + \cdots$. Replace x by $-x^2$ to get $e^{-x^2} = 1 - \frac{x^2}{1} + \frac{x^4}{2!} - \frac{x^6}{3!} + \cdots$. Integrate each term of this series from $x = 0$ to $x = 1$:

$$\begin{aligned}\int_0^1 e^{-x^2}\,dx &= [x - \frac{x^3}{3} + \frac{x^5}{5\times 2!} - \frac{x^7}{7\times 3!} + \cdots]_0^1 \\ &= 1 - \frac{1}{3} + \frac{1}{5\times 2!} - \frac{1}{7\times 3!} + \frac{1}{9\times 4!} - \frac{1}{11\times 5!} \approx 0.747.\end{aligned}$$

For three-place accuracy, we need an error less than 5×10^{-4}. Since the series is alternating with decreasing terms, the actual error is less than the first unused term $\frac{1}{13 \times 6!} \approx 10^{-4}$.

If $|x| < |X|$ and $\sum a_n X^n$ converges, then the series $\sum a_n x^n$ also **converges.** There is convergence in a **symmetric** interval around the **origin.** For $\sum (2x)^n$ the convergence radius is $r = \mathbf{\frac{1}{2}}$. For $\sum x^n/n!$ the radius is $r = \infty$. For $\sum (x-3)^n$ there is convergence for $|x - 3| < \mathbf{1}$. Then x is between **2** and **4.**

Starting with $f(x)$, its Taylor series $\sum a_n x^n$ has $a_n = \mathbf{f^{(n)}(0)/n!}$. With basepoint a, the coefficient of $(x-a)^n$ is $\mathbf{f^{(n)}(a)/n!}$. The error after the x^n term is called the **remainder** $R_n(x)$. It is equal to $\mathbf{f^{(n+1)}(c)(x-a)^{n+1}/(n+1)!}$ where the unknown point c is between **a and x**. Thus the error is controlled by the $\mathbf{(n+1)}$**st** derivative.

The circle of convergence reaches out to the first point where $f(x)$ fails. For $f = 4/(2-x)$, that point is $x = \mathbf{2}$. Around the basepoint $a = 5$, the convergence radius would be $r = \mathbf{3}$. For $\sin x$ and $\cos x$ the radius is $r = \infty$.

The series for $\sqrt{1+x}$ is the **binomial** series with $p = \frac{1}{2}$. Its coefficients are $a_n = \mathbf{(\frac{1}{2})(-\frac{1}{2})(-\frac{3}{2}) \cdots /n!}$. Its convergence radius is **1**. Its square is the very short series $1 + x$.

2 In the geometric series $\frac{1}{1-x} = 1 + x + x^2 + \cdots$ change x to $4x^2$: $\frac{1}{1-4x^2} = \mathbf{1 + 4x^2 + 16x^4} + \cdots$. Convergence fails when $4x^2$ reaches 1(thus $\mathbf{x = \frac{1}{2}}$ or $\mathbf{x = -\frac{1}{2}}$). The radius of convergence is $\mathbf{r = \frac{1}{2}}$.

4 $\tan x$ has derivatives $\sec^2 x, 2 \sec^2 x \tan x, 2 \sec^4 x + 4 \sec^2 x \tan^2 x$. At $x = 0$ the series is $1 + 0x + \frac{2}{2}x^2 + 0x^3 = \mathbf{1 + x^2}$. The function $\tan x = \frac{\sin x}{\cos x}$ is infinite when $\cos x = 0$, at $\mathbf{x = \frac{\pi}{2}}$ and $\mathbf{x = -\frac{\pi}{2}}$. Then $\mathbf{r = \frac{\pi}{2}}$.

6 In the geometric series replace x by $-4x^2$. Then $\frac{1}{1+4x^2} = \mathbf{1 - 4x^2 + 16x^4} - \cdots$. Convergence fails when $|4x^2|$ reaches 1. The function blows up when $4x^2 = -1$, at $\mathbf{x = \frac{i}{2}}$ and $\mathbf{x = -\frac{i}{2}}$. The radius of convergence is $\mathbf{r = \frac{1}{2}}$.

14 (a) Combine $x + x^4 + x^7 + \cdots = \frac{x}{1-x^3}$ and $x^2 + x^5 + x^8 + \cdots = \frac{x^2}{1-x^3}$ and $-(x^3 + x^6 + \cdots) = -\frac{x^3}{1-x^3}$ to get $\mathbf{\frac{x+x^2-x^3}{1-x^3}}$. (b) Adding the series for $\cos x$ and $\cosh x$ leads to $1 + \frac{x^4}{4!} + \frac{x^8}{8!} + \cdots = \mathbf{\frac{1}{2}(\cos x + \cosh x)}$. (c) $\ln(x-1) = x - \frac{1}{2}x^2 + \frac{1}{3}x^3 \cdots$ so changing x to $x - 1$ gives the series for $\mathbf{\ln(x-2)}$ around $a = 1$.

22 The remainder after n terms of the series for e^x around $a = 1$ is $R_n(x) = e^c \frac{(x-1)^{n+1}}{(n+1)!}$. The factor e^c is between 1 and e^x. As $n \to \infty$ the factorial assures that $R_n(x) \to 0$ and the series converges to e^x.

26 The derivatives of $(1-x)^{-1/2}$ are $\frac{1}{2}(1-x)^{-3/2}, \frac{1 \cdot 3}{2 \cdot 2}(1-x)^{-5/2}, \ldots, \frac{1 \cdot 3 \cdot 5 \cdots (2n-1)}{2^n}(1-x)^{-(2n+1)/2}$. At $x = 0$ this nth derivative divided by $n!$ is the coefficient a_n.

50 If $a_n^{1/n}$ approaches L then $(a_n x^n)^{1/n}$ approaches $\frac{x}{L}$. By the root test the series $\sum a_n x^n$ converges when $|\frac{x}{L}| < 1$ and diverges when $|\frac{x}{L}| > 1$. So the radius of convergence is $r = L$.

10 Chapter Review Problems

Review Problems

R1 Define the convergence of a sequence $s_1, s_2, s_3, \cdots$ and the convergence of a series $\sum a_n$.

R2 Find the sums of (a) $\frac{2}{5} - \frac{4}{25} + \frac{8}{125} - \cdots$ (b) $3.646464\cdots$ (c) $\frac{1}{e} + \frac{1}{e^2} + \frac{1}{e^3} + \cdots$

R3 When is $a_0 + a_1 + a_2 + \cdots$ a geometric series? When does it converge?

R4 Explain these tests for the convergence (or divergence!) of the infinite series $\sum a_n$ with all $a_n \geq 0$: Comparison test, Integral test, Ratio test, Root test.

R5 When can you be sure that an *alternating series* converges? Give examples of absolute convergence, conditional convergence, and divergence.

R6 Find the Taylor Series for $y = \cos x$ around the basepoint $a = \frac{\pi}{3}$.

R7 If $a_n \to 0$ show by comparison that $\sum a_n x^n$ converges at least for $-1 < x < 1$.

R8 What is a p-series? Which p-series converge and which diverge?

R9 In using the Taylor series to evaluate $f(x)$, why should the basepoint a be close to the evaluation point?

R10 If the root test for $\sum a_n (x-4)^n$ yields $L = 3|x-4|$, what is the convergence interval $|x-4| < r$?

Drill Problems

Determine whether the positive series **D1** – **D8** converge or diverge. Name your test.

D1 $\sum_1^\infty \frac{1}{\sqrt{2n+1}}$ **D2** $\sum_1^\infty \frac{1}{2n+1}$ **D3** $\sum_1^\infty \frac{1}{3^n+1}$ **D4** $\sum \frac{n - \ln n}{n^2}$

D5 $\sum \frac{(\ln n)^3}{n^2}$ **D6** $\sum \frac{2^n}{n^2}$ **D7** $\sum \frac{n^2}{3n^3+7n}$ **D8** $\frac{e}{3} + \frac{e^2}{9} + \frac{e^3}{27} + \cdots$

Decide whether the alternating series **D9** – **D12** diverge, converge absolutely, or converge conditionally.

D9 $\sum \frac{(-1)^n}{\sqrt{n}}$ **D10** $1.1 - 1.01 + 1.001 - \cdots$ **D11** $\sum \frac{(-1)^n}{n^2}$ **D12** $\sum \frac{1}{n} \sin \frac{2\pi}{n}$

D13 Find the first four terms of the Taylor series for $f(x) = \ln(\sin x)$ about $a = \frac{\pi}{4}$.

D14 Write out three non-zero terms of the Taylor series for $f(x) = \tan x$ about $a = 0$.

D15 Use your answer from **D14** to find the first three terms for $f(x) = \sec^2 x$.

D16 Use your answer from **D14** to find the first three terms for $\ln \cos x$.

D17 Find the geometric series for $f(x) = \frac{1}{1-2x}$. What is the interval of convergence?

D18 Find three terms of the binomial series for $(1+x)^{2/3}$.

Find the interval of convergence for the power series **D19** to **D23**.

D19 $\sum_1^\infty \frac{(2x-3)^n}{n^2}$ **D20** $\sum_2^\infty \frac{x^n}{\ln n}$ **D21** $\sum \frac{n^2}{n!} x^n$ **D22** $\sum \frac{(-1)^n (x+4)^n}{3^n}$ **D23** $\sum_1^\infty n! x^n$

D24 Show that $y = \sum_{n=0}^\infty \frac{x^{2n}}{n!}$ solves the differential equation $y' = 2xy$ with $y(0) = 1$. Explain $y = e^{x^2}$.

D25 Estimate the error when $1 - \frac{1}{2} + \frac{1}{3} - \frac{1}{4} + \cdots - \frac{1}{10}$ approximates $\sum \frac{(-1)^{n+1}}{n} =$ ______.

D26 We know that $e^{-1} = 1 - \frac{1}{1!} + \frac{1}{2!} - \frac{1}{3!} + \frac{1}{4!} \cdots$ How many terms to give six correct decimals?

D27 Without a calculator $\int_0^\pi \sin x \, dx$ equals 2. With a calculator $\int_0^\pi (x - \frac{x^3}{3!} + \frac{x^5}{5!} - \frac{x^7}{7!}) dx =$ ______?

CHAPTER 11 VECTORS AND MATRICES

11.1 Vectors and Dot Products (page 405)

One side of the brain thinks of a vector as a pair of numbers (x, y) or a triple of numbers (x, y, z) or a string of n numbers $(x_1, x_2, \cdots, x_n)$. This is the algebra side. The other hemisphere of the brain deals with geometry. It thinks of a vector as a point (or maybe an arrow.) For (x, y) the point is in a plane and for (x, y, z) the point is in 3-dimensional space. The picture gets fuzzy in n dimensions but our intuition is backed up by algebra.

The big operation on vectors is to take ***linear combinations*** $a\mathbf{u} + b\mathbf{v}$. The second biggest operation is to take *dot products* $\mathbf{u} \cdot \mathbf{v}$. Those are the key steps in matrix multiplication and they are best seen by example!

$$3\begin{bmatrix}1\\2\end{bmatrix} + 1\begin{bmatrix}4\\5\end{bmatrix} = \begin{bmatrix}7\\11\end{bmatrix} \text{ and } \begin{bmatrix}1\\2\end{bmatrix} \cdot \begin{bmatrix}4\\5\end{bmatrix} = 14.$$

Linear combinations produce vectors. Dot products yield numbers. The dot product of a vector with itself is the length squared. The dot product of a vector with another vector reveals the angle between them:

$$\mathbf{v} = \begin{bmatrix}1\\2\end{bmatrix} \text{ has } \mathbf{v} \cdot \mathbf{v} = |\mathbf{v}|^2 = 5. \text{ The length is } |\mathbf{v}| = \sqrt{5}.$$

$$\mathbf{v} = \begin{bmatrix}1\\2\end{bmatrix} \text{ and } \mathbf{w} = \begin{bmatrix}4\\5\end{bmatrix} \text{ have } \cos\theta = \frac{\mathbf{v} \cdot \mathbf{w}}{|\mathbf{v}||\mathbf{w}|} = \frac{14}{\sqrt{5}\sqrt{41}} = .9778. \text{ The angle is } \theta = \cos^{-1} .9778 = .21$$

Since cosines are never above 1, dot products $\mathbf{v} \cdot \mathbf{w}$ are never above $|\mathbf{v}|$ times $|\mathbf{w}|$. This is the *Cauchy-Schwarz inequality.*

1. Find the lengths of $\mathbf{v} = \begin{bmatrix}3\\4\end{bmatrix}$ and $\mathbf{w} = \begin{bmatrix}4\\-3\end{bmatrix}$. What is the angle between them?

   - Since $\mathbf{v} \cdot \mathbf{v} = 9 + 16 = 25$, the length is $|\mathbf{v}| = \sqrt{25} = 5$. Similarly $\mathbf{w} \cdot \mathbf{w} = 25$ and the length is $|\mathbf{w}| = 5$. The dot product is $\mathbf{v} \cdot \mathbf{w} = 12 - 12 = 0$. *The cosine of θ is zero!* Therefore $\theta = 90°$ and these two vectors are ***perpendicular***.

The arrow from the origin to the point $\mathbf{v} = (3, 4)$ has slope $\frac{4}{3}$. The arrow to the other point $\mathbf{w} = (4, -3)$ has slope $-\frac{3}{4}$. The slopes multiply to give -1. This is our old test for perpendicular lines. Dot products give a new and better test $\mathbf{v} \cdot \mathbf{w} = 0$. This applies in three dimensions or in n dimensions.

Read-throughs and selected even-numbered solutions :

A vector has length and **direction**. If $\mathbf{v}$ has components 6 and -8, its length is $|\mathbf{v}| = \mathbf{10}$ and its direction vector is $\mathbf{u} = \mathbf{.6i - .8j}$. The product of $|\mathbf{v}|$ with $\mathbf{u}$ is $\mathbf{v}$. This vector goes from (0,0) to the point $x = \mathbf{6}, y = \mathbf{-8}$. A combination of the coordinate vectors $\mathbf{i} = \mathbf{(1, 0)}$ and $\mathbf{j} = \mathbf{(0, 1)}$ produces $\mathbf{v} = \mathbf{x\,i + y\,j}$.

To add vectors we add their **components**. The sum of $(6, -8)$ and (1,0) is $\mathbf{(7, -8)}$. To see $\mathbf{v} + \mathbf{i}$ geometrically, put the **tail** of $\mathbf{i}$ at the **head** of $\mathbf{v}$. The vectors form a **parallelogram with diagonal $\mathbf{v} + \mathbf{i}$**. (The other diagonal is $\mathbf{v} - \mathbf{i}$). The vectors $2\mathbf{v}$ and $-\mathbf{v}$ are $\mathbf{(12, -16)}$ and $\mathbf{(-6, 8)}$. Their lengths are **20** and **10**.

In a space without axes and coordinates, the tail of $\mathbf{V}$ can be placed **anywhere**. Two vectors with the same **components or the same length and direction** are the same. If a triangle starts with $\mathbf{V}$ and continues with $\mathbf{W}$, the third side is $\mathbf{V} + \mathbf{W}$. The vector connecting the midpoint of $\mathbf{V}$ to the midpoint of $\mathbf{W}$ is $\mathbf{\frac{1}{2}(V + W)}$. That vector is **half of** the third side. In this coordinate-free form the dot product is $\mathbf{V} \cdot \mathbf{W} = \mathbf{|V||W|\cos\theta}$.

Using components, $\mathbf{V} \cdot \mathbf{W} = \mathbf{V_1W_1 + V_2W_2 + V_3W_3}$ and $(1, 2, 1) \cdot (2, -3, 7) = \mathbf{3}$. The vectors are perpendicular if $\mathbf{V} \cdot \mathbf{W} = \mathbf{0}$. The vectors are parallel if **$\mathbf{V}$ is a multiple of $\mathbf{W}$**. $\mathbf{V} \cdot \mathbf{V}$ is the same as $\mathbf{|V|^2}$. The dot product of $\mathbf{U}+\mathbf{V}$ with $\mathbf{W}$ equals $\mathbf{U}\cdot\mathbf{W}+\mathbf{V}\cdot\mathbf{W}$. The angle between $\mathbf{V}$ and $\mathbf{W}$ has $\cos\theta = \mathbf{V}\cdot\mathbf{W}/|\mathbf{V}||\mathbf{W}|$. When $\mathbf{V}\cdot\mathbf{W}$ is negative then θ is **greater than 90°**. The angle between $\mathbf{i}+\mathbf{j}$ and $\mathbf{i}+\mathbf{k}$ is $\pi/3$ **with cosine** $\frac{1}{2}$. The Cauchy-Schwarz inequality is $|\mathbf{V}\cdot\mathbf{W}| \le |\mathbf{V}||\mathbf{W}|$, and for $\mathbf{V}=\mathbf{i}+\mathbf{j}$ and $\mathbf{W}=\mathbf{i}+\mathbf{k}$ it becomes $\mathbf{1}\le\mathbf{2}$.

12 We want $\mathbf{V}\cdot(\mathbf{W}-c\mathbf{V})=0$ or $\mathbf{V}\cdot\mathbf{W}=c\mathbf{V}\cdot\mathbf{V}$. Then $c=\frac{6}{3}=2$ and $\mathbf{W}-c\mathbf{V}=(\mathbf{-1},\mathbf{0},\mathbf{1})$.

14 (a) Try two possibilities: keep clock vectors 1 through 5 or 1 through 6. The five add to $1+2\cos 30^\circ+2\cos 60^\circ = 2\sqrt{3} = 3.73$ (in the direction of 3:00). The six add to $2\cos 15^\circ + 2\cos 45^\circ + 2\cos 75^\circ = \mathbf{3.86}$ which is longer (in the direction of 3:30). (b) The 12 o'clock vector (call it $\mathbf{j}$ because it is vertical) is subtracted from all twelve clock vectors. So the sum changes from $\mathbf{V}=0$ to $\mathbf{V}^*=-12\mathbf{j}$.

18 (a) The points $t\mathbf{B}$ form a **line from the origin in the direction of B.** (b) $\mathbf{A}+t\mathbf{B}$ forms a **line from A in the direction of B.** (c) $s\mathbf{A}+t\mathbf{B}$ forms a **plane containing A and B.** (d) $\mathbf{v}\cdot\mathbf{A}=\mathbf{v}\cdot\mathbf{B}$ means $\frac{\cos\theta_1}{\cos\theta_2}$ = fixed number $\frac{|\mathbf{B}|}{|\mathbf{A}|}$ where θ_1 and θ_2 are the angles from $\mathbf{v}$ to $\mathbf{A}$ and $\mathbf{B}$. Then $\mathbf{v}$ is on the plane through the origin that gives this fixed number. (If $|\mathbf{A}|=|\mathbf{B}|$ the plane bisects the angle between those vectors.)

28 $\mathbf{I}\cdot\mathbf{J}=\frac{\mathbf{i}+\mathbf{j}}{\sqrt{2}}\cdot\frac{\mathbf{i}-\mathbf{j}}{\sqrt{2}}=\frac{1-1}{2}=0$. Add $\mathbf{i}+\mathbf{j}=\sqrt{2}\mathbf{I}$ to $\mathbf{i}-\mathbf{j}=\sqrt{2}\mathbf{J}$ to find $\mathbf{i}=\frac{\sqrt{2}}{2}(\mathbf{I}+\mathbf{J})$. Substitute back to find $\mathbf{j}=\frac{\sqrt{2}}{2}(\mathbf{I}-\mathbf{J})$. Then $\mathbf{A}=2\mathbf{i}+3\mathbf{j}=\sqrt{2}(\mathbf{I}+\mathbf{J})+\frac{3\sqrt{2}}{2}(\mathbf{I}-\mathbf{J})=a\mathbf{I}+b\mathbf{J}$ with $a=\sqrt{2}+\frac{3\sqrt{2}}{2}$ and $b=\sqrt{2}-\frac{3\sqrt{2}}{2}$.

34 The diagonals are $\mathbf{A}+\mathbf{B}$ and $\mathbf{B}-\mathbf{A}$. Suppose $|\mathbf{A}+\mathbf{B}|^2=\mathbf{A}\cdot\mathbf{A}+\mathbf{A}\cdot\mathbf{B}+\mathbf{B}\cdot\mathbf{A}+\mathbf{B}\cdot\mathbf{B}$ equals $|\mathbf{B}-\mathbf{A}|^2 = \mathbf{B}\cdot\mathbf{B}-\mathbf{A}\cdot\mathbf{B}-\mathbf{B}\cdot\mathbf{A}+\mathbf{B}\cdot\mathbf{B}$. After cancelling this is $4\mathbf{A}\cdot\mathbf{B}=0$ (note that $\mathbf{A}\cdot\mathbf{B}$ is the same as $\mathbf{B}\cdot\mathbf{A}$). The region is a **rectangle.**

40 Choose $\mathbf{W}=(\mathbf{1},\mathbf{1},\mathbf{1})$. Then $\mathbf{V}\cdot\mathbf{W}=V_1+V_2+V_3$. The Schwarz inequality $|\mathbf{V}\cdot\mathbf{W}|^2\le|\mathbf{V}|^2|\mathbf{W}|^2$ is $(V_1+V_2+V_3)^2\le 3(V_1^2+V_2^2+V_3^2)$.

42 $|\mathbf{A}+\mathbf{B}|\le|\mathbf{A}|+|\mathbf{B}|$ or $|\mathbf{C}|\le|\mathbf{A}|+|\mathbf{B}|$ says that any side length is less than **the sum of the other two side lengths.** Proof: $|\mathbf{A}+\mathbf{B}|^2\le$ (using Schwarz for $\mathbf{A}\cdot\mathbf{B}$)$|\mathbf{A}|^2+2|\mathbf{A}||\mathbf{B}|+|\mathbf{B}|^2=(|\mathbf{A}|+|\mathbf{B}|)^2$.

11.2 Planes and Projections (page 414)

The main point is to understand the equation of a plane (in three dimensions). In general it is $ax+by+cz=d$. Specific examples are $2x+4y+z=14$ and $2x+4y+z=0$. *Those planes are parallel. They have the same normal vector* $\mathbf{N}=(2,4,1)$. The equations of the planes can be written as $\mathbf{N}\cdot\mathbf{P}=d$. A typical point in $\mathbf{N}\cdot\mathbf{P}=14$ is $(x,y,z)=(2,2,2)$ because then $2x+4y+z=14$. A typical point in $\mathbf{N}\cdot\mathbf{P}=0$ is $(2,-1,0)$ because then $2x+4y+z=0$.

From $\mathbf{N}\cdot\mathbf{P}=0$ we see that $\mathbf{N}$ is perpendicular (= *normal*) to every vector $\mathbf{P}$ in the plane. The components of $\mathbf{P}$ are (x,y,z) and the vector is $\mathbf{P}=x\mathbf{i}+y\mathbf{j}+z\mathbf{k}$. This plane goes through the origin, because $x=0$, $y=0$, $z=0$ satisfies the equation $2x+4y+z=0$. The other plane $\mathbf{N}\cdot\mathbf{P}=14$ does not go through the origin. Its equation can be written as $\mathbf{N}\cdot\mathbf{P}=\mathbf{N}\cdot\mathbf{P}_0$ where $\mathbf{P}_0$ is any particular point on the plane (since every one of those points has $\mathbf{N}\cdot\mathbf{P}_0=14$). Then the plane equation is $\mathbf{N}\cdot(\mathbf{P}-\mathbf{P}_0)=0$, which shows that the normal vector $\mathbf{N}=(2,4,1)$ is perpendicular to every vector $\mathbf{P}-\mathbf{P}_0$ lying in the plane.

To repeat: $\mathbf{P}$ and $\mathbf{P}_0$ are vectors *to* the plane, starting at (0,0,0). $\mathbf{P}-\mathbf{P}_0$ is a vector *in* the plane.

The distance *to* the plane is the shortest vector $\mathbf{P}$ with $\mathbf{N}\cdot\mathbf{P} = 14$. The shortest $\mathbf{P}$ is in the same direction as $\mathbf{N}$. Call it $\mathbf{P} = t\,\mathbf{N}$. Then $\mathbf{N}\cdot\mathbf{P} = 14$ becomes $t\,\mathbf{N}\cdot\mathbf{N} = 14$. In this example $\mathbf{N}\cdot\mathbf{N} = 2^2 + 4^2 + 1^2 = 21$. Therefore $21t = 14$ and $t = \frac{2}{3}$. The shortest $\mathbf{P}$ is $\frac{2}{3}\,\mathbf{N}$.

The general formula for distance from $(0,0,0)$ ***to*** $\mathbf{N}\cdot\mathbf{P} = d$ ***is*** $|d|$ ***divided by*** $|\mathbf{N}|$. The distance to $\mathbf{N}\cdot\mathbf{P} = 0$ is 0.

1. Find the plane through $\mathbf{P}_0 = (1,2,2)$ perpendicular to $\mathbf{N} = (3,1,4)$.

   - Because of $\mathbf{N}$ the equation must be $3x+y+4z = d$. Because of $\mathbf{P}_0$ the number d must be $3\cdot1+2+4\cdot2 = 13$. The plane is $3x + y + 4z = 13$ or $\mathbf{N}\cdot\mathbf{P} = \mathbf{N}\cdot\mathbf{P}_0$.

2. Find the plane parallel to $x + y + z = 4$ but going through $(1,4,5)$.

   - To be parallel the equation must be $x + y + z = d$. To go through $(1,4,5)$ we need $d = 10$. The plane $x + y + z = 10$ has the desired normal vector $\mathbf{N} = (1,1,1)$. The distance from the origin is $\frac{|d|}{|\mathbf{N}|} = \frac{10}{\sqrt{3}}$.

Projection onto a line We look for the projection of a vector $\mathbf{B}$ in the direction of $\mathbf{A}$. The projection is an unknown number x times $\mathbf{A}$. Geometry says that the vector from $\mathbf{B}$ down to $x\mathbf{A}$ should be perpendicular to $\mathbf{A}$:

$$\mathbf{A}\cdot(\mathbf{B} - x\mathbf{A}) = 0 \text{ or } \mathbf{A}\cdot\mathbf{B} = x\mathbf{A}\cdot\mathbf{A}. \text{ Therefore } x = \frac{\mathbf{A}\cdot\mathbf{B}}{\mathbf{A}\cdot\mathbf{A}} \textbf{ and } \mathbf{P} = x\mathbf{A} = \frac{\mathbf{A}\cdot\mathbf{B}}{\mathbf{A}\cdot\mathbf{B}}\mathbf{A}.$$

The projection is $\mathbf{P}$ (*a new use for this letter*). Its length is $|\mathbf{B}||\cos\theta|$. This is the length of $\mathbf{B}$ "*along* $\mathbf{A}$."

3. In terms of $\mathbf{A}$ and $\mathbf{B}$, find the length of the projection $\mathbf{P}$.

   - The length is $|\mathbf{B}||\cos\theta| = |\mathbf{B}|\frac{|\mathbf{A}\cdot\mathbf{B}|}{|\mathbf{A}||\mathbf{B}|} = \frac{|\mathbf{A}\cdot\mathbf{B}|}{|\mathbf{A}|}$.
   - From the formula $\mathbf{P} = \frac{\mathbf{A}\cdot\mathbf{B}}{\mathbf{A}\cdot\mathbf{A}}\mathbf{A}$ the length is again $\frac{|\mathbf{A}\cdot\mathbf{B}|}{|\mathbf{A}|}$.

4. Project the vector $\mathbf{B} = (1,3,1)$ onto the vector $\mathbf{A} = (2,1,2)$. Find the component $\mathbf{P}$ along $\mathbf{A}$ and also find the component $\mathbf{B}-\mathbf{P}$ perpendicular to $\mathbf{A}$.

   - The dot product is $\mathbf{A}\cdot\mathbf{B} = 7$. Also $\mathbf{A}\cdot\mathbf{A} = 9$. Therefore $\mathbf{P} = \frac{7}{9}\mathbf{A} = (\frac{14}{9}, \frac{7}{9}, \frac{14}{9})$.

   The perpendicular component is $\mathbf{B}-\mathbf{P} = (1,3,1)-\mathbf{P} = (-\frac{5}{9}, \frac{20}{9}, -\frac{5}{9})$. Check that it really is perpendicular: $\mathbf{A}\cdot(\mathbf{B}-\mathbf{P}) = -\frac{10}{9} + \frac{20}{9} - \frac{10}{9} = 0$.

Read-throughs and selected even-numbered solutions :

A plane in space is determined by a point $P_0 = (x_0, y_0, z_0)$ and a **normal** vector $\mathbf{N}$ with components (a,b,c). The point $P = (x,y,z)$ is on the plane if the dot product of $\mathbf{N}$ with $\mathbf{P} - \mathbf{P_0}$ is zero. (*That answer was not P!*) The equation of this plane is $a(\mathbf{x} - \mathbf{x_0}) + b(\mathbf{y} - \mathbf{y_0}) + c(\mathbf{z} - \mathbf{z_0}) = 0$. The equation is also written as $ax + by + cz = d$, where d equals $\mathbf{ax_0} + \mathbf{by_0} + \mathbf{cz_0}$ or $\mathbf{N}\cdot\mathbf{P_0}$. A parallel plane has the same $\mathbf{N}$ and a different $\mathbf{d}$. A plane through the origin has $d = \mathbf{0}$.

The equation of the plane through $P_0 = (2,1,0)$ perpendicular to $\mathrm{N} = (3,4,5)$ is $\mathbf{3x + 4y + 5z = 10}$. A second point in the plane is $P = (0,0,\mathbf{2})$. The vector from P_0 to P is $(\mathbf{-2}, \mathbf{-1}, \mathbf{2})$, and it is **perpendicular** to N. (Check by dot product). The plane through $P_0 = (2,1,0)$ perpendicular to the z axis has $\mathrm{N} = (\mathbf{0},\mathbf{0},\mathbf{1})$ and equation $\mathbf{z = 0}$.

The component of **B** in the direction of **A** is $|\mathbf{B}|\cos\theta$, where θ is the angle between the vectors. This is $\mathbf{A}\cdot\mathbf{B}$ divided by $|\mathbf{A}|$. The projection vector **P** is $|\mathbf{B}|\cos\theta$ times a **unit** vector in the direction of **A**. Then $\mathbf{P} = (|\mathbf{B}|\cos\theta)(\mathbf{A}/|\mathbf{A}|)$ simplifies to $(\mathbf{A}\cdot\mathbf{B})\mathbf{A}/|\mathbf{A}|^2$. When **B** is doubled, **P** is **doubled**. When **A** is doubled, **P** is **not changed**. If **B** reverses direction, so does **P**. If **A** reverses direction, then **P stays the same**.

When **B** is a velocity vector, **P** represents the **velocity in the A direction**. When **B** is a force vector, **P is the force component along A**. The component of **B** perpendicular to **A** equals $\mathbf{B}-\mathbf{P}$. The shortest distance from (0,0,0) to the plane $ax+by+cz=d$ is along the **normal** vector. The distance from the origin is $|\mathbf{d}|/\sqrt{\mathbf{a^2+b^2+c^2}}$ and the point on the plane closest to the origin is $P = (\mathbf{da,db,dc})/(\mathbf{a^2+b^2+c^2})$. The distance from $\mathbf{Q}=(x_1,y_1,z_1)$ to the plane is $|\mathbf{d-ax_1-by_1-cz_1}|/\sqrt{\mathbf{a^2+b^2+c^2}}$.

6 The plane $y-z=0$ contains the given points (0,0,0) and (1,0,0) and (0,1,1). The normal vector is $\mathbf{N}=\mathbf{j}-\mathbf{k}$. (Certainly $P=(0,1,1)$ and $P_0=(0,0,0)$ give $\mathbf{N}\cdot(P-P_0)=0$.)

12 (a) **No**: the line where the planes (or walls) meet is not perpendicular to itself. (b) A third plane perpendicular to the first plane could make **any angle** with the second plane.

22 If **B** makes a 60° angle with **A** then the length of **P** is $|\mathbf{B}|\cos 60° = 2\cdot\frac{1}{2}=1$. Since **P** is in the direction of **A** it must be $\frac{\mathbf{A}}{|\mathbf{A}|}$.

32 The points at distance 1 from the plane $x+2y+2z=3$ fill two parallel planes $\mathbf{x+2y+2z=6}$ and $\mathbf{x+2y+2z=0}$. Check: The point (0,0,0) on the last plane is a distance $\frac{|d|}{|\mathbf{N}|}=\frac{3}{3}=1$ from the plane $x+2y+2z=3$.

38 The point $P=Q+t\mathbf{N}=(3+t,3+2t)$ lies on the line $x+2y=4$ if $(3+t)+2(3+2t)=4$ or $9+5t=4$ or $t=\mathbf{-1}$. Then $P=(\mathbf{2},\mathbf{1})$.

40 The drug runner takes $\frac{1}{2}$ second to go the 4 meters. You have 5 meters to travel in the same $\frac{1}{2}$ second. Your speed must be **10 meters per second**. The projection of your velocity (a vector) onto the drug runner's velocity equals **the drug runner's velocity.**

11.3 Cross Products and Determinants (page 423)

This section is mostly in three dimensions. We take the cross product of vectors **A** and **B**. That produces a third vector $\mathbf{A}\times\mathbf{B}$ perpendicular to the first two. The length of this cross product is adjusted to equal $|\mathbf{A}||\mathbf{B}|\sin\theta$. This leads to a (fairly) neat formula

$$\mathbf{A}\times\mathbf{B} = \begin{vmatrix} \mathbf{i} & \mathbf{j} & \mathbf{k} \\ a_1 & a_2 & a_3 \\ b_1 & b_2 & b_3 \end{vmatrix} = (a_2b_3-a_3b_2)\mathbf{i} + (a_3b_1-a_1b_3)\mathbf{j} + (a_1b_2-a_2b_1)\mathbf{k}.$$

1. Find the cross product of $\mathbf{A}=\mathbf{i}$ and $\mathbf{B}=\mathbf{k}$.

   - The perpendicular direction is **j**. But the cross product $\mathbf{i}\times\mathbf{k}$ is $-\mathbf{j}$. This comes from the formula above, when you substitute 1,0,0 in the second row and 0,0,1 in the third row. The minus sign also comes from the right hand rule. The length of $\mathbf{A}\times\mathbf{B}$ is $|\mathbf{A}||\mathbf{B}|\sin\theta$. The vectors $\mathbf{A}=\mathbf{i}$ and $\mathbf{B}=\mathbf{k}$ are perpendicular so $\theta=90°$ and $\sin\theta=1$.

2. Find the cross product of **k** and **i**.

   - In this opposite order the cross product changes sign: $\mathbf{k} \times \mathbf{i} = +\mathbf{j}$.

3. Find $(\mathbf{i} + \mathbf{k}) \times (\mathbf{i} + \mathbf{k})$. Also split this into $\mathbf{i} \times \mathbf{i} + \mathbf{i} \times \mathbf{k} + \mathbf{k} \times \mathbf{i} + \mathbf{k} \times \mathbf{k}$.

   - $\mathbf{i} \times \mathbf{i}$ is the zero vector. So is $\mathbf{k} \times \mathbf{k}$. So is any $\mathbf{A} \times \mathbf{A}$ because the angle is $\theta = 0$ and its sine is zero. The splitting also gives $\mathbf{0} - \mathbf{j} + \mathbf{j} + \mathbf{0} = \mathbf{0}$.

Most determinants do not involve **i, j, k**. They are determinants of ordinary matrices. For a 2 by 2 matrix we get the ***area*** of a parallelogram. For a 3 by 3 matrix we get the ***volume*** of a parallelepiped (a box):

$$\begin{vmatrix} a & b \\ c & d \end{vmatrix} = ad - bc = \text{ area with corners } (0,0), (a,b), (c,d), (a+c, b+d)$$

$$\begin{vmatrix} a_1 & a_2 & a_3 \\ b_1 & b_2 & b_3 \\ c_1 & c_2 & c_3 \end{vmatrix} = \mathbf{A} \cdot (\mathbf{B} \times \mathbf{C}) = a_1 b_2 c_3 + \text{ five other terms } = \text{ volume of box.}$$

A 3 by 3 determinant has six terms. An n by n determinant has $n!$ terms. Half the signs are minus.

4. Compute these determinants. Interpret as areas or volumes:

$$\text{(a)} \begin{vmatrix} 2 & 5 \\ 4 & 10 \end{vmatrix} \quad \text{(b)} \begin{vmatrix} 0 & 1 & 0 \\ 0 & 0 & 1 \\ 1 & 0 & 0 \end{vmatrix} \quad \text{(c)} \begin{vmatrix} 1 & 2 & 3 \\ 1 & 4 & 5 \\ 2 & 6 & 8 \end{vmatrix}$$

   - (a) The determinant is $(2)(10) - (-5)(4) = 0$. Because (2,5) is parallel to (4,10), the parallelogram is crushed. It has no area. The matrix has no inverse.
   - (b) The determinant is $(1)(1)(1) = 1$ with a *plus sign.* This is also $\mathbf{j} \cdot (\mathbf{k} \times \mathbf{i})$ which is $\mathbf{j} \cdot \mathbf{j} = 1$. It is the volume of a *unit cube.* Exchange two rows and the determinant is -1 (left-handed cube).
   - (c) The six terms in the determinant are $+32 + 18 + 20 - 24 - 30 - 16$. This gives zero! The volume is zero because the 3-dimension box is completely squashed into a plane.
     The third row of this matrix is the sum of the first two rows. In such a case the determinant is zero. We never leave the plane of the first two rows. $\mathbf{A} \times \mathbf{B}$ is perpendicular to that plane, **C** is in that plane, so the determinant $(\mathbf{A} \times \mathbf{B}) \cdot \mathbf{C}$ is zero.

Read-throughs and selected even-numbered solutions :

The cross product $\mathbf{A} \times \mathbf{B}$ is a **vector** whose length is $|\mathbf{A}||\mathbf{B}| \sin \theta$. Its direction is **perpendicular** to **A** and **B**. That length is the area of a **parallelogram**, whose base is $|\mathbf{A}|$ and whose height is $|\mathbf{B}| \sin \theta$. When $\mathbf{A} = a_1 \mathbf{i} + a_2 \mathbf{j}$ and $\mathbf{B} = b_1 \mathbf{i} + b_2 \mathbf{j}$, the area is $|\mathbf{a_1 b_2 - a_2 b_1}|$. This equals a 2 by 2 **determinant**. In general $|\mathbf{A} \cdot \mathbf{B}|^2 + |\mathbf{A} \times \mathbf{B}|^2 = |\mathbf{A}|^{\mathbf{2}} |\mathbf{B}|^{\mathbf{2}}$.

The rules for cross products are $\mathbf{A} \times \mathbf{A} = \mathbf{0}$ and $\mathbf{A} \times \mathbf{B} = -(\mathbf{B} \times \mathbf{A})$ and $\mathbf{A} \times (\mathbf{B} + \mathbf{C}) = \mathbf{A} \times \mathbf{B} + \mathbf{A} \times \mathbf{C}$. In particular $\mathbf{A} \times \mathbf{B}$ needs the **right**-hand rule to decide its direction. If the fingers curl from **A** towards **B** (not more than 180°), then $\mathbf{A} \times \mathbf{B}$ points **along the right thumb**. By this rule $\mathbf{i} \times \mathbf{j} = \mathbf{k}$ and $\mathbf{i} \times \mathbf{k} = -\mathbf{j}$ and $\mathbf{j} \times \mathbf{k} = \mathbf{i}$.

The vectors $a_1 \mathbf{i} + a_2 \mathbf{j} + a_3 \mathbf{k}$ and $b_1 \mathbf{i} + b_2 \mathbf{j} + b_3 \mathbf{k}$ have cross product $(\mathbf{a_2 b_3} - \mathbf{a_3 b_2})\mathbf{i} + (\mathbf{a_3 b_1} - \mathbf{a_1 b_3})\mathbf{j} + (\mathbf{a_1 b_2} - \mathbf{a_2 b_1})\mathbf{k}$. The vectors $\mathbf{A} = \mathbf{i} + \mathbf{j} + \mathbf{k}$ and $\mathbf{B} = \mathbf{i} + \mathbf{j}$ have $\mathbf{A} \times \mathbf{B} = -\mathbf{i} + \mathbf{j}$. (*This is also the 3 by 3 determinant* $\begin{vmatrix} \mathbf{i} & \mathbf{j} & \mathbf{k} \\ 1 & 1 & 1 \\ 1 & 1 & 0 \end{vmatrix}$.) Perpendicular to the plane containing (0,0,0), (1,1,1), (1,1,0) is the normal vector **N** $= -\mathbf{i} + \mathbf{j}$. The area of the triangle with those three vertices is $\frac{1}{2}\sqrt{2}$, which is half the area of the parallelogram with fourth vertex at $(\mathbf{2}, \mathbf{2}, \mathbf{1})$.

Vectors **A, B, C** from the origin determine a **box**. Its volume $|\mathbf{A} \cdot (\mathbf{B} \times \mathbf{C})|$ comes from a 3 by 3 **determinant.** There are six terms, **three** with a plus sign and **three** with minus. In every term each row and **column** is represented once. The rows (1,0,0), (0,0,1), and (0,1,0) have determinant $=\mathbf{-1}$. That box is a **cube**, but its sides form a **left**-handed triple in the order given.

If **A, B, C** lie in the same plane then $\mathbf{A} \cdot (\mathbf{B} \times \mathbf{C})$ is **zero**. For $\mathbf{A} = x\mathbf{i} + y\mathbf{j} + z\mathbf{k}$ the first row contains the letters **x,y,z**. So the plane containing **B** and **C** has the equation $\mathbf{A} \cdot \mathbf{B} \times \mathbf{C} = 0$. When $\mathbf{B} = \mathbf{i} + \mathbf{j}$ and $\mathbf{C} = \mathbf{k}$ that equation is $\mathbf{x} - \mathbf{y} = \mathbf{0}$. $\mathbf{B} \times \mathbf{C}$ is $\mathbf{i} - \mathbf{j}$.

A 3 by 3 determinant splits into **three** 2 by 2 determinants. They come from rows 2 and 3, and are multiplied by the entries in row 1. With **i, j, k** in row 1, this determinant equals the **cross** product. Its **j** component is $-(\mathbf{a_1 b_3} - \mathbf{a_3 b_1})$, including the **minus** sign which is easy to forget.

10 (a) **True** ($\mathbf{A} \times \mathbf{B}$ is a vector, $\mathbf{A} \cdot \mathbf{B}$ is a number) (b) **True** (Equation (1) becomes $0 = |\mathbf{A}|^2|\mathbf{B}|^2$ so $\mathbf{A} = \mathbf{0}$ or $\mathbf{B} = \mathbf{0}$) (c) **False:** $\mathbf{i} \times (\mathbf{j}) = \mathbf{i} \times (\mathbf{i} + \mathbf{j})$

16 $|\mathbf{A} \times \mathbf{B}|^2 = |\mathbf{A}|^2|\mathbf{B}|^2 - (\mathbf{A} \cdot \mathbf{B})^2$ which is $(a_1^2 + a_2^2 + a_3^2)(b_1^2 + b_2^2 + b_3^2) - (a_1 b_1 + a_2 b_2 + a_3 b_3)^2$. Multiplying and simplifying leads to $(a_1 b_2 - a_2 b_1)^2 + (a_1 b_3 - a_3 b_1)^2 + (a_2 b_3 - a_3 b_2)^2$ which confirms $|\mathbf{A} \times \mathbf{B}|$ in eq. (6).

26 The plane has normal $\mathbf{N} = (\mathbf{i} + \mathbf{j}) \times \mathbf{k} = \mathbf{i} \times \mathbf{k} + \mathbf{j} \times \mathbf{k} = -\mathbf{j} + \mathbf{i}$. So the plane is $x - y = d$. If the plane goes through the origin, its equation is $x - y = 0$.

42 The two sides going out from (a_1, b_1) are $(a_2 - a_1)\mathbf{i} + (b_2 - b_1)\mathbf{j}$ and $(a_3 - a_1)\mathbf{i} + (b_3 - b_1)\mathbf{j}$. The cross product gives the area of the parallelogram as $|(a_2 - a_1)(b_3 - b_1) - (a_3 - a_1)(b_2 - b_1)|$. Divide by 2 for triangle.

48 The triple vector product in Problem 47 is $(\mathbf{A} \times \mathbf{B}) \times \mathbf{C} = (\mathbf{A} \cdot \mathbf{C})\mathbf{B} - (\mathbf{B} \cdot \mathbf{C})\mathbf{A}$. Take the dot product with **D**. The right side is easy: $(\mathbf{A} \cdot \mathbf{C})(\mathbf{B} \cdot \mathbf{D}) - (\mathbf{B} \cdot \mathbf{C})(\mathbf{A} \cdot \mathbf{D})$. The left side is $((\mathbf{A} \times \mathbf{B}) \times \mathbf{C}) \cdot \mathbf{D}$ and the vectors $\mathbf{A} \times \mathbf{B}, \mathbf{C}, \mathbf{D}$ can be put in any cyclic order (see "useful facts" about volume of a box, after Theorem 11G). We choose $(\mathbf{A} \times \mathbf{B}) \cdot (\mathbf{C} \times \mathbf{D})$.

11.4 Matrices and Linear Equations (page 433)

A system of linear equations is written $\mathrm{A}\mathbf{x} = \mathbf{b}$ or in this book $\mathrm{A}\mathbf{u} = \mathbf{d}$. The coefficient matrix A multiplies the unknown vector **u**. This matrix-vector multiplication is set up so that it reproduces the given linear equations:

$$\begin{aligned} 2x + 5y &= 9 \\ x + 3y &= 5 \end{aligned} \quad \text{becomes} \begin{bmatrix} 2 & 5 \\ 1 & 3 \end{bmatrix} \begin{bmatrix} x \\ y \end{bmatrix} = \begin{bmatrix} 9 \\ 5 \end{bmatrix} = \mathbf{d}$$

Then the solution is $\mathbf{u} = \mathrm{A}^{-1}\mathbf{d}$. It can be found in three or more ways:

(a) Find A^{-1} explicitly. (This matrix satisfies $\mathrm{A}\ \mathrm{A}^{-1} = \mathbf{I}$.) Then multiply $\mathbf{A}^{-1}$ times **d**.

(b) Use Cramer's Rule. Then x and y are ratios of determinants.

(c) Eliminate x from the second equation by subtracting $\frac{1}{2}$ of the first equation. From $3-\frac{1}{2}(5)$ and $5-\frac{1}{2}(9)$ you get $\frac{1}{2}y=\frac{1}{2}$. Then $y=1$. Back substitution gives $x=2$.

Elimination (c) is preferred for larger problems. For a 2 by 2 problem it is reasonable to know A^{-1}:

$$\begin{bmatrix} a & b \\ c & d \end{bmatrix}^{-1} = \frac{1}{ad-bc}\begin{bmatrix} d & -b \\ -c & a \end{bmatrix} \text{ and } \begin{bmatrix} 2 & 5 \\ 1 & 3 \end{bmatrix}^{-1} = \frac{1}{1}\begin{bmatrix} 3 & -5 \\ -1 & 2 \end{bmatrix}.$$

The multiplication $A^{-1}\mathbf{d} = \begin{bmatrix} 3 & -5 \\ -1 & 2 \end{bmatrix}\begin{bmatrix} 9 \\ 5 \end{bmatrix}$ gives $\mathbf{u} = \begin{bmatrix} x \\ y \end{bmatrix} = \begin{bmatrix} 2 \\ 1 \end{bmatrix}$ which agrees with elimination.

Multiply by dot products $3\cdot 9 - 5\cdot 5 = 2$ and $-1\cdot 9 + 2\cdot 5 = 1$ or by linear combination of columns :

$$9\begin{bmatrix} 3 \\ -1 \end{bmatrix} + 5\begin{bmatrix} -5 \\ 2 \end{bmatrix} = \begin{bmatrix} 2 \\ 1 \end{bmatrix}.$$

1. Write the equations $\begin{aligned} x+3y &= 0 \\ 2x+4y &= 10 \end{aligned}$ as $A\mathbf{u}=\mathbf{d}$. Solve by A^{-1} and also by elimination.

- $A = \begin{bmatrix} 1 & 3 \\ 2 & 4 \end{bmatrix}$ has $A^{-1} = \frac{1}{-2}\begin{bmatrix} 4 & -3 \\ -2 & 1 \end{bmatrix}$. Multiply by $\mathbf{d} = \begin{bmatrix} 0 \\ 10 \end{bmatrix}$ to get $\mathbf{u} = \begin{bmatrix} 15 \\ -5 \end{bmatrix}$.
- Eliminate x by subtracting two times the first equation from the second equation. This leaves $-2y=10$. Then $y=-5$. Back substitution gives $x=15$.

Projection onto a plane We look for the combination $\mathbf{p} = x\,\mathbf{a} + y\,\mathbf{b}$ that is closest to a given vector $\mathbf{d}$. It is the projection of $\mathbf{d}$ onto the plane of $\mathbf{a}$ and $\mathbf{b}$. The numbers x and y come from the "normal equations"

$$\begin{aligned} (\mathbf{a}\cdot\mathbf{a})\,x + (\mathbf{a}\cdot\mathbf{b})\,y &= \mathbf{a}\cdot\mathbf{d} \\ (\mathbf{b}\cdot\mathbf{a})\,x + (\mathbf{b}\cdot\mathbf{b})\,y &= \mathbf{b}\cdot\mathbf{d} \end{aligned}$$

2. Project the vector $\mathbf{d}$ = (1,2,4) onto the plane of $\mathbf{a}$ = (1,1,1) and $\mathbf{b}$ = (1,2,3). Interpret this projection as **least squares fitting by a straight line.**

- After computing dot products, the normal equations are $3x+6y=7$ and $6x+14y=17$. Subtract 2 times the first to get $2y=3$. Then $y=\frac{3}{2}$ and $x=-\frac{2}{3}$. The projection is $\mathbf{p} = -\frac{2}{3}(1,1,1)+\frac{3}{2}(1,2,3)$.
- To see the line-fitting problem, write down $x\,\mathbf{a} + y\,\mathbf{b} \approx \mathbf{d}$ (this has no solution!):

$$x\begin{bmatrix} 1 \\ 1 \\ 1 \end{bmatrix} + y\begin{bmatrix} 1 \\ 2 \\ 3 \end{bmatrix} \approx \begin{bmatrix} 1 \\ 2 \\ 4 \end{bmatrix} \quad \text{or} \quad \begin{aligned} x+y &\approx 1 \\ x+2y &\approx 2 \\ x+3y &\approx 4. \end{aligned}$$

The straight line $f(t)=x+yt$ has intercept x and slope y. We are trying to make it go through the three points (1,1), (2,2), and (3,4). This is doomed to failure. If those points were on a line, we could solve our three equations. We can't. The best solution (the least squares solution) has $x=-\frac{2}{3}$ and $y=\frac{2}{3}$. That line $f(t) = -\frac{2}{3}+\frac{3}{2}t$ comes closest to the three points. It minimizes the sum of squares of the three errors.

Read-throughs and selected even-numbered solutions :

The equations $3x+y=8$ and $x+y=6$ combine into the vector equation $\mathbf{x}\begin{bmatrix} 3 \\ 1 \end{bmatrix} + \mathbf{y}\begin{bmatrix} 1 \\ 1 \end{bmatrix} = \begin{bmatrix} 8 \\ 6 \end{bmatrix} = \mathbf{d}$. The left side is $A\mathbf{u}$ with coefficient matrix $A = \begin{bmatrix} 3 & 1 \\ 1 & 1 \end{bmatrix}$ and unknown vector $\mathbf{u} = \begin{bmatrix} x \\ y \end{bmatrix}$. The determinant of **A** is **2**, so this problem is not **singular**. The row picture shows two intersecting **lines**. The column picture shows

$x\mathbf{a} + y\mathbf{b} = \mathbf{d}$, where $\mathbf{a} = \begin{bmatrix} 3 \\ 1 \end{bmatrix}$ and $\mathbf{b} = \begin{bmatrix} 1 \\ 1 \end{bmatrix}$. The inverse matrix is $A^{-1} = \frac{1}{2}\begin{bmatrix} 1 & -1 \\ -1 & 3 \end{bmatrix}$. The solution is $\mathbf{u} = A^{-1}\mathbf{d} = \begin{bmatrix} 1 \\ 5 \end{bmatrix}$.

A matrix-vector multiplication produces a vector of dot **products** from the rows, and also a combination of the **columns:**

$$\begin{bmatrix} \mathbf{A} \\ \mathbf{B} \end{bmatrix}\begin{bmatrix} \mathbf{u} \end{bmatrix} = \begin{bmatrix} \mathbf{A}\cdot\mathbf{u} \\ \mathbf{B}\cdot\mathbf{u} \end{bmatrix} \text{ and } \begin{bmatrix} \mathbf{a} & \mathbf{b} \end{bmatrix}\begin{bmatrix} x \\ y \end{bmatrix} = \begin{bmatrix} x\mathbf{a} + y\mathbf{b} \end{bmatrix} \text{ and } \begin{bmatrix} 3 & 1 \\ 1 & 1 \end{bmatrix}\begin{bmatrix} 1 \\ 5 \end{bmatrix} = \begin{bmatrix} 8 \\ 6 \end{bmatrix}.$$

If the entries are a, b, c, d, the determinant is $D = \mathbf{ad} - \mathbf{bc}$. A^{-1} is $\begin{bmatrix} d & -b \\ -c & a \end{bmatrix}$ divided by D. Cramer's Rule shows components of $\mathbf{u} = A^{-1}\mathbf{d}$ as ratios of determinants: $x = (\mathbf{b_2 d_1} - \mathbf{b_1 d_2})/D$ and $y = (\mathbf{a_1 d_2} - \mathbf{a_2 d_1})/D$.

A matrix-matrix multiplication MV yields a matrix of dot products, from the rows of M and columns of $\mathbf{V}$:

$$\begin{bmatrix} \mathbf{A} \\ \mathbf{B} \end{bmatrix}\begin{bmatrix} \mathbf{v}_1 & \mathbf{v}_2 \end{bmatrix} = \begin{bmatrix} \mathbf{A}\cdot\mathbf{v}_1 & \mathbf{A}\cdot\mathbf{v}_2 \\ \mathbf{B}\cdot\mathbf{v}_1 & \mathbf{B}\cdot\mathbf{v}_2 \end{bmatrix} \qquad \begin{bmatrix} 3 & 1 \\ 1 & 1 \end{bmatrix}\begin{bmatrix} 1 & 2 \\ 5 & 6 \end{bmatrix} = \begin{bmatrix} 8 & 12 \\ 6 & 8 \end{bmatrix}$$

$$\begin{bmatrix} 3 & 1 \\ 1 & 1 \end{bmatrix}\begin{bmatrix} 1/2 & -1/2 \\ -1/2 & 3/2 \end{bmatrix} = \begin{bmatrix} 1 & 0 \\ 0 & 1 \end{bmatrix}\begin{bmatrix} 1 & 0 \\ 0 & 1 \end{bmatrix}\begin{bmatrix} A \end{bmatrix} = \begin{bmatrix} A \end{bmatrix}.$$

The last line contains the **identity** matrix, denoted by I. It has the property that $IA = AI = \mathbf{A}$ for every matrix A, and $I\mathbf{u} = \mathbf{u}$ for every vector $\mathbf{u}$. The inverse matrix satisfies $A^{-1}A = \mathbf{I}$. Then $A\mathbf{u} = \mathbf{d}$ is solved by multiplying both sides by $\mathbf{A^{-1}}$, to give $\mathbf{u} = \mathbf{A^{-1}d}$. There is no inverse matrix when $\mathbf{det\, A = 0}$.

The combination $x\mathbf{a} + y\mathbf{b}$ is the projection of $\mathbf{d}$ when the error $\mathbf{d} - \mathbf{xa} - \mathbf{yb}$ is perpendicular to $\mathbf{a}$ and $\mathbf{b}$. If $\mathbf{a} = (1,1,1)$, $\mathbf{b} = (1,2,3)$, and $\mathbf{d} = (0,8,4)$, the equations for x and y are $\mathbf{3x + 6y = 12}$ **and** $\mathbf{6x + 14y = 28}$. Solving them also gives the closest **line** to the data points (1,0), **(2,8)**, and (3,4). The solution is $x = 0, y = 2$, which means the best line is **horizontal**. The projection is $0\mathbf{a} + 2\mathbf{b} = \mathbf{(2,4,6)}$. The three error components are $\mathbf{(-2,4,-2)}$. Check perpendicularity: $\mathbf{(1,1,1)}\cdot\mathbf{(-2,4,-2)} = 0$ and $\mathbf{(1,2,3)}\cdot\mathbf{(-2,4,-2)} = 0$. Applying calculus to this problem, x and y minimize the sum of squares $E = (\mathbf{-x-y})^2 + (\mathbf{8-x-2y})^2 + (\mathbf{4-x-3y})^2$.

8 The solution is $x = \frac{d-b}{ad-bc}, y = \frac{a-c}{ad-bc}$ (ok to use Cramer's Rule). The solution breaks down if $ad = bc$.

$$\frac{d-b}{ad-bc}\begin{bmatrix} a \\ c \end{bmatrix} + \frac{a-c}{ad-bc}\begin{bmatrix} b \\ d \end{bmatrix} = \begin{bmatrix} 1 \\ 1 \end{bmatrix}; \begin{vmatrix} a & b \\ c & d \end{vmatrix} = ad - bc.$$

12 With $A = I$ the equations are $\begin{matrix} 1x + 0y = d_1 \\ 0x + 1y = d_2 \end{matrix}$. Then $x = \dfrac{\begin{vmatrix} d_1 & 0 \\ d_2 & 1 \end{vmatrix}}{\begin{vmatrix} 1 & 0 \\ 0 & 1 \end{vmatrix}} = d_1$ and $y = \dfrac{\begin{vmatrix} 1 & d_1 \\ 0 & d_2 \end{vmatrix}}{\begin{vmatrix} 1 & 0 \\ 0 & 1 \end{vmatrix}} = d_2$.

22 Problem 21 is $\begin{matrix} .96x + .02y = d_1 \\ .04x + .98y = d_2. \end{matrix}$ The sums down the columns of $\mathbf{A}$ are $.96 + .04 = 1$ and $.02 + .98 = 1$.
Reason: Everybody has to be accounted for. Nobody is lost or gained. Then $x + y$ (total population before move) equals $d_1 + d_2$ (total population after move).

34 $AB = \begin{bmatrix} 2 & 6 \\ 0 & 2 \end{bmatrix}$ has $(AB)^{-1} = \frac{1}{4}\begin{bmatrix} 2 & -6 \\ 0 & 2 \end{bmatrix}$. Check that this is $B^{-1}A^{-1} = \frac{1}{2}\begin{bmatrix} 1 & -2 \\ 0 & 2 \end{bmatrix}\frac{1}{2}\begin{bmatrix} 2 & -4 \\ 0 & 1 \end{bmatrix}$.

40 Compute $\mathbf{a}\cdot\mathbf{a}=3$ and $\mathbf{a}\cdot\mathbf{b}=\mathbf{b}\cdot\mathbf{a}=2$ and $\mathbf{b}\cdot\mathbf{b}=6$ and $\mathbf{a}\cdot\mathbf{d}=5$ and $\mathbf{b}\cdot\mathbf{d}=6$. The normal equation (14) is $\begin{matrix}3x+2y=5\\2x+6y=6\end{matrix}$ with solution $x=\frac{\mathbf{18}}{\mathbf{14}}=\frac{\mathbf{9}}{\mathbf{7}}$ and $y=\frac{\mathbf{8}}{\mathbf{14}}=\frac{\mathbf{4}}{\mathbf{7}}$. The nearest combination $x\mathbf{a}+y\mathbf{b}$ is $\mathbf{p}=(\frac{5}{7},\frac{13}{7},\frac{17}{7})$. The vector of three errors is $\mathbf{d}-\mathbf{p}=(\frac{2}{7},-\frac{6}{7},\frac{4}{7})$. It is perpendicular to $\mathbf{a}$ and $\mathbf{b}$. The best straight line is $f=x+yt=\frac{9}{7}+\frac{4}{7}t$.

11.5 Linear Algebra (page 443)

This section is about 3 by 3 matrices A, leading to three equations $A\mathbf{u}=\mathbf{d}$ in three unknowns. The ideas of A^{-1} and determinants and elimination still apply – but the formulas are beginning to get complicated. The determinant formula (six terms) we know from Section 11.3. The inverse divides by this determinant D. ***When*** $D=0$ ***the matrix A has no inverse***. Such a matrix is called ***singular***. Otherwise see pages 438-439 for A^{-1}.

Again we can solve $A\mathbf{u}=\mathbf{d}$ in three ways: use A^{-1} which has D in the denominator, or use Cramer's Rule which has D in *every* denominator, or use elimination which will fail if $D=0$.

1. Solve the three equations $\begin{matrix}x&+&y&+z&=&5\\x&+&2y&+2z&=&9\\&&y&-z&=&0\end{matrix}$

- My first choice is elimination. Subtract equation 1 from equation 2 to get $y+z=4$. Subtract this from equation 3 to get $-2z=-4$. Then $z=2$. Then $y=2$. Then $x=1$. The solution is $\mathbf{u}=(1,2,2)$.

The determinant of $\mathbf{A}$ is $D=-2+1+0-2+1+0=-2$. Find A^{-1} from the formulas:

$$A^{-1}=\frac{1}{-2}\begin{bmatrix}-4&2&0\\1&-1&-1\\1&-1&1\end{bmatrix}.\ \text{Check } AA^{-1}=I.\ \text{Then multiply } A^{-1}\mathbf{d}=\begin{bmatrix}1\\2\\2\end{bmatrix}.$$

The 1993 textbook "Introduction to Linear Algebra" by Gilbert Strang goes beyond these formulas. In calculus we are seeing lines and planes as special cases of curves and surfaces. In linear algebra we see lines and planes as special cases of n-dimensional vector spaces. Please take that course – it is truly useful.

Read-throughs and selected even-numbered solutions :

Three equations in three unknowns can be written as $A\mathbf{u}=\mathbf{d}$. The **vector u** has components x,y,z and $\mathbf{A}$ is a **3 by 3 matrix.** The row picture has a **plane** for each equation. The first two planes intersect in a **line,** and all three planes intersect in a **point,** which is **u**. The column picture starts with vectors **a,b,c** from the columns of $\mathbf{A}$ and combines them to produce $x\mathbf{a}+y\mathbf{b}+z\mathbf{c}$. The vector equation is $x\mathbf{a}+y\mathbf{b}+z\mathbf{c}=\mathbf{d}$.

The determinant of $\mathbf{A}$ is the triple product $\mathbf{a}\cdot\mathbf{b}\times\mathbf{c}$. This is the volume of a box, whose edges from the origin are $\mathbf{a},\mathbf{b},\mathbf{c}$. If det $A=\mathbf{0}$ then the system is **singular**. Otherwise there is an **inverse** matrix such that $A^{-1}A=\mathbf{I}$ (the **identity** matrix). In this case the solution to $A\mathbf{u}=\mathbf{d}$ is $\mathbf{u}=\mathbf{A}^{-1}\mathbf{d}$.

The rows of A^{-1} are the cross products $\mathbf{b}\times\mathbf{c},\mathbf{c}\times\mathbf{a},\mathbf{a}\times\mathbf{b}$, divided by D. The entries of A^{-1} are 2 by 2 **determinants**, divided by D. The upper left entry equals $(\mathbf{b_2c_3}-\mathbf{b_3c_2})/\mathbf{D}$. The 2 by 2 determinants needed for a row of A^{-1} do not use the corresponding **column** of A.

The solution is $\mathbf{u} = A^{-1}\mathbf{d}$. Its first component x is a ratio of determinants, $|\mathbf{d}\ \mathbf{b}\ \mathbf{c}|$ divided by $|\mathbf{a}\ \mathbf{b}\ \mathbf{c}|$. Cramer's Rule breaks down when det $A = \mathbf{0}$. Then the columns **a,b,c** lie in the same **plane.** There is no solution to $x\mathbf{a} + y\mathbf{b} + z\mathbf{c} = \mathbf{d}$, if $\mathbf{d}$ is not on that **plane.** In a singular row picture, the intersection of planes 1 and 2 is **parallel** to the third plane.

In practice $\mathbf{u}$ is computed by **elimination.** The algorithm starts by subtracting a multiple of row 1 to eliminate x from **the second equation.** If the first two equations are $x - y = 1$ and $3x + z = 7$, this elimination step leaves $\mathbf{3y + z = 4}$. Similarly x is eliminated from the third equation, and then **y** is eliminated. The equations are solved by back **substitution.** When the system has no solution, we reach an impossible equation like $\mathbf{1 = 0}$. The example $x - y = 1, 3x + z = 7$ has no solution if the third equation is $\mathbf{3y + z = 5}$.

8 $\begin{matrix} x + 2y + 2z = 0 \\ 2x + 3y + 5z = 0 \\ 2y + 2z = 8 \end{matrix} \rightarrow \begin{matrix} x + 2y + 2z = 0 \\ -y + z = 0 \\ 2y + 2z = 8 \end{matrix} \rightarrow \begin{matrix} x + 2y + 2z = 0 \\ -y + z = 0 \\ 4z = 8 \end{matrix} \rightarrow \begin{matrix} x = -8 \\ y = 2 \\ z = 2 \end{matrix}$

18 Choose $\mathbf{d} = \begin{bmatrix} 0 \\ 0 \\ 1 \end{bmatrix}$ as right side. The same steps as in Problem 16 end with $y - 3z = 0$ and $-y + 3z = 1$. Addition leaves $\mathbf{0 = 1}$. *No solution. Note:* The left sides of the three equations add to zero. There is a solution only if the right sides (components of **d**) also add to zero.

20 $BC = \begin{bmatrix} 0 & -1 & 3 \\ 1 & 0 & -6 \\ 2 & 2 & -18 \end{bmatrix}$ and $CB = \begin{bmatrix} -20 & -13 & 1 \\ 4 & 2 & -1 \\ 16 & 11 & 0 \end{bmatrix}$. The columns of CB add to zero (they are combinations of columns of C, and those add to zero). BC and CB are singular because C is.

22 $2A = \begin{bmatrix} 2 & 8 & 0 \\ 0 & 4 & 12 \\ 0 & 0 & 6 \end{bmatrix}$ has determinant 48 which is **8** times det A. If an n by n matrix is multiplied by 2, the determinant is multiplied by 2^n. Here $2^3 = 8$.

28 $\begin{bmatrix} 1 & 0 & 0 \\ 0 & 1 & 0 \\ 0 & 0 & 1 \end{bmatrix} \begin{bmatrix} 0 & 1 & 0 \\ 0 & 0 & 1 \\ 1 & 0 & 0 \end{bmatrix} \begin{bmatrix} 0 & 0 & 1 \\ 1 & 0 & 0 \\ 0 & 1 & 0 \end{bmatrix} \quad \begin{bmatrix} 1 & 0 & 0 \\ 0 & 0 & 1 \\ 0 & 1 & 0 \end{bmatrix} \begin{bmatrix} 0 & 1 & 0 \\ 1 & 0 & 0 \\ 0 & 0 & 1 \end{bmatrix} \begin{bmatrix} 0 & 0 & 1 \\ 0 & 1 & 0 \\ 1 & 0 & 0 \end{bmatrix}$

These are **"even"** These are **"odd"**

11 Chapter Review Problems

Review Problems

R1 If the vectors **v** and **w** form two sides of a triangle, the third side has squared length $(\mathbf{v}-\mathbf{w})\cdot(\mathbf{v}-\mathbf{w}) = \mathbf{v}\cdot\mathbf{v} + \mathbf{w}\cdot\mathbf{w} - 2\mathbf{v}\cdot\mathbf{w}$. How is this connected to the Law of Cosines?

R2 Find unit vectors in the direction of $\mathbf{i} + \mathbf{j} + \mathbf{k}$ and $2\mathbf{i} + 2\mathbf{j} + \mathbf{k}$.

R3 Explain why the derivative of a dot product $\mathbf{v}(t)\cdot\mathbf{w}(t)$ is $\mathbf{v}(t)\cdot\mathbf{w}'(t) + \mathbf{v}'(t)\cdot\mathbf{w}(t)$.

R4 Find a plane through (1,1,1) that is perpendicular to the plane $2x + 2y - z = 3$.

R5 How far is the origin from the plane through (1,0,0), (0,2,0), (0,0,4)? What is the nearest point?

R6 Project $\mathbf{B} = (1,4,4)$ onto the vector $\mathbf{A} = (2,2,-1)$. Then project the projection back onto $\mathbf{B}$.

R7 Find the dot product and cross product of $\mathbf{A} = (3,1,2)$ and $\mathbf{B} = (1,-1,-1)$.

R8 Starting from a 2 by 2 matrix with entries a, b, c, d, fill out the other 5 entries of a 3 by 3 matrix that has the same determinant.

R9 Multiply $AB = \begin{bmatrix} a & b \\ c & d \end{bmatrix} \begin{bmatrix} 3 & 1 \\ 2 & 4 \end{bmatrix}$. Compute the determinant of AB and check that it equals the determinant of A times the determinant of B.

R10 Fit the three points (0,1), (1,3), (2,3) by the closest line $f = x + yt$ (least squares).

Drill Problems

D1 Find the lengths and dot products of $\mathbf{A} = (1,2,1)$ and $\mathbf{B} = (1,1,2)$. Check the inequality $|\mathbf{A} \cdot \mathbf{B}| \leq |\mathbf{A}||\mathbf{B}|$.

D2 Find the normal vector to the plane $3x - y - z = 0$ and find a vector *in* the plane.

D3 Multiply $A = \begin{bmatrix} 3 & 2 \\ 4 & 1 \end{bmatrix}$ times A^{-1} after computing A^{-1}.

D4 For the same matrix compute A^2 and $2A$ and their determinants.

D5 Find the area of the parallelogram with $\mathbf{u} = \mathbf{i} + 3\mathbf{j}$ and $\mathbf{v} = 3\mathbf{i} + \mathbf{j}$ as two sides.

D6 Find the volume of the box with those sides $\mathbf{u}$ and $\mathbf{v}$ and third side $\mathbf{w} = \mathbf{j} + 2\mathbf{k}$.

D7 Write down two equations in x and y that have no solution. Change the right sides of those equations to produce infinitely many solutions. Change the left side to produce one unique solution and find it.

CHAPTER 12 MOTION ALONG A CURVE

12.1 The Position Vector (page 452)

This section explains the key vectors that describe motion. They are functions of t. In other words, the time is a "parameter." We know more than the path that the point travels along. We also know *where the point is at every time t*. So we can find the *velocity* (by taking a derivative).

In Chapter 1 the velocity $v(t)$ was the derivative of the distance $f(t)$. That was in one direction, along a line. Now $f(t)$ is replaced by the ***position vector*** $\mathbf{R}(t)$. Its derivative is the ***velocity vector*** $\mathbf{v}(t)$:

$$\textbf{position } \mathbf{R}(t) = x(t)\mathbf{i} + y(t)\mathbf{j} + z(t)\mathbf{k} \qquad \textbf{velocity } \mathbf{v}(t) = \frac{d\mathbf{R}}{dt} = \frac{dx}{dt}\mathbf{i} + \frac{dy}{dt}\mathbf{j} + \frac{dz}{dt}\mathbf{k}.$$

The next derivative is the ***acceleration vector*** $\mathbf{a} = d\mathbf{v}/dt$.

The direction of $\mathbf{v}$ is a unit vector $\frac{\mathbf{v}}{|\mathbf{v}|}$. This is the ***tangent vector*** $\mathbf{T}$. It is tangent to the path because $\mathbf{v}$ is tangent to the path. But it does not tell us the speed because we divide by $|\mathbf{v}|$; the length is always $|\mathbf{T}| = 1$. The length $|\mathbf{v}|$ is the speed ds/dt. Integrate to find the distance traveled. Remember: $\mathbf{R}$, $\mathbf{v}$, $\mathbf{a}$, $\mathbf{T}$ are vectors.

Read-throughs and selected even-numbered solutions :

The position vector $\mathbf{R(t)}$ along the curve changes with the parameter t. The velocity is $\mathbf{dR/dt}$. The acceleration is $\mathbf{d^2R/dt^2}$. If the position is $\mathbf{i} + t\mathbf{j} + t^2\mathbf{k}$, then $\mathbf{v} = \mathbf{j + 2t\,k}$ and $\mathbf{a} = \mathbf{2k}$. In that example the speed is $|\mathbf{v}| = \mathbf{\sqrt{1+4t^2}}$. This equals $\mathbf{ds/dt}$, where s measures **the distance along the curve.** Then $s = \int \mathbf{(ds/dt)dt}$.

The tangent vector is in the same direction as the **velocity,** but $\mathbf{T}$ is a **unit vector.** In general $\mathbf{T} = \mathbf{v}/|\mathbf{v}|$ and in the example $\mathbf{T} = \mathbf{(j + 2t\,k)/\sqrt{1+4t^2}}$.

Steady motion along a line has $\mathbf{a} = \mathbf{zero}$. If the line is $x = y = z$, the unit tangent vector is $\mathbf{T} = \mathbf{(i + j + k)/\sqrt{3}}$. If the speed is $|\mathbf{v}| = \sqrt{3}$, the velocity vector is $\mathbf{v} = \mathbf{i + j + k}$. If the initial position is (1,0,0), the position vector is $\mathbf{R}(t) = \mathbf{(1+t)i + t\,j + t\,k}$. The general equation of a line is $x = x_0 + tv_1$, $y = \mathbf{y_0 + tv_2}$, $z = \mathbf{z_0 + t\,v_3}$. In vector notation this is $\mathbf{R}(t) = \mathbf{R_0 + t\,v}$. Eliminating t leaves the equations $(x - x_0)/v_1 = (y - y_0)/v_2 = \mathbf{(z - z_0)/v_3}$. A line in space needs **two** equations where a plane needs **one.** A line has one parameter where a plane has **two.** The line from $\mathbf{R}_0 = (1,0,0)$ to (2,2,2) with $|\mathbf{v}| = 3$ is $\mathbf{R}(t) = \mathbf{(1+t)i + 2t\,j + 2t\,k}$.

Steady motion around a circle (radius r, angular velocity ω) has $x = \mathbf{r\cos\omega t}$, $y = \mathbf{r\sin\omega t}$, $z = 0$. The velocity is $\mathbf{v} = \mathbf{-r\omega\sin\omega t\,i + r\,\omega\cos\omega t\,j}$. The speed is $|\mathbf{v}| = \mathbf{r\omega}$. The acceleration is $\mathbf{a} = \mathbf{-r\omega^2(\cos\omega t\,i + \sin\omega t\,j)}$, which has magnitude $\mathbf{r\omega^2}$ and direction **toward (0,0).** Combining upward motion $\mathbf{R} = t\mathbf{k}$ with this circular motion produces motion around a **helix.** Then $\mathbf{v} = \mathbf{-r\omega\sin\omega t\,i + r\omega\cos\omega t\,j + k}$ and $|\mathbf{v}| = \mathbf{\sqrt{1+r^2\omega^2}}$.

2 The path is the line $x + y = 2$. The speed is $\sqrt{(dx/dt)^2 + (dy/dt)^2} = \mathbf{\sqrt{2}}$.

10 The line is $(x, y, z) = (3, 1, -2) + \mathbf{t(-1, -\frac{1}{3}, \frac{2}{3})}$. Then at $t = 3$ this gives $(0, 0, 0)$. The speed is $\frac{\text{distance}}{\text{time}} = \frac{\sqrt{9+1+4}}{3} = \mathbf{\frac{\sqrt{14}}{3}}$. For speed e^t choose $(x, y, z) = \mathbf{(3, 1, -2) + \frac{e^t}{\sqrt{14}}(-3, -1, 2)}$.

14 $x^2 + y^2 = (1+t)^2 + (2-t)^2$ is a minimum when $2(1+t) - 2(2-t) = 0$ or $4t = 2$ or $\mathbf{t = \frac{1}{2}}$. The path crosses $y = x$ when $1 + t = 2 - t$ or $t = \frac{1}{2}$ (again) at $\mathbf{x = y = \frac{3}{2}}$. The line never crosses a **parallel line like** $\mathbf{x = 2 + t, y = 2 - t}$.

20 If $x\frac{dx}{dt} + y\frac{dy}{dt} = 0$ along a path then $\mathbf{\frac{d}{dt}(x^2 + y^2) = 0}$ and $x^2 + y^2 = \text{constant}$.

32 Given only the path $y = f(x)$, it is impossible to find the **velocity** but still possible to find the **tangent vector** (or the *slope*).

40 The first particle has speed 1 and arrives at $t = \frac{\pi}{2}$. The second particle arrives when $v_2 t = 1$ and $-v_1 t = 1$, so $t = \frac{1}{v_2}$ and $v_1 = -v_2$. Its speed is $\sqrt{v_1^2 + v_2^2} = \sqrt{2}v_2$. So it should have $\sqrt{2}\mathbf{v_2} < \mathbf{1}$ (to go slower) and $\frac{\mathbf{1}}{\mathbf{v_2}} < \frac{\pi}{2}$ (to win). OK to take $\mathbf{v_2} = \frac{\mathbf{2}}{\mathbf{3}}$.

42 $\mathbf{v} \times \mathbf{w}$ is perpendicular to both lines, so the distance between lines is the length of the projection of $\mathbf{u} = Q - P$ onto $\mathbf{v} \times \mathbf{w}$. The formula for the distance is $\frac{|\mathbf{u}\cdot(\mathbf{v}\times\mathbf{w})|}{|\mathbf{v}\times\mathbf{w}|}$.

12.2 Plane Motion: Projectiles and Cycloids (page 457)

This section discusses two particularly important motions. One is from a ball in free flight (a projectile). The other is from a point on a rolling wheel. The first travels on a parabola and the second on a cycloid. In both cases we have to figure out the position vector $\mathbf{R}(t)$ from the situation:

$$\textbf{projectile motion} \qquad x(t) = (\mathrm{v}_0 \cos\alpha)t \quad \text{and} \quad y(t) = (\mathrm{v}_0 \sin\alpha)t - \frac{1}{2}gt^2.$$

Actually this came from two integrations. In free flight the only acceleration is gravity: $\mathbf{a} = -g\mathbf{k}$ (simple but important). One integration gives the velocity $\mathbf{v} = -gt\mathbf{k} + \mathbf{v}(0)$. Another integration gives the position $\mathbf{R}(t) = -\frac{1}{2}gt^2\ \mathbf{k} + \mathbf{v}(0)t + \mathbf{R}(0)$. Assume that the starting point $\mathbf{R}(0)$ is the origin. Split $\mathbf{v}(0)$ into its x and y components $\mathrm{v}_0 \cos x$ and $\mathrm{v}_0 \sin\alpha$ (where α is the *launch angle*). The component $-\frac{1}{2}gt^2$ is all in the y direction. Now you have the components $x(t)$ and $y(t)$ given above. The constant g is 9.8 meters/sec^2.

1. If a ball is launched at $\alpha = 60°$ with speed 50 meters/second, how high does it go and how far does it go?

   - To find how *far* it goes before hitting the ground, solve $y = 0$ to find the flight time $\mathbf{T} = 2\ \mathrm{v}_0(\sin\alpha)/g = 8.8$ seconds. Compute $x(\mathbf{T}) = (50)(\frac{1}{2})(8.8) = 220$ meters for the distance (the range). The highest point is at the half-time $t = 4.4$. Then $y_{\max} = (50)(\frac{\sqrt{3}}{2})(4.4) - \frac{1}{2}(9.8)(4.4)^2 = 96$ meters.

The **cycloid** is in Figure 12.6. The parameter is not t but the rotation angle θ:

$$\textbf{position} \quad x(\theta) = a(\theta - \sin\theta) \text{ and } y(\theta) = a(1 - \cos\theta) \text{ and slope} \frac{dy}{dx} = \frac{dy/d\theta}{dx/d\theta} = \frac{a\sin\theta}{a(1-\cos\theta)}.$$

The text computes the area $3\pi a^2$ under the cycloid and the length $8a$ of the curve.

2. Where is the cycloid's slope zero? Where is the slope infinite?

   - The slope is zero at $\theta = \pi$, because then $\sin\theta = 0$. Top of cycloid is flat.
   - The slope is infinite at $\theta = 0$, even though $\sin\theta = 0$ again. *Reason:* $1 - \cos\theta$ is also zero. We have $\frac{0}{0}$ and go to l'Hôpital. The ratio of θ derivatives is $\frac{a\cos\theta}{a\sin\theta} \to \infty$ at $\theta = 0$. Bottom of cycloid is a cusp.

Read-throughs and selected even-numbered solutions :

A projectile starts with speed v_0 and angle α. At time t its velocity is $dx/dt = \mathbf{v_0}\ \mathbf{cos}\ \alpha$, $dy/dt = \mathbf{v_0}\ \mathbf{sin}\ \alpha - \mathbf{gt}$

(the downward acceleration is g). Starting from (0,0), the position is $x = \mathbf{v_0} \cos\alpha\, t, y = \mathbf{v_0} \sin\alpha\, t - \frac{1}{2}\mathbf{g}t^2$. The flight time back to $y = 0$ is $T = \mathbf{2v_0(\sin\alpha)/g}$. At that time the horizontal range is $R = \mathbf{(v_0^2 \sin 2\alpha)/g}$. The flight path is a **parabola**.

The three quantities v_0, α, t determine the projectile's motion. Knowing v_0 and the position of the target, we cannot solve for α. Knowing α and the position of the target, we can solve for v_0.

A cycloid is traced out by a point on a rolling circle. If the radius is a and the turning angle is θ, the center of the circle is at $x = \mathbf{a}\theta, y = \mathbf{a}$. The point is at $x = \mathbf{a}(\theta - \sin\theta), y = \mathbf{a}(1 - \cos\theta)$, starting from (0,0). It travels a distance **8a** in a full turn of the circle. The curve has a **cusp** at the end of every turn. An upside-down cycloid gives the **fastest** slide between two points.

6 If the maximum height is $\frac{(v_0 \sin\alpha)^2}{2g} = 6$ meters, then $\sin^2\alpha = \frac{12(9.8)}{30^2} \approx .13$ gives $\alpha \approx .37$ or $21°$.

10 Substitute into $(gx/v_0)^2 + 2gy = g^2t^2\cos^2\alpha + 2gv_0t\sin\alpha - t^2 = 2gv_0t\sin\alpha - g^2t^2\sin^2\alpha$. This is less than v_0^2 because $(\mathbf{v_0} - \mathbf{g\,t}\sin\alpha)^2 \geq \mathbf{0}$. For $y = H$ the largest x is when equality holds: $v_0^2 = (gx/v_0)^2 + 2gH$ or $\mathbf{x} = \sqrt{\mathbf{v_0^2} - \mathbf{2gH}}(\frac{\mathbf{v_0}}{\mathbf{g}})$. If $2gH$ is larger than v_0, the height H can't be reached.

20 $\frac{dy}{dx} = \frac{\sin\theta}{1-\cos\theta}$ becomes $\frac{0}{0}$ at $\theta = 0$, so use **l'Hôpital's Rule**: The ratio of derivatives is $\frac{\cos\theta}{\sin\theta}$ which becomes infinite. $\frac{\sin\theta}{1-\cos\theta} \approx \frac{\theta}{\theta^2/2} = \frac{2}{\theta}$ equals **20** at $\theta = \frac{1}{10}$ and **−20** at $\theta = -\frac{1}{10}$. The slope is 1 when $\sin\theta = 1 - \cos\theta$ which happens at $\theta = \frac{\pi}{2}$.

38 On the line $x = \frac{\pi}{2}y$ the distance is $ds = \sqrt{(dx)^2 + (dy)^2} = \sqrt{(\pi/2)^2 + 1}\,dy$. The last step in equation (5) integrates $\frac{\text{constant}}{\sqrt{y}}$ to give $\frac{\sqrt{\pi^2+4}}{2\sqrt{2g}}[2\sqrt{y}]_0^{2a} = \sqrt{\pi^2+4}\,\frac{2\sqrt{2a}}{2\sqrt{2g}} = \sqrt{\pi^2+4}\,\sqrt{\frac{a}{g}}$.

40 I have read (but don't believe) that the rolling circle jumps as the weight descends.

12.3 Curvature and Normal Vector (page 463)

The tangent vector $\mathbf{T}$ is a unit vector along the path. When the path is a straight line, $\mathbf{T}$ is constant. When the path curves, $\mathbf{T}$ changes. *The rate of change gives the curvature* κ(kappa). This comes from the shape of the path but not the speed. The curvature is $|\frac{d\mathbf{T}}{ds}|$ not $|\frac{d\mathbf{T}}{dt}|$. Since we know the position $\mathbf{R}$ and velocity $\mathbf{v}$ and tangent $\mathbf{T}$ and distance s as functions of t, the available derivatives are time derivatives. So use the chain rule:

$$\kappa = \left|\frac{d\mathbf{T}}{ds}\right| = \frac{|d\mathbf{T}/dt|}{|ds/dt|} = \frac{|d\mathbf{T}/dt|}{|\mathbf{v}|}.$$

1. Find the curvature $\kappa(t)$ if $x = 3\cos 2t$ and $y = 3\sin 2t$ (a ***circle***). Thus $\mathbf{R} = 3\cos 2t\,\mathbf{i} + 3\sin 2t\,\mathbf{j}$.

   - The velocity is $\mathbf{v} = \frac{d\mathbf{R}}{dt} = -6\sin 2t\,\mathbf{i} + 6\cos 2t\,\mathbf{j}$. The speed is $|\mathbf{v}| = 6$. The tangent vector is $\mathbf{T} = \frac{\mathbf{v}}{|\mathbf{v}|} = -\sin 2t\,\mathbf{i} + \cos 2t\,\mathbf{j}$. $\mathbf{T}$ is a unit vector but it changes direction. Its time derivative is

$$\frac{d\mathbf{T}}{dt} = -2\cos 2t\,\mathbf{i} - 2\sin 2t\,\mathbf{j} \text{ and } \left|\frac{d\mathbf{T}}{dt}\right| = 2 \text{ and } \kappa = \frac{2}{6} = \frac{1}{3}.$$

   Since the path is a circle of radius 3, we were expecting its curvature to be $\frac{1}{3}$.

There is also a new vector involved as $\mathbf{T}$ changes direction. It is the ***normal vector*** $\mathbf{N}$, which tells which way $\mathbf{T}$ is changing. The rule is that $\mathbf{N}$ is always a unit vector perpendicular to $\mathbf{T}$. In the xy plane this leaves practically no choice (once we know $\mathbf{T}$). In xyz space we need a formula:

$$\mathbf{N} = \frac{d\mathbf{T}/dt}{|d\mathbf{T}/dt|} \quad \text{or} \quad \mathbf{N} = \frac{d\mathbf{T}/ds}{|d\mathbf{T}/ds|} = \frac{1}{\kappa}\frac{d\mathbf{T}}{ds}.$$

2. Find the normal vector $\mathbf{N}$ to the circle in Problem 1. The derivative $d\mathbf{T}/dt$ is already computed above.

- The length of $|\frac{d\mathbf{T}}{dt}|$ is 2. Divide by 2 to get $\mathbf{N} = \frac{1}{2}\frac{d\mathbf{T}}{dt} = -\cos 2t\ \mathbf{i} - \sin 2t\ \mathbf{j}$. Check that $\mathbf{N}$ is perpendicular to $\mathbf{T} = -\sin 2t\ \mathbf{i} + \cos 2t\ \mathbf{j}$.

3. From the fact that $\mathbf{T}\cdot\mathbf{T} = 1$ show that $\frac{d\mathbf{T}}{dt}$ is perpendicular to $\mathbf{T}$.

- The derivative of the dot product $\mathbf{T}\cdot\mathbf{T}$ is like the ordinary product rule (first vector dot derivative of second vector plus second vector dot derivative of first vector). Here that gives $2\mathbf{T}\cdot\frac{d\mathbf{T}}{dt}$ = derivative of $1 = 0$. Since $\mathbf{N}$ is in the direction of $\frac{d\mathbf{T}}{dt}$, this means $\mathbf{T}\cdot\mathbf{N} = 0$.

The third unit vector, perpendicular to $\mathbf{T}$ and $\mathbf{N}$, is $\mathbf{B} = \mathbf{T}\times\mathbf{N}$ in Problem 25.

Read-throughs and selected even-numbered solutions :

The curvature tells how fast the curve **turns.** For a circle of radius a, the direction changes by 2π in a distance $\mathbf{2\pi a}$, so $\kappa = \mathbf{1/a}$. For a plane curve $y = f(x)$ the formula is $\kappa = |y''|/(\mathbf{1} + (\mathbf{y}')^{\mathbf{2}})^{\mathbf{3/2}}$. The curvature of $y = \sin x$ is $|\mathbf{\sin x}|/(\mathbf{1} + \mathbf{\cos^2 x})^{\mathbf{3/2}}$. At a point where $y'' = 0$ (an **inflection** point) the curve is momentarily straight and $\kappa =$ **zero**. For a space curve $\kappa = |\mathbf{v}\times\mathbf{a}|/|\mathbf{v}|^{\mathbf{3}}$.

The normal vector $\mathbf{N}$ is perpendicular to **the curve (and therefore to v and T).** It is a **unit** vector along the derivative of $\mathbf{T}$, so $\mathbf{N} = \mathbf{T}'/|\mathbf{T}'|$. For motion around a circle $\mathbf{N}$ points **inward.** Up a helix $\mathbf{N}$ also points **inward.** Moving at unit speed on any curve, the time t is the same as the **distance** s. Then $|\mathbf{v}| = \mathbf{1}$ and $d^2s/dt^2 = \mathbf{0}$ and $\mathbf{a}$ is in the direction of $\mathbf{N}$. Acceleration equals $\mathbf{d^2s/dt^2}\ \mathbf{T} + \kappa|\mathbf{v}|^{\mathbf{2}}\ \mathbf{N}$. At unit speed around a unit circle, those components are **zero and one.** An astronaut who spins once a second in a radius of one meter has $|\mathbf{a}| = \omega^{\mathbf{2}} = (\mathbf{2\pi})^{\mathbf{2}}$ meters/sec^2, which is about **4g.**

2 $y = \ln x$ has $\kappa = \frac{|y''|}{(1+y'^2)^{3/2}} = \frac{1/x^2}{(1+\frac{1}{x^2})^{3/2}} = \frac{\mathbf{x}}{(\mathbf{x^2+1})^{\mathbf{3/2}}}$. Maximum of κ when its derivative is zero: $(x^2+1)^{3/2} = x\frac{3}{2}(x^2+1)^{1/2}(2x)$ or $x^2+1 = 3x^2$ or $\mathbf{x^2} = \mathbf{\frac{1}{2}}$.

14 When $\kappa = 0$ the path is a **straight line.** This happens when $\mathbf{v}$ and $\mathbf{a}$ are **parallel.** Then $\mathbf{v}\times\mathbf{a} = \mathbf{0}$.

18 Using equation (8), $\mathbf{v}\times\mathbf{a} = |\mathbf{v}|\mathbf{T}\times(\frac{d^2s}{dt^2}\mathbf{T} + \kappa(\frac{ds}{dt})^2\mathbf{N}) = \kappa|\mathbf{v}|^3\mathbf{T}\times\mathbf{N}$ because $\mathbf{T}\times\mathbf{T} = \mathbf{0}$ and $|\mathbf{v}|$ is the same as $|\frac{ds}{dt}|$. Since $|\mathbf{T}\times\mathbf{N}| = 1$ this gives $|\mathbf{v}\times\mathbf{a}| = \kappa|\mathbf{v}|^3$ or $\kappa = \frac{|\mathbf{v}\times\mathbf{a}|}{|\mathbf{v}|^3}$.

22 The parabola through the three points is $y = x^2 - 2x$ which has a constant second derivative $\frac{d^2y}{dx^2} = 2$. The circle through the three points has radius $= 1$ and $\kappa = \frac{1}{\textbf{radius}} = 1$. These are the smallest possible (Proof?)

32 $\mathbf{T} = \cos\theta\ \mathbf{i} + \sin\theta\ \mathbf{j}$ gives $\frac{d\mathbf{T}}{d\theta} = -\sin\theta\ \mathbf{i} + \cos\theta\ \mathbf{j}$ so $|\frac{d\mathbf{T}}{d\theta}| = 1$. Then $\kappa = |\frac{d\mathbf{T}}{ds}| = |\frac{d\mathbf{T}}{d\theta}||\frac{d\theta}{ds}| = |\frac{d\theta}{ds}|$. Curvature is **rate of change of slope of path.**

12.4 Polar Coordinates and Planetary Motion (page 468)

This section has two main ideas, one from mathematics and the other from physics and astronomy. The mathematics idea is to switch from **i** and **j** (unit vectors across and up) to a polar coordinate system $\mathbf{u}_r$ and $\mathbf{u}_\theta$ (unit vectors out from and around the origin). Very suitable for travel in circles – the tangent vector **T** is just $\mathbf{u}_\theta$. Very suitable for central forces – the force vector is a multiple of $\mathbf{u}_r$.

Just one difficulty: $\mathbf{u}_r$ and $\mathbf{u}_\theta$ change direction as you move, which didn't happen for **i** and **j**:

$$\mathbf{u}_r = \cos\theta\ \mathbf{i} + \sin\theta\ \mathbf{j} \qquad \text{and} \qquad \mathbf{u}_\theta = \frac{d\mathbf{u}_r}{d\theta} = -\sin\theta\ \mathbf{i} + \cos\theta\ \mathbf{j}.$$

The idea from physics is ***gravity***. This is a central force from the sun. (We ignore planets that are smaller and stars that are farther away.) By using polar coordinates we derive the path of the planets around the sun. Kepler found ellipses in the measurements. Newton found ellipses by using $\mathbf{F} = m\mathbf{a}$ and $\mathbf{F} = \text{constant}/r^2$. We find ellipses by the chain rule (page 467).

An important step in mathematical physics is to write the velocity and acceleration in polar coordinates:

$$\mathbf{v} = \frac{dr}{dt}\mathbf{u}_r + r\frac{d\theta}{dt}\mathbf{u}_\theta \qquad \text{and} \qquad \mathbf{a} = \left(\frac{d^2r}{dt^2} - r\left(\frac{d\theta}{dt}\right)^2\right)\mathbf{u}_r + \left(r\frac{d^2\theta}{dt^2} + 2\frac{dr}{dt}\frac{d\theta}{dt}\right)\mathbf{u}_\theta$$

Physicists tend to like those formulas. Other people tend to hate them. It was not until writing this book that I understood what Newton did in Calculus III. He solved Problems 12.4.34 and 14.4.26 and so can you.

Read-throughs and selected even-numbered solutions :

A central force points toward **the origin**. Then $\mathbf{R} \times d^2\mathbf{R}/dt^2 = \mathbf{0}$ because **these vectors are parallel**. Therefore $\mathbf{R} \times d\mathbf{R}/dt$ is a **constant** (called **H**).

In polar coordinates, the outward unit vector is $\mathbf{u}_r = \cos\theta\ \mathbf{i} + \mathbf{sin}\ \theta\ \mathbf{j}$. Rotated by 90° this becomes $\mathbf{u}_\theta = -\mathbf{sin}\ \theta\ \mathbf{i} + \mathbf{cos}\ \theta\ \mathbf{j}$. The position vector **R** is the distance r times $\mathbf{u_r}$. The velocity $\mathbf{v} = d\mathbf{R}/dt$ is $(\mathbf{dr/dt})\mathbf{u}_r + (\mathbf{r}\ \mathbf{d}\theta/\mathbf{dt})\mathbf{u}_\theta$. For steady motion around the circle $r = 5$ with $\theta = 4t$, $\mathbf{v}$ is $-\mathbf{20\ sin\ 4t\ i} + \mathbf{20\ cos\ 4t\ j}$ and $|\mathbf{v}|$ is **20** and **a** is $-\mathbf{80\ cos\ 4t\ i} - \mathbf{80\ sin\ 4t\ j}$.

For motion under a circular force, r^2 times $\mathbf{d}\theta/\mathbf{dt}$ is constant. Dividing by 2 gives Kepler's second law $dA/dt = \frac{1}{2}\mathbf{r}^2\mathbf{d}\theta/\mathbf{dt} =$ **constant**. The first law says that the orbit is an **ellipse** with the sun at **a focus**. The polar equation for a conic section is $\mathbf{1/r} = C - D\cos\theta$. Using $\mathbf{F} = m\mathbf{a}$ we found $q_{\theta\theta} + \mathbf{q} = C$. So the path is a conic section; it must be an ellipse because **planets come around again**. The properties of an ellipse lead to the period $T = \mathbf{2\pi a^{3/2}}/\sqrt{\mathbf{GM}}$, which is Kepler's third law.

8 The distance $r\theta$ around the circle is the integral of the speed $8t$: thus $4\theta = 4t^2$ and $\theta = t^2$. The circle is complete at $\mathbf{t} = \sqrt{\mathbf{2}\pi}$. At that time $\mathbf{v} = r\frac{d\theta}{dt}\mathbf{u}_\theta = 4(2\sqrt{2\pi})\mathbf{j}$ and $\mathbf{a} = -\mathbf{4(8\pi)i} + \mathbf{4(2)j}$.

12 Since $\mathbf{u_r}$ has constant length, its derivatives are **perpendicular** to itself. In fact $\frac{du_r}{dr} = 0$ and $\frac{d\mathbf{u_r}}{d\theta} = \mathbf{u}_\theta$.

14 $R = re^{i\theta}$ has $\frac{d^2R}{dt^2} = \frac{d^2r}{dt^2}e^{i\theta} + 2\frac{dr}{dt}(ie^{i\theta}\frac{d\theta}{dt}) + ir\frac{d^2\theta}{dt^2}e^{i\theta} + i^2r(\frac{d\theta}{dt})^2e^{i\theta}$. (Note repeated term gives factor 2.) The coefficient of $e^{i\theta}$ is $\frac{d^2r}{dt^2} - r(\frac{d\theta}{dt})^2$. The coefficient of $ie^{i\theta}$ is $2\frac{dr}{dt}\frac{d\theta}{dt} + r\frac{d^2\theta}{dt^2}$. These are the $\mathbf{u_r}$ and $\mathbf{u}_\theta$ components of **a**.

20 (a) **False:** The paths are conics but they could be hyperbolas and possibly parabolas.

(b) **True:** A circle has $r =$ constant and $r^2 \frac{d\theta}{dt} =$ constant so $\frac{d\theta}{dt} =$ constant.

(c) **False:** The central force might not be proportional to $\frac{1}{r^2}$.

32 $T = \frac{2\pi}{\sqrt{GM}}(1.6 \cdot 10^9)^{3/2} \approx 71$ years. So the comet will return in the year $1986 + 71 = 2057$.

34 First derivative: $\frac{dr}{dt} = \frac{d}{dt}\left(\frac{1}{C - D\cos\theta}\right) = \frac{-D\sin\theta \frac{d\theta}{dt}}{(C - D\cos\theta)^2} = -D\sin\theta \, r^2 \frac{d\theta}{dt} = -Dh\sin\theta$.

Next derivative: $\frac{d^2r}{dt^2} = -Dh\cos\theta \frac{d\theta}{dt} = \frac{-Dh^2\cos\theta}{r^2}$. But $C - D\cos\theta = \frac{1}{r}$ so $-D\cos\theta = (\frac{1}{r} - C)$.

The acceleration terms $\frac{d^2r}{dt^2} - r(\frac{d\theta}{dt})^2$ combine into $(\frac{1}{r} - C)\frac{h^2}{r^2} - \frac{h^2}{r^3} = -C\frac{h^2}{r^2}$. Conclusion by Newton:

The elliptical orbit $r = \frac{1}{C - D\cos\theta}$ requires acceleration $= \frac{\text{constant}}{r^2}$: *the inverse square law.*

12 Chapter Review Problems

Graph Problems

G1 Draw these three curves as t goes from 0 to 2π. Mark distances on the x and y axes.

(a) $x = 2t, y = \sin t$ (b) $x = 2\sin t, y = \sin t$ (c) $x = 2\cos t, y = \sin t$

G2 Draw a helix $x = \cos t$, $y = \sin t$, $z = t$. Also draw its projections on the xy and yz planes.

Review Problems

R1 Find $x(t)$ and $y(t)$ for travel around the unit circle at speed 2. Start at $(x, y) = (-1, 0)$.

R2 Find $x(t)$ and $y(t)$ along the circle $(x - 3)^2 + (y - 4)^2 = 5^2$.

R3 Find $x(t)$ and $y(t)$ to produce travel along a cycloid. At what time do you repeat?

R4 Find the tangent vector $\mathbf{T}$ and the speed $|\mathbf{v}|$ for your cycloid travel.

R5 Find the velocity vector and the speed along the curve $x = t^2, y = \sqrt{t}$. What curve is it?

R6 Explain why the velocity vector $\mathbf{v}(t) = d\mathbf{R}/dt$ is tangent to the curve. $\mathbf{R}(t)$ gives the position.

R7 Is the acceleration vector $\mathbf{a}(t) = d\mathbf{v}/dt = d^2\mathbf{R}/dt^2$ always perpendicular to $\mathbf{v}$ and the curve?

R8 Choose $x(t), y(t)$, and $z(t)$ for travel from (2,0,0) at $t = 0$ to (8,3,6) at $t = 3$ with no acceleration.

R9 Figure 12.5a shows flights at 30° and 60°. The range R is the same. One flight goes ___ times as high as the other.

R10 Why do flights at $\alpha = 10°$ and $80°$ have the same range? Are the flight times T the same?

R11 If $\mathbf{T}(t)$ is always a unit vector, why isn't $d\mathbf{T}/dt = 0$? Prove that $\mathbf{T} \cdot d\mathbf{T}/dt = 0$.

R12 Define the curvature $\kappa(t)$. Compute it for $x = y = z = \sin t$.

R13 If a satellite stays above Hawaii, why does that determine its distance from Earth?

Drill Problems

D1 With position vector $\mathbf{R} = 3t\,\mathbf{i} + 4t\,\mathbf{j}$ find the velocity $\mathbf{v}$ and speed $|\mathbf{v}|$ and unit tangent $\mathbf{T}$.

D2 True or false: A projectile that goes from (0,0) to (1,1) has starting angle $\alpha = 45°$.

D3 State the r, θ equation for an ellipse. Where is its center point? At the sun?

D4 Give the equations (using t) for the line through (1,4,5) in the direction of $\mathbf{i} + 2\mathbf{j}$.

D5 Give the equations for the same line without the parameter t. (Eliminate t.).

D6 State the equations $x =$ ___ and $y =$ ___ for a ball thrown down at a 30° angle with speed $\mathbf{v}_0$.

D7 What are the rules for the derivatives of $\mathbf{v}(t) \cdot \mathbf{w}(t)$ and $\mathbf{v}(t) \times \mathbf{w}(t)$?

D8 Give the formula for the curvature κ of $y = f(x)$. Compute κ for $y = \sqrt{1 - x^2}$ and explain.

D9 At $(x, y) = (3, 4)$ what are the unit vectors $\mathbf{u}_r$ and $\mathbf{u}_\theta$? Why are they the same at (6,8)?

D10 Motion in a central force field always stays in a (circle, ellipse, plane). Why is $\mathbf{R} \times \mathbf{a} = \mathbf{0}$?

D11 What integral gives the distance traveled back to $y = 0$ on the curved flight $x = t, y = t - \frac{1}{2}gt^2$?

D12 What is the velocity $\mathbf{v} =$ ___$\mathbf{u}_r +$ ___$\mathbf{u}_\theta$ along the spiral $r = t, \theta = t$?

CHAPTER 13 PARTIAL DERIVATIVES

13.1 Surfaces and Level Curves (page 475)

The graph of $z = f(x, y)$ is a surface in xyz space. When f is a linear function, the surface is flat (a plane). When $f = x^2 + y^2$ the surface is curved (a parabola is revolved to make a bowl). When $f = \sqrt{x^2 + y^2}$ the surface is pointed (a cone resting on the origin). These three examples carry you a long way.

To visualize a surface we cut through it by planes. Often the cutting planes are horizontal, with the simple equation $z = c$ (a constant). This plane meets the surface in a ***level curve***, and the equation of that curve is $c = f(x, y)$. The cutting is up at all different heights c, but we move all the level curves down to the xy plane. For the bowl $z = x^2 + y^2$ the level curves are $c = x^2 + y^2$ (circles). For the cone $z = \sqrt{x^2 + y^2}$ the level curves are $c = \sqrt{x^2 + y^2}$ (again circles – just square both sides). For the plane $z = x + y$ the level curves are straight lines $c = x + y$ (parallel to each other as c changes).

The collection of level curves in the xy plane is a ***contour map***. If you are climbing on the surface, the map tells you two important things:

1. Which way is up: Perpendicular to the level curve is the steepest direction.
2. How steep the surface is: Divide the change in c by the distance between level curves.

A climbing map shows the curves at equal steps of c. The mountain is steeper when the level curves are closer.

1. Describe the level curves for the saddle surface $z = xy$.

   - The curve $xy = 1$ is a *hyperbola*. One branch is in the first quadrant through (1,1). The other branch is in the third quadrant through $(-1, -1)$. At these points the saddle surface has $z = 1$.

   The curve $xy = -1$ is also a hyperbola. Its two pieces go through $(1, -1)$ and $(-1, 1)$. At these points the surface has $z = xy = -1$ and it is below the plane $z = 0$.

2. How does a maximum of $f(x, y)$ show up on the contour map of level curves?

   - Think about the top point of the surface. The highest cutting plane just touches that top point. The level curve is only a point! When the plane moves lower, it cuts out a curve that goes around the top point. So the contour map shows "near-circles" closing in on a single maximum point. A minimum looks just the same, but the c's decrease as the contour lines close in.

Read-throughs and selected even-numbered solutions :

The graph of $z = f(x, y)$ is a **surface** in **three**-dimensional space. The **level** curve $f(x, y) = 7$ lies down in the base plane. Above this level curve are all points at height **7** in the surface. The **plane** $z = 7$ cuts through the surface at those points. The level curves $f(x, y) = \mathbf{c}$ are drawn in the xy plane and labeled by **c**. The family of labeled curves is a **contour** map.

For $z = f(x, y) = x^2 - y^2$, the equation for a level curve is $\mathbf{x^2 - y^2 = c}$. This curve is a **hyperbola.** For $z = x - y$ the curves are **straight lines.** *Level curves never cross because* **f(x,y) cannot equal two numbers c and c′.** They crowd together when the surface is **steep.** The curves tighten to a point when **f reaches a maximum or minimum.** The steepest direction on a mountain is **perpendicular** to the **level curve.**

6 $(x + y)^2 = 0$ gives the **line** $\mathbf{y = -x}$; $(x + y)^2 = 1$ gives the **pair of lines** $x + y = 1$ and $x + y = -1$; similarly $\mathbf{x + y = \sqrt{2}}$ and $\mathbf{x + y = -\sqrt{2}}$; **no level curve** $(x + y)^2 = -4$.

16 $f(x,y) = \{$ maximum of x^2+y^2-1 and zero $\}$ is zero inside the unit circle.

18 $\sqrt{4x^2+y^2} = c+2x$ gives $4x^2+y^2 = c^2+4cx+4x^2$ or $\mathbf{y^2 = c^2 + 4cx}$. This is a **parabola** opening to the left or right.

30 Direct approach: $xy = (\frac{x_1+x_2}{2})(\frac{y_1+y_2}{2}) = \frac{1}{4}(x_1y_1+x_2y_2+x_1y_2+x_2y_1) = \frac{1}{4}(1+1+\frac{x_1}{x_2}+\frac{x_2}{x_1})$ $= 1+\frac{(x_1-x_2)^2}{4x_1x_2} \geq 1$. **Quicker approach:** $y = \frac{1}{x}$ is concave up (or convex) because $y'' = \frac{2}{x^3}$ is positive. *Note for convex functions:* Tangent lines below curve, secant line segments above curve!

13.2 Partial Derivatives (page 479)

I am sure you are good at taking partial derivatives. They are like ordinary derivatives, when you close your eyes to the other variables. As the text says, "***Do not treat y as zero!*** Treat it as a constant." Just pretend that $y = 5$. That applies to $\frac{\partial}{\partial x}e^{xy} = y\,e^{xy}$ and $\frac{\partial}{\partial x}(x^2+xy^2) = 2x+y^2$.

Remember that $\frac{\partial f}{\partial x}$ is also written f_x. The y-derivative of this function is $\frac{\partial^2 f}{\partial y\,\partial x}$ or f_{xy}. *A major point is that* $f_{xy} = f_{yx}$. The y-derivative of $\frac{\partial f}{\partial x}$ equals the x-derivative of $\frac{\partial f}{\partial y}$. Take $f = x^2+xy^2$ with $\frac{\partial f}{\partial y} = 2xy$:

$$\frac{\partial^2 f}{\partial y\,\partial x} = \frac{\partial}{\partial y}(2x+y^2) = 2y \qquad \text{and} \qquad \frac{\partial^2 f}{\partial x\,\partial y} = \frac{\partial}{\partial x}(2xy) = 2y.$$

Problem 43 proves this rule $f_{xy} = f_{yx}$, assuming that both functions are continuous. Here is another example:

1. The partial derivatives of $f(x,y) = e^{xy}$ are $f_x = ye^{xy}$ and $f_y = xe^{xy}$. Find f_{xx}, f_{xy}, f_{yx}, and f_{yy}.

   - f_{xx} is $\frac{\partial^2 f}{\partial x^2}$ or $\frac{\partial}{\partial x}(\frac{\partial f}{\partial x})$. This is $\frac{\partial}{\partial x}(ye^{xy}) = y^2e^{xy}$. Similarly f_{yy} is $\frac{\partial}{\partial y}(xe^{xy}) = x^2e^{xy}$. The mixed derivatives are equal as usual:

   $$\frac{\partial}{\partial y}(\frac{\partial f}{\partial x}) = \frac{\partial}{\partial y}(ye^{xy}) = y(xe^{xy}) + 1(e^{xy}) \text{ by the product rule}$$

   $$\frac{\partial}{\partial x}(\frac{\partial f}{\partial y}) = \frac{\partial}{\partial x}(xe^{xy}) = x(ye^{xy}) + 1(e^{xy}) \text{ by the product rule}$$

   You *must* notice that it is $\partial^2 f$ above and ∂x^2 below. We divide $\Delta(\Delta f)$ by $(\Delta x)^2$.

2. What does that mean? How is $\Delta(\Delta f)$ different from $(\Delta f)^2$?

   - Start with $f(x)$. The forward difference Δf is $f(x+\Delta x) - f(x)$. This is a function of x. So we can take *its* forward difference:

   $$\Delta(\Delta f) = \Delta f(x+\Delta x) - \Delta f(x) = [f(x+2\Delta x) - f(x+\Delta x)] - [f(x+\Delta x) - f(x)]$$

   This is totally different from $(\Delta f)^2 = [f(x+\Delta x) - f(x)]^2$. In the limit $\frac{\partial^2 f}{\partial x^2}$ is totally different from $(\frac{\partial f}{\partial x})^2$.

3. Which third derivatives are equal to f_{xxy}? This is $\frac{\partial}{\partial y}(f_{xx})$ or $\frac{\partial^3 f}{\partial y\,\partial x^2}$.

   - We are taking *one* y-derivative and *two* x-derivatives. The order does not matter (for a smooth function). Therefore $f_{xxy} = f_{xyx} = f_{yxx}$.

Notice Problems 45 – 52 about limits and continuity for functions $f(x, y)$. This two-variable case is more subtle than limits and continuity of $f(x)$. In a course on mathematical analysis this topic would be expanded. In a calculus course I believe in completing the definitions and applying them.

More important in practice are ***partial differential equations*** like $\frac{\partial f}{\partial t} = \frac{\partial f}{\partial x}$ and $\frac{\partial^2 f}{\partial t^2} = \frac{\partial^2 f}{\partial x^2}$ and $\frac{\partial f}{\partial t} = \frac{\partial^2 f}{\partial x^2}$. Those are the *one-way* wave equation and the *two-way* wave equation and the *heat* equation. Problem 42 says that if $\frac{\partial f}{\partial t} = \frac{\partial f}{\partial x}$ then automatically $\frac{\partial^2 f}{\partial t^2} = \frac{\partial^2 f}{\partial x^2}$. A one-way wave is a special case of a two-way wave.

4. Solve Problem 42. Then find $f(x, t)$ that satisfies the 2-way equation but not the 1-way equation.

- Suppose a particular function satisfies $f_t = f_x$. Take t-derivatives to get $f_{tt} = f_{xt}$. Take x-derivatives to get $f_{tx} = f_{xx}$. The mixed derivatives agree for *any* smooth function: $f_{xt} = f_{tx}$. Therefore $f_{tt} = f_{xx}$.

Example of a 1-way wave: $f = (x+t)^2$. The function $f = (x-t)^2$ does *not* satisfy the 1-way equation, because $f_x = 2(x-t)$ and $f_t = -2(x-t)$. It satisfies the ***other-way*** wave equation $f_t = -f_x$ with a minus sign. But this is enough for the 2-way equation because $f_{xx} = 2$ and $f_{tt} = 2$.

In general $F(x+t)$ solves the one-way equation, $G(x-t)$ solves the other-way equation, and their sum $F + G$ solves the two-way equation.

Read-throughs and selected even-numbered solutions :

The **partial** derivative $\partial f/\partial y$ comes from fixing **x** and moving **y**. It is the limit of $(\mathbf{f(x, y+\Delta y) - f(x, y)})/\mathbf{\Delta y}$. If $f = e^{2x}\sin y$ then $\partial f/\partial x = \mathbf{2e^{2x}\sin y}$ and $\partial f/\partial y = \mathbf{e^{2x}\cos y}$. If $f = (x^2+y^2)^{1/2}$ then $f_x = \mathbf{x/(x^2+y^2)^{1/2}}$ and $f_y = \mathbf{y/(x^2+y^2)^{1/2}}$. At (x_0, y_0) the partial derivative f_x is the ordinary derivative of the **partial** function $f(x, y_0)$. Similarly f_y comes from $\mathbf{f(x_0, y)}$. Those functions are cut out by vertical planes $x = x_0$ and $\mathbf{y = y_0}$, while the level curves are cut out by **horizontal** planes.

The four second derivatives are $f_{xx}, \mathbf{f_{xy}, f_{yx}, f_{yy}}$. For $f = xy$ they are $\mathbf{0, 1, 1, 0}$. For $f = \cos 2x \cos 3y$ they are $\mathbf{-4\cos 2x \cos y, 6\sin 2x}\sin \mathbf{3y}, \mathbf{-9\cos 2x \cos 3y}$. In those examples the derivatives $\mathbf{f_{xy}}$ and $\mathbf{f_{yx}}$ are the same. That is always true when the second derivatives are **continuous.** At the origin, $\cos 2x \cos 3y$ is curving **down** in the x and y directions, while xy goes **up** in the 45° direction and **down** in the −45° direction.

8 $\frac{\partial f}{\partial x} = \frac{1}{x+2y}, \frac{\partial f}{\partial y} = \frac{2}{x+2y}$

18 $f_{xx} = n(n-1)(x+y)^{n-2} = f_{xy} = f_{yx} = f_{yy}!$

20 $f_{xx} = \frac{2}{(x+iy)^3}, f_{xy} = f_{yx} = \frac{2i}{(x+iy)^3}, f_{yy} = \frac{2i^2}{(x+iy)^3} = \frac{-2}{(x+iy)^3}$ Note $f_{xx} + f_{yy} = 0$.

28 $\frac{\partial f}{\partial x} = -v(x)$ and $\frac{\partial f}{\partial y} = v(y)$.

36 $f_x = \frac{1}{\sqrt{t}}(\frac{-2x}{4t})e^{-x^2/4t}$. Then $f_{xx} = f_t = \frac{-1}{2t^{3/2}}e^{-x^2/4t} + \frac{x^2}{4t^{5/2}}e^{-x^2/4t}$.

38 $e^{-m^2t-n^2t}\sin mx \cos ny$ solves $f_t = f_{xx} + f_{yy}$. Also $f = \frac{1}{t}e^{-(x^2+y^2)/4t}$ has $f_t = f_{xx} + f_{yy} = (-\frac{1}{t^2} + \frac{x^2+y^2}{4t^3})e^{-(x^2+y^2)/4t}$.

50 Along $y = mx$ the function is $\frac{mx^3}{x^4+m^2x^2} \to 0$ (the ratio is near $\frac{mx^3}{m^2x^2}$ for small x). But on the parabola $y = x^2$ the function is $\frac{x^4}{2x^4} = \mathbf{\frac{1}{2}}$. So this function $f(x, y)$ has **no limit:** not continuous at (0,0).

13.3 Tangent Planes and Linear Approximations (page 488)

A smooth curve has tangent lines. The equation of the line uses the derivative (the slope). *A smooth surface has tangent planes.* The equation of the plane uses *two partial derivatives* f_x and f_y (two slopes). Compare

$$\textbf{line} \quad y - f(a) = f'(a)(x-a) \qquad \text{with} \qquad z - f(a,b) = f_x(a,b)(x-a) + f_y(a,b)(y-b) \quad \textbf{plane.}$$

These are *linear equations.* On the left is $y = mx$ + constant. On the right is $z = Mx + Ny$ + constant. Linear equations give lines in the xy plane, and they give planes in xyz space. The nice thing is that the first slope $M = \partial f/\partial x$ stays completely separate from the second slope $N = \partial f/\partial y$.

I will follow up that last sentence. Suppose we change a by Δx and b by Δy. The basepoint is (a,b) and the movement is to $(a+\Delta x,\ b+\Delta y)$. Knowing the function f and its derivatives at the basepoint, we can predict the function (*linear approximation*) at the nearby point. In one variable we follow the tangent line to $f(a)+f'(a)\Delta x$. In two variables we follow the tangent plane to the nearby point:

$$f(a+\Delta x, b+\Delta y) \ \approx\ f(a,b) + \Delta x\ f_x(a,b) + \Delta y\ f_y(a,b).$$

We add on *two linear corrections,* in the x and y directions. Often these formulas are written with x instead of a and y instead of b. The movement is from $f(x,y)$ to $f(x+\Delta x, y+\Delta y)$. The change is $\Delta x\ f_x + \Delta y\ f_y$.

1. Estimate the change in $f(x,y) = x^3y^4$ when you move from (1,1) to $(1+\Delta x, 1+\Delta y)$.

   - The x-derivative is $f_x = 3x^2y^4 = 3$ at the basepoint (1,1). The y-derivative is $f_y = 4x^3y^3 = 4$ at the basepoint. The change Δf is approximately $f_x\Delta x + f_y\Delta y$. This is $3\Delta x + 4\Delta y$:

   $$f(x,y) = (1+\Delta x)^3(1+\Delta y)^4 \approx 1 + 3\Delta x + 4\Delta y.$$

   On the left, the high powers $(\Delta x)^3(\Delta y)^4$ would multiply. But the lowest powers Δx and Δy just add. You can see that if you write out $(1+\Delta x)^3$ and $(1+\Delta y)^4$ and start multiplying:

   $$(1+3\Delta x + 3(\Delta x)^2 + (\Delta x)^3)(1+4\Delta y+\cdots) = 1 + 3\Delta x + 4\Delta y + \text{ higher terms.}$$

   These higher terms come into the complete Taylor series. The constant and linear terms are the start of that series. They give the linear approximation.

2. Find the equation of the *tangent plane* to the surface $z = x^3y^4$ at $(x,y) = (1,1)$.

   The plane is $z - 1 = 3(x-1) + 4(y-1)$. If $x-1$ is Δx and $y-1$ is Δy, this is $z = 1 + 3\Delta x + 4\Delta y$. Same as Question 1. The tangent plane gives the linear approximation!

Some surfaces do not have "explicit equations" $z = f(x,y)$. That gives one z for each x and y. A more general equation is $F(x,y,z) = 0$. An example is the sphere $F = x^2+y^2+z^2-4 = 0$. We could solve to find $z = \sqrt{4-x^2-y^2}$ and also $z = -\sqrt{4-x^2-y^2}$. These are *two* surfaces of the type $z = f(x,y)$, to give the top half and bottom half of the sphere. In other examples it is difficult or impossible to solve for z and we really want to stay with the *"implicit equation"* $F(x,y,z) = 0$.

How do you find tangent planes and linear approximations for $F(x,y,z) = 0$? Problem 3 shows by example.

3. The surface $xz + 2yz - 10 = 0$ goes through the point $(x_0, y_0, z_0) = (1,2,2)$. Find the tangent plane and normal vector. Estimate z when $x = 1.1$ and $y = 1.9$.

- Main idea: *Go ahead and differentiate* $F = xz + 2yz - 10$. Not only x and y derivatives, also z :

$$\frac{\partial F}{\partial x} = z = 2 \quad \text{and} \quad \frac{\partial F}{\partial y} = 2z = 4 \quad \text{and} \quad \frac{\partial F}{\partial z} = x + 2y = 5 \quad \text{at the basepoint } (1,2,2).$$

The tangent plane is $2(x-1) + 4(y-2) + 5(z-2) = 0$. The normal vector is $N = (2,4,5)$. Notice how F_x, F_y, and F_z multiply Δx and Δy and Δz. The total change is ΔF which is zero (because F is constant: the surface is $F = 0$). A linear approximation stays on the tangent plane! So if you know $x = 1.1$ and $y = 1.9$ you can solve for z on the plane:

$$2(1.1-1) + 4(1.9-2) + 5(z-2) = 0 \text{ gives } z = 2 - \frac{2(.1)}{5} - \frac{4(-.1)}{5}. \text{ This is } z = z_0 - \frac{F_x}{F_z}\Delta x - \frac{F_y}{F_z}\Delta y.$$

I would memorize the tangent plane formula, which is $(F_x)(x - x_0) + (F_y)(y - y_0) + (F_z)(z - z_0) = 0$.

In this example you *could* solve $F = xz + 2yz - 10 = 0$ to find z. The explicit equation $z = f(x,y)$ is $z = \frac{10}{x+2y}$. Its x and y derivatives give the same tangent plane as the x, y, z derivatives of F.

The last topic in this important section is ***Newton's method.*** It deals with *two* functions $g(x,y)$ and $h(x,y)$. Solving $g(x,y) = 0$ should give a curve, solving $h(x,y) = 0$ should give another curve, and solving *both* equations should give the point (or points) where the two curves meet. When the functions are complicated – they usually are – we "linearize." Instead of $g(x,y) = 0$ and $h(x,y) = 0$ Newton solves

$$g(x_0, y_0) + (\frac{\partial g}{\partial x})_0(\Delta x) + (\frac{\partial g}{\partial y})_0(\Delta y) = 0$$

$$h(x_0, y_0) + (\frac{\partial h}{\partial x})_0(\Delta x) + (\frac{\partial h}{\partial y})_0(\Delta y) = 0.$$

Those are linear equations for Δx *and* Δy. We move to the new basepoint $(x_1, y_1) = (x_0 + \Delta x, y_0 + \Delta y)$ and start again. Newton's method solves many linear equations instead of $g(x,y) = 0$ and $h(x,y) = 0$.

4. Take one Newton step from $(x_0, y_0) = (1,2)$ toward the solution of $g = xy - 3 = 0$ and $h = x + y - 2 = 0$.

   - The partial derivatives at the basepoint (1,2) are $g_x = y = 2$ and $g_y = x = 1$ and $h_x = 1$ and $h_y = 1$. The functions themselves are $g = -1$ and $h = 1$. Newton solves the two linear equations above (tangent equations) for Δx and Δy:

$$\begin{aligned} -1 + 2\Delta x + \Delta y &= 0 \\ 1 + \Delta x + \Delta y &= 0 \end{aligned} \quad \text{give} \quad \begin{aligned} \Delta x &= 2 \\ \Delta y &= -3 \end{aligned} \quad \text{The new guess is} \quad \begin{aligned} x_1 &= x_0 + \Delta x = 3 \\ y_1 &= y_0 + \Delta y = -1. \end{aligned}$$

The new point $(3,-1)$ exactly solves $h = x + y - 2 = 0$. It misses badly on $g = xy - 3 = 0$. This surprised me because the method is usually terrific. Then I tried to solve the equations exactly by algebra.

Substituting $y = 2 - x$ from the second equation into the first gave $x(2-x) - 3 = 0$. This is a quadratic $x^2 - 2x + 3 = 0$. But it has no real solutions! Both roots are complex numbers. Newton never had a chance.

Read-throughs and selected even-numbered solutions :

The tangent line to $y = f(x)$ is $y - y_0 = \mathbf{f'(x_0)(x - x_0)}$. The tangent plane to $w = f(x,y)$ is $w - w_0 = \mathbf{(\partial f/\partial x)_0(x - x_0) + (\partial f/\partial y)_0(y - y_0)}$. The normal vector is $\mathbf{N} = \mathbf{(f_x, f_y, -1)}$. For $w = x^3 + y^3$ the tangent equation at (1,1,2) is $\mathbf{w - 2 = 3(x-1) + 3(y-1)}$. The normal vector is $\mathbf{N} = \mathbf{(3,3,-1)}$. For a sphere, the direction of $\mathbf{N}$ is **out from the origin.**

The surface given implicitly by $F(x,y,z) = c$ has tangent plane with equation $(\partial F/\partial x)_0(x - x_0) + \mathbf{(\partial F/\partial y)(y - y_0) + (\partial F/\partial z)_0(z - z_0) = 0}$. For $xyz = 6$ at (1,2,3) the tangent plane has the equation

$6(x-1)+3(y-2)+2(z-3)=0$. On that plane the differentials satisfy $6dx+3dy+2dz=0$. The differential of $z=f(x,y)$ is $dz=f_x dx+f_y dy$. This holds exactly on the tangent plane, while $\Delta z\approx f_x\Delta x+f_y\Delta y$ holds approximately on the **surface.** The height $z=3x+7y$ is more sensitive to a change in **y** than in x, because the partial derivative $\partial z/\partial y=7$ is larger than $\partial z/\partial x=3$.

The linear approximation to $f(x,y)$ is $f(x_0,y_0)+(\partial f/\partial x)_0(x-x_0)+(\partial f/\partial y)_0(y-y_0)$. This is the same as $\Delta f\approx(\partial f/\partial x)\Delta x+(\partial f/\partial y)\Delta y$. The error is of order $(\Delta x)^2+(\Delta y)^2$. For $f=\sin xy$ the linear approximation around (0,0) is $f_L=0$. We are moving along the **tangent plane** instead of the **surface.** When the equation is given as $F(x,y,z)=c$, the linear approximation is $F_x\Delta x+F_y\Delta y+F_z\Delta z=0$.

Newton's method solves $g(x,y)=0$ and $h(x,y)=0$ by a **linear** approximation. Starting from x_n,y_n the equations are replaced by $g_x\Delta x+g_y\Delta y=-g(x_n,y_n)$ and $h_x\Delta x+h_y\Delta y=-h(x_n,y_n)$. The steps Δx and Δy go to the next point (x_{n+1},y_{n+1}). Each solution has a basin of **attraction.** Those basins are likely to be **fractals.**

8 $N=8\pi i+4\pi j-k$; $8\pi(r-2)+4\pi(h-2)=V-8\pi$

12 $N_1=2i+4j-k$ and $N_2=2i+6j-k$ give $v=\begin{vmatrix} i & j & k\\ 2 & 4 & -1\\ 2 & 6 & -1\end{vmatrix}=2i+4k$ tangent to both surfaces

14 The direction of N is $2xy^2 i+2x^2y j-k=8i+4j-k$. So the line through (1,2,4) has $x=1+8t, y=2+4t, z=4-t$.

18 $df=yz\,dx+xz\,dy+xy\,dz$.

32 $\frac{3}{4}\Delta x-\Delta y=\frac{3}{8}$ and $-\Delta x+\frac{3}{4}\Delta y=\frac{3}{8}$ give $\Delta x=\Delta y=-\frac{3}{2}$. The new point is $(-1,-1)$, an exact solution. The point $(\frac{1}{2},\frac{1}{2})$ is in the gray band (upper right in Figure 13.11a) or the blue band on the front cover.

38 A famous fractal shows the three basins of attraction – see almost any book on fractals. Remarkable property of the boundaries points between basins: **they touch all three basins!** Try to draw 3 regions with this property.

13.4 Directional Derivatives and Gradients (page 495)

The partial derivatives $\partial f/\partial x$ and $\partial f/\partial y$ are directional derivatives, in special directions. They give the slope in directions $u=(1,0)$ and $u=(0,1)$, parallel to the x and y axes. From those two partial derivatives we can quickly find the derivative in any other direction $u=(\cos\theta,\sin\theta)$:

$$\textbf{directional derivative}\quad D_u f=\left(\frac{\partial f}{\partial x}\right)\cos\theta+\left(\frac{\partial f}{\partial y}\right)\sin\theta.$$

It makes sense that the slope of the surface $z=f(x,y)$, climbing at an angle between the x direction and y direction, should be a combination of slopes $\partial f/\partial x$ and $\partial f/\partial y$. That slope formula is really a dot product between the ***direction vector*** u and the ***derivative vector*** (called the ***gradient***):

$$\text{gradient}=\left(\frac{\partial f}{\partial x},\frac{\partial f}{\partial y}\right)=\nabla f\qquad\text{direction}=(\cos\theta,\sin\theta)=u\qquad\text{directional derivative}=\nabla f\cdot u.$$

1. Find the gradient of $f(x,y) = 4x + y - 7$. Find the derivative in the 45° direction, along the line $y = x$.

   - The partial derivatives are $f_x = 4$ and $f_y = 1$. So the gradient is the vector $\nabla f = (4, 1)$.
   - Along the 45° line $y = x$, the direction vector is $\mathbf{u} = (\cos\frac{\pi}{4}, \sin\frac{\pi}{4})$. This is $\mathbf{u} = (\frac{\sqrt{2}}{2}, \frac{\sqrt{2}}{2})$. The dot product $\nabla f \cdot \mathbf{u} = 4\frac{\sqrt{2}}{2} + 1\frac{\sqrt{2}}{2} = 5\frac{\sqrt{2}}{2}$ is $D_{\mathbf{u}}f$, the directional derivative.

2. Which direction gives the largest value of $D_{\mathbf{u}}f$? This is the ***steepest direction***.

   - The derivative is the dot product of $\nabla f = (4,1)$ with $\mathbf{u} = (\cos\theta, \sin\theta)$. A dot product equals the length $|\nabla f| = \sqrt{4^2 + 1^2} = \sqrt{17}$ times the length $|\mathbf{u}| = 1$ times the *cosine of the angle* between ∇f and $\mathbf{u}$. To maximize the dot product and maximize that cosine, ***choose* u *in the same direction as*** ∇f. Make $\mathbf{u}$ a unit vector:

$$\mathbf{u} = \frac{\nabla f}{|\nabla f|} = (\frac{4}{\sqrt{17}}, \frac{1}{\sqrt{17}}) \quad \text{and} \quad \nabla f \cdot \mathbf{u} = 4(\frac{4}{\sqrt{17}}) + 1(\frac{1}{\sqrt{17}}) = \frac{17}{\sqrt{17}} = \sqrt{17}.$$

   This is the general rule: The steepest direction is parallel to the gradient $\nabla f = (f_x, f_y)$. The steepness (the slope) is $|\nabla f| = \sqrt{f_x^2 + f_y^2}$. This is the largest value of $D_{\mathbf{u}}f$.

3. Find a function $f(x,y)$ for which the steepest direction is the x direction.

   - The question is asking for $\frac{\partial f}{\partial y} = 0$. Then the gradient is $(\frac{\partial f}{\partial x}, 0)$. It points in the x-direction. The maximum slope is $\sqrt{(\frac{\partial f}{\partial x})^2 + 0^2}$ which is just $|\frac{\partial f}{\partial x}|$.

   The answer is: Don't let f depend on y. Choose $f = x$ or $f = e^x$ or any $f(x)$. *The slope in the y-direction is zero!* The steepest slope is in the pure x-direction. At every in-between direction the slope is a mixture of $\frac{\partial f}{\partial x}$ and 0. The steepest slope is $|\frac{\partial f}{\partial x}|$ with no zero in the mixture.

$\nabla f \cdot \mathbf{u}$ is the directional derivative along a straight line (in the direction $\mathbf{u}$). What if we travel along a curve? The value of $f(x,y)$ changes as we travel, and calculus asks how fast it changes. This is an "instantaneous" question, at a single point on the curved path. *At each point the path direction is the tangent direction.* So replace the fixed vector $\mathbf{u}$ by the tangent vector $\mathbf{T}$ at that point: Slope of $f(x,y)$ going along path $= \nabla f \cdot \mathbf{T}$.

The tangent vector $\mathbf{T}$ was in Section 12.1. We are given $x(t)$ and $y(t)$, the position as we move along the path. The derivative $(\frac{dx}{dt}, \frac{dy}{dt})$ is the velocity vector $\mathbf{v}$. This is along the tangent direction (parallel to $\mathbf{T}$), but $\mathbf{T}$ is required to be a unit vector. So divide $\mathbf{v}$ by its length which is the speed $|\mathbf{v}| = \sqrt{(\frac{dx}{dt})^2 + (\frac{dy}{dt})^2} = ds/dt$:

$$\mathbf{v} = (\tfrac{dx}{dt}, \tfrac{dy}{dt}) \text{ gives } \nabla f \cdot \mathbf{v} = \frac{\partial f}{\partial x}\frac{\partial x}{\partial t} + \frac{\partial f}{\partial y}\frac{\partial y}{\partial t}. \text{ This is } \frac{df}{dt}.$$

$$\frac{\mathbf{v}}{|\mathbf{v}|} = \frac{\mathbf{v}}{ds/dt} = \mathbf{T} \text{ gives } \nabla f \cdot \mathbf{T}. \text{ This is } \frac{df/dt}{ds/dt} = \frac{df}{ds}.$$

The speed is divided out of the slope df/ds. The speed is *not* divided out of the rate of change df/dt. One says how steeply you climb. The other says how *fast* you climb.

4. How steeply do you climb and how fast do you climb on a roller-coaster of height $f(x,y) = 2x + y$? You travel around the circle $x = \cos 4t$, $y = \sin 4t$ with velocity $\mathbf{v} = (-4\sin 4t, 4\cos 4t)$ and speed $|\mathbf{v}| = 4$.

   - The gradient of $f = 2x + y$ is $\nabla f = (2,1)$. The tangent vector is $\mathbf{T} = \frac{\mathbf{v}}{|\mathbf{v}|} = (-\sin 4t, \ \cos 4t)$.

$$\text{Slope of path} = \nabla f \cdot \mathbf{T} = -2\sin 4t + \cos 4t \qquad \text{Maximum slope } \sqrt{5}.$$

$$\text{Climbing rate} = \nabla f \cdot \mathbf{v} = -8\sin 4t + 4\cos 4t \qquad \text{Maximum rate } 4\sqrt{5}.$$

   How fast you climb = (how steeply you climb) × (how fast you travel).

Read-throughs and selected even-numbered solutions :

$D_{\mathbf{u}}f$ gives the rate of change of $\mathbf{f(x,y)}$ in the direction $\mathbf{u}$. It can be computed from the two derivatives $\partial\mathbf{f}/\partial\mathbf{x}$ **and** $\partial\mathbf{f}/\partial\mathbf{y}$ in the special directions **(1,0) and (0,1).** In terms of u_1, u_2 the formula is $D_{\mathbf{u}}f = \mathbf{f_x u_1 + f_y u_2}$. This is a **dot** product of $\mathbf{u}$ with the vector $(\mathbf{f_x}, \mathbf{f_y})$, which is called the **gradient.** For the linear function $f = ax + by$, the gradient is grad $f = (\mathbf{a}, \mathbf{b})$ and the directional derivative is $D_{\mathbf{u}}f = (\mathbf{a}, \mathbf{b}) \cdot \mathbf{u}$.

The gradient $\nabla f = (f_x, f_y)$ is not a vector in **three** dimensions, it is a vector in the **base plane.** It is perpendicular to the **level** lines. It points in the direction of **steepest** climb. Its magnitude $|\text{grad} f|$ is **the steepness** $\sqrt{\mathbf{f_x^2 + f_y^2}}$. For $f = x^2 + y^2$ the gradient points **out from the origin** and the slope in that steepest direction is $|(\mathbf{2x}, \mathbf{2y})| = \mathbf{2r}$.

The gradient of $f(x,y,z)$ is $(\mathbf{f_x}, \mathbf{f_y}, \mathbf{f_z})$. This is different from the gradient on the surface $F(x,y,z) = 0$, which is $-(F_x/F_z)\mathbf{i} - (F_y/F_z)\mathbf{j}$. Traveling with velocity $\mathbf{v}$ on a curved path, the rate of change of f is $df/dt = (\mathbf{grad\ f}) \cdot \mathbf{v}$. When the tangent direction is $\mathbf{T}$, the slope of f is $df/ds = (\mathbf{grad\ f}) \cdot \mathbf{T}$. In a straight direction $\mathbf{u}$, df/ds is the same as **the directional derivative** $D_{\mathbf{u}}f$.

12 In one dimension the gradient of $f(x)$ is $\frac{df}{dx}\mathbf{i}$. The two possible directions are $\mathbf{u} = \mathbf{i}$ and $\mathbf{u} = -\mathbf{i}$. The two directional derivatives are $+\frac{\mathbf{df}}{\mathbf{dx}}$ and $-\frac{\mathbf{df}}{\mathbf{dx}}$. The normal vector $\mathbf{N}$ is $\frac{df}{dx}\mathbf{i} - \mathbf{j}$.

14 Here $f = 2x$ above the line $y = 2x$ and $f = y$ below that line. The two pieces agree on the line. Then grad $f = 2\mathbf{i}$ above and grad $f = \mathbf{j}$ below. Surprisingly f increases fastest *along* the line, which is the direction $\mathbf{u} = \frac{1}{\sqrt{5}}(\mathbf{i} + 2\mathbf{j})$ and gives $D_{\mathbf{u}}f = \frac{2}{\sqrt{5}}$.

28 (a) **False** because $f + C$ has the same gradient as f (b) **True** because the line direction $(1,1,-1)$ is also the normal direction $\mathbf{N}$ (c) **False** because the gradient is in 2 dimensions.

30 $\theta = \tan^{-1}\frac{y}{x}$ has grad $\theta = \left(\frac{-y/x^2}{1+(y/x)^2}, \frac{1/x}{1+(y/x)^2}\right) = \frac{(-y,x)}{x^2+y^2}$. The unit vector in this direction is $\mathbf{T} = \left(\frac{-y}{\sqrt{x^2+y^2}}, \frac{x}{\sqrt{x^2+y^2}}\right)$. Then grad $\theta \cdot T = \frac{y^2+x^2}{(x^2+y^2)^{3/2}} = \frac{1}{r}$.

34 The gradient is $(2ax + c)\mathbf{i} + (2by + d)\mathbf{j}$. The figure shows $c = 0$ and $d \approx \frac{1}{3}$ at the origin. Then $b \approx \frac{1}{3}$ from the gradient at (0,1). Then $a \approx -\frac{1}{4}$ from the gradient at (2,0). The function $-\frac{1}{4}x^2 + \frac{1}{3}y^2 + \frac{1}{3}y$ has hyperbolas opening upwards as level curves.

44 $\mathbf{v} = (2t, 0)$ and $\mathbf{T} = (1,0)$; grad $f = (y,x)$ so $\frac{df}{dt} = 2ty = 6t$ and $\frac{df}{ds} = y = 3$.

48 $D^2 = (x-1)^2 + (y-2)^2$ has $2D\frac{\partial D}{\partial x} = 2(x-1)$ or $\frac{\partial D}{\partial x} = \frac{x-1}{D}$. Similarly $2D\frac{\partial D}{\partial y} = 2(y-2)$ and $\frac{\partial D}{\partial y} = \frac{y-2}{D}$. Then $|\text{grad } D| = (\frac{x-1}{D})^2 + (\frac{y-2}{D})^2 = 1$. The graph of D is a 45° cone with its vertex at (1,2).

13.5 The Chain Rule (page 503)

Chain Rule 1 On the surface $z = g(x, y)$ the partial derivatives of $f(z)$ are $\frac{\partial f}{\partial x} = \frac{df}{dz}\frac{\partial z}{\partial x}$ and $\frac{\partial f}{\partial y} = \frac{df}{dz}\frac{\partial z}{\partial y}$. $z = x^2 + y^2$ gives a bowl. Then $f(z) = \sqrt{z} = \sqrt{x^2+y^2}$ gives a sharp-pointed cone. The slope of the cone in the x-direction is

$$\frac{\partial f}{\partial x} = \frac{df}{dz}\frac{\partial z}{\partial x} = \left(\frac{1}{2}z^{-1/2}\right)(2x) = \frac{x}{\sqrt{z}} = \frac{x}{\sqrt{x^2+y^2}}.$$

Check that by directly taking the x-derivative of $f(g(x, y)) = \sqrt{x^2+y^2}$.

Chain Rule 2 For $z = f(x, y)$ on the curve $x = x(t)$ and $y = y(t)$ the t-derivative is $\frac{\partial z}{\partial t} = \frac{\partial f}{\partial x}\frac{dx}{dt} + \frac{\partial f}{\partial y}\frac{dy}{dt}$. This is exactly the climbing rate from the previous section 13.4.

Chain Rule 3 For $z = f(x, y)$ when $x = x(t, u)$ and $y = y(t, u)$ the t-derivative is $\frac{\partial z}{\partial t} = \frac{\partial f}{\partial x}\frac{\partial x}{\partial t} + \frac{\partial f}{\partial y}\frac{\partial y}{\partial t}$. This combines Rule 1 and Rule 2. The outer function f has two variables x, y as in Rule 2. The inner functions x and y have two variables as in Rule 1. So all derivatives are partial derivatives. But notice:

$$\frac{\partial z}{\partial u} \text{ is not } \frac{\partial z}{\partial x}\frac{\partial x}{\partial u}. \text{ The correct rule is } \frac{\partial z}{\partial x}\frac{\partial x}{\partial u} + \frac{\partial z}{\partial y}\frac{\partial y}{\partial u}$$

1. A change in u produces a change in $x = tu$ and $y = t/u$. These produce a change in $z = 3x + 2y$. Find $\partial z/\partial u$.

$$\frac{\partial z}{\partial u} \text{ is } \frac{\partial z}{\partial x}\frac{\partial x}{\partial u} + \frac{\partial z}{\partial y}\frac{\partial y}{\partial u} = (3)(t) + (2)\left(\frac{-t}{u^2}\right).$$

2. When would Rule 3 reduce to Rule 2? • The inner functions x and y depend only on t, not u.

Please read the paradox on page 501. Its main point is: For partial derivatives you must know which variable is moving and also which variable is *not* moving.

Read-throughs and selected even-numbered solutions :

The chain rule applies to a function of a **function.** The x derivative of $f(g(x, y))$ is $\partial f/\partial x = (\partial \mathbf{f}/\partial \mathbf{g})(\partial \mathbf{g}/\partial \mathbf{x})$. The y derivative is $\partial f/\partial y = (\partial \mathbf{f}/\partial \mathbf{g})(\partial \mathbf{g}/\partial \mathbf{y})$. The example $f = (x+y)^n$ has $g = \mathbf{x} + \mathbf{y}$. Because $\partial g/\partial x = \partial g/\partial y$ we know that $\partial \mathbf{f}/\partial \mathbf{x} = \partial \mathbf{f}/\partial \mathbf{y}$. This **partial** differential equation is satisfied by any function of $x + y$.

Along a path, the derivative of $f(x(t), y(t))$ is $df/dt = (\partial \mathbf{f}/\partial \mathbf{x})(\mathbf{dx/dt}) + (\partial \mathbf{f}/\partial \mathbf{y})(\mathbf{dy/dt})$. The derivative of $f(x(t), y(t), z(t))$ is $\mathbf{f_x x_t} + \mathbf{f_y y_t} + \mathbf{f_z z_t}$. If $f = xy$ then the chain rule gives $df/dt = \mathbf{y\,dx/dt} + \mathbf{x\,dy/dt}$. That is the same as the **product** rule! When $x = u_1 t$ and $y = u_2 t$ the path is **a straight line.** The chain rule for $f(x, y)$ gives $df/dt = \mathbf{f_x u_1} + \mathbf{f_y u_2}$. That is the **directional** derivative $D_{\mathbf{u}} f$.

The chain rule for $f(x(t, u), y(t, u))$ is $\partial f/\partial t = (\partial \mathbf{f}/\partial \mathbf{x})(\partial \mathbf{x}/\partial \mathbf{t}) + (\partial \mathbf{f}/\partial \mathbf{y})(\partial \mathbf{y}/\partial \mathbf{t})$. We don't write df/dt because **f also depends on u.** If $x = r\cos\theta$ and $y = r\sin\theta$, the variables t, u change to **r and** θ. In this case $\partial f/\partial r = (\partial \mathbf{f}/\partial \mathbf{x})\cos\theta + (\partial \mathbf{f}/\partial \mathbf{y})\,\mathbf{sin}\,\theta$ and $\partial f/\partial\theta = (\partial \mathbf{f}/\partial \mathbf{x})(-\mathbf{r}\,\mathbf{sin}\,\theta) + (\mathbf{df/dy})(\mathbf{r}\,\mathbf{cos}\,\theta)$. That connects the derivatives in **rectangular** and **polar** coordinates. The difference between $\partial r/\partial x = x/r$ and $\partial r/\partial x = 1/\cos\theta$ is because **y** is constant in the first and θ is constant in the second.

With a relation like $xyz = 1$, the three variables are **not** independent. The derivatives $(\partial f/\partial x)_y$ and $(\partial f/\partial x)_z$ and $(\partial f/\partial x)$ mean **that y is held constant, and z is constant, and both are constant.** For

$f = x^2 + y^2 + z^2$ with $xyz = 1$, we compute $(\partial f/\partial x)_z$ from the chain rule $\partial \mathbf{f}/\partial \mathbf{x} + (\partial \mathbf{f}/\partial \mathbf{y})(\partial \mathbf{y}/\partial \mathbf{x})$. In that rule $\partial z/\partial x = \mathbf{-1/x^2y}$ from the relation $xyz = 1$.

4 $f_x = \frac{1}{x+7y}$ and $f_y = \frac{7}{x+7y}$; $7f_x = f_y$.

6 $\frac{df}{dt} = \frac{\partial f}{\partial x}\frac{dx}{dt} + \frac{\partial f}{\partial y}\frac{dy}{dt}$ is the **product rule** $\mathbf{y\frac{dx}{dt} + x\frac{dy}{dt}}$. In terms of u and v this is $\frac{d}{dt}(uv) = v\frac{du}{dt} + u\frac{dv}{dt}$.

12 (a) $f_r = 2re^{2i\theta}$, $f_{rr} = 2e^{2i\theta}$, $f_{\theta\theta} = r^2(2i)^2e^{2i\theta}$ and $f_{rr} + \frac{f_r}{r} + \frac{f_{\theta\theta}}{r^2} = 0$. Take real parts *throughout* to find the same for $r^2 \cos 2\theta$ (and imaginary parts for $r^2 \sin 2\theta$). (b) Any function $f(re^{i\theta})$ has $f_r = e^{i\theta}f'(re^{i\theta})$ and $f_{rr} = (e^{i\theta})^2 f''(re^{i\theta})$ and $f_\theta = ire^{i\theta}f'(re^{i\theta})$ and $f_{\theta\theta} = i^2re^{i\theta}f' + (ire^{i\theta})^2 f''$. Any $f(re^{i\theta})$ or any $f(x+iy)$ will satisfy the polar or rectangular form of Laplace's equation.

16 Since $\frac{x}{y} = \frac{1}{2}$ we must find $\frac{df}{dt} = 0$. The chain rule gives $\frac{1}{y}\frac{dx}{dt} - \frac{x}{y^2}\frac{dy}{dt} = \frac{1}{2e^t}(e^t) - \frac{e^t}{4e^{2t}}(2e^t) = 0$.

32 $\frac{\partial r}{\partial x} = \frac{x}{r}$ and then $\frac{\partial^2 r}{\partial y\, \partial x} = -\frac{x}{r^2}\frac{\partial r}{\partial y} = -\frac{x}{r^2}\frac{y}{r} = -\frac{xy}{r^3}$.

40 (a) $\frac{\partial f}{\partial x} = \mathbf{2x}$ (b) $f = x^2 + y^2 + (x^2+y^2)^2$ so $\frac{\partial f}{\partial x} = \mathbf{2x + 4x(x^2+y^2)}$ (c) $\frac{\partial f}{\partial x} + \frac{\partial f}{\partial z}\frac{\partial z}{\partial x} = 2x + 2z(2x) = \mathbf{2x + 4x(x^2+y^2)}$ (d) y is constant for $(\frac{\partial f}{\partial x})_y$.

13.6 Maxima, Minima, and Saddle Points (page 512)

A one-variable function $f(x)$ reaches its maximum and minimum at three types of critical points:

1. Stationary points where $\dfrac{df}{dx} = 0$ **2.** Rough points **3.** Endpoints (possibly at ∞ or $-\infty$).

A two-variable function $f(x, y)$ has the same three possible types of critical points:

1. Stationary points where $\dfrac{\partial f}{\partial x} = 0$ and $\dfrac{\partial f}{\partial y} = 0$ **2.** Rough points **3.** Boundary points.

The stationary points come first. Notice that they involve two equations (both partial derivatives are zero). There are two unknowns (the coordinates x and y of the stationary point). The tangent is horizontal as usual, but it is a tangent *plane* to the surface $z = f(x, y)$.

It is harder to solve two equations than one. And the second derivative test (which was previously $f'' > 0$ for a minimum and $f'' < 0$ for a maximum) now involves all three derivatives f_{xx}, f_{yy}, and $f_{xy} = f_{yx}$:

$$\textit{Minimum}\ \begin{array}{rcl} f_{xx} & > & 0 \\ f_{xx}f_{yy} & > & (f_{xy})^2 \end{array} \qquad \textit{Maximum}\ \begin{array}{rcl} f_{xx} & < & 0 \\ f_{xx}f_{yy} & > & (f_{xy})^2 \end{array} \qquad \textit{Saddle}\ f_{xx}f_{yy} < (f_{xy})^2$$

When $f_{xx}f_{yy} = (f_{xy})^2$ the test gives no answer. This is like $f'' = 0$ for a one-variable function $f(x)$.

Our two-variable case really has a 2 by 2 matrix of second derivatives. Its determinant is the critical quantity $f_{xx}f_{yy} - (f_{xy})^2$. This pattern continues on to $f(x,y,z)$ or $f(x,y,z,t)$. Those have 3 by 3 and 4 by 4 matrices of second derivatives and we check 3 or 4 determinants. In linear algebra, a *positive definite second-derivative matrix* indicates that the stationary point is a minimum.

1. (13.6.26) Find the stationary points of $f(x, y) = xy - \frac{1}{4}x^4 - \frac{1}{4}y^4$ and decide between **min**, **max**, and **saddle**.

- The partial derivatives are $f_x = y - x^3$ and $f_y = x - y^3$. Set both derivatives to zero: $y = x^3$ and $x = y^3$ lead to $y = y^9$. This gives $y = 0,\ 1,$ or -1. Then $x = y^3$ gives $x = 0,\ 1,$ or -1. The stationary points are (0,0) and (1,1) and $(-1,-1)$. The second derivatives are $f_{xx} = -3x^2$ and $f_{yy} = -3y^2$ and $f_{xy} = 1$:

$(0,0)$ is a **saddle point** because $f_{xx}f_{yy} = (0)(0)$ is less than $(1)^2$

$(1,1)$ and $(-1,-1)$ are **maxima** because $f_{xx}f_{yy} = (-3)(-3)$ is greater than $(1)^2$ and $f_{xx} = -3$.

Solving $x = y^3$ and $y = x^3$ is our example of the ***two – variable Newton method*** in Section 11.3. This is really important in practice. For this function $xy - \frac{1}{4}x^4 - \frac{1}{4}y^4$ we found a saddle point and two maximum points. The *minimum* is at infinity. This counts as a "boundary point".

2. (This is Problem 13.6.56) Show that a solution to Laplace's equation $f_{xx} + f_{yy} = 0$ has no maximum or minimum stationary points. So where are the maximum and minimum of $f(x, y)$?

   - A maximum requires $f_{xx} < 0$. ***It also requires*** $f_{yy} < 0$. We didn't say that, but it follows from the requirement $f_{xx}f_{yy} > (f_{xy})^2$. The left side has to be positive, so f_{xx} and f_{yy} must have the same sign. If $f_{xx} + f_{yy} = 0$ this can't happen; stationary points must be saddle points (or f = constant). A *max* or *min* is impossible. Those must occur at ***rough points*** or ***boundary points***.

Example A $\quad f(x, y) = \ln(x^2 + y^2)$ has a minimum of $-\infty$ at $(x, y) = (0,0)$, since $\ln 0 = -\infty$. This is a rough point because $f_x = \frac{2x}{x^2+y^2}$ is unbounded. You could check Laplace's equation two ways. One is to compute $f_{xx} = \frac{2}{x^2+y^2} - (\frac{2x}{x^2+y^2})^2$. Also $f_{yy} = \frac{2}{x^2+y^2} - (\frac{2y}{x^2+y^2})^2$. Add to get zero. The other way is to write $f = \ln r^2 = 2\ln r$ in polar coordinates. Then $f_r = \frac{2}{r}$ and $f_{rr} = -\frac{2}{r^2}$. Substitute into the *polar* Laplace equation to get $f_{rr} + \frac{1}{r}f_r + \frac{1}{r^2}f_{\theta\theta} = 0$.

Example B $\quad f(x, y) = xy$ satisfies Laplace's equation because $f_{xx} + f_{yy} = 0 + 0$. The stationary point at the origin cannot be a max or min. It is a typical and famous saddle point: We find $f_{xy} = 1$ and then $f_{xx}f_{yy} = (0)(0) < (1)^2$. There are no rough points. The min and max must be at ***boundary points***.

Note: Possibly there are no restrictions on x and y. The boundary is *at infinity*. Then the max and min occur out *at infinity*. Maximum when x and y go to $+\infty$. Minimum when $x \to +\infty$ and $y \to -\infty$, because then $xy \to -\infty$. (Also max when x and y go to $-\infty$. Also min when $x \to -\infty$ and $y \to +\infty$).

Suppose x and y are restricted to stay in the square $1 \le x \le 2$ and $1 \le y \le 2$. Then the max and min of xy occur on the ***boundary of the square***. Maximum at $x = y = 2$. Minimum at $x = y = 1$. In a way those are "rough points of the boundary," because they are sharp corners.

Suppose x and y are restricted to stay in the unit circle $x = \cos t$ and $y = \sin t$. The maximum of xy is on the boundary (where $xy = \cos t \sin t$). The circle has no rough points. The maximum is at the 45° angle $t = \frac{\pi}{4}$ (also at $t = \frac{5\pi}{4}$). At those points $xy = \cos\frac{\pi}{4}\sin\frac{\pi}{4} = \frac{1}{2}$. To emphasize again: This maximum occurred on the *boundary* of the circle.

Finally we call attention to the ***Taylor Series*** for a function $f(x, y)$. The text chose (0,0) as basepoint. The whole idea is to match each derivative $(\frac{\partial}{\partial x})^n(\frac{\partial}{\partial y})^m f(x, y)$ at the basepoint by one term in the Taylor series. Since $\frac{x^n y^m}{n!m!}$ has derivative equal to 1, multiply this standard power by the required derivative to find the correct term in the Taylor Series.

When the basepoint moves to (x_0, y_0), change from $x^n y^m$ to $(x - x_0)^n (y - y_0)^m$. Divide by the same $n!m!$

3. Find the Taylor series of $f(x, y) = e^{x-y}$ with (0,0) as the basepoint. Notice $f(x, y) = e^x$ *times* e^{-y}.

   - Method 1: Multiply the series for e^x and e^{-y} to get $e^{x-y} = (1 + x + \frac{1}{2!}x^2 + \cdots)(1 - y + \frac{1}{2!}y^2 - \cdots) = 1 + x - y + \frac{1}{2}x^2 - xy + \cdots$
   - Method 2: Substitute $x - y$ directly into the series to get $e^{x-y} = 1 + (x - y) + \frac{1}{2!}(x - y)^2 + \cdots$.
   - Method 3: (general method): Find all the derivatives of $f(x, y) = e^{x-y}$ at the basepoint (0,0):

   $$f(0,0) = 1 \quad f_x(0,0) = 1 \quad f_y(0,0) = -1 \quad f_{xx}(0,0) = 1 \quad f_{xy}(0,0) = -1 \quad f_{yy}(0,0) = 1 \quad \cdots$$

   Then the Taylor Series is $\frac{1}{0!0!} + \frac{1}{1!0!}x + \frac{-1}{0!1!}y + \frac{1}{2!0!}x^2 + \frac{-1}{1!1!}xy + \frac{1}{0!2!}y^2 + \cdots$. Remember that $0! = 1$.

Read-throughs and selected even-numbered solutions :

A minimum occurs at a **stationary** point (where $f_x = f_y = 0$) or a **rough** point (no derivative) or a **boundary** point. Since $f = x^2 - xy + 2y$ has $f_x = \mathbf{2x - y}$ and $f_y = \mathbf{2 - x}$, the stationary point is $x = \mathbf{2}, y = \mathbf{4}$. This is not a minimum, because f decreases when $y = 2x$ **increases**.

The minimum of $d^2 = (x - x_1)^2 + (y - y_1)^2$ occurs at the rough point $(\mathbf{x_1}, \mathbf{y_1})$. The graph of d is a **cone** and grad d is a **unit** vector that points **out from** $(\mathbf{x_1}, \mathbf{y_1})$. The graph of $f = |xy|$ touches bottom along the lines $\mathbf{x = 0}$ **and** $\mathbf{y = 0}$. Those are "rough lines" because the derivative **does not exist**. The maximum of d and f must occur on the **boundary** of the allowed region because it doesn't occur **inside.**

When the boundary curve is $x = x(t)$, $y = y(t)$, the derivative of $f(x, y)$ along the boundary is $\mathbf{f_x x_t + f_y y_t}$ (chain rule). If $f = x^2 + 2y^2$ and the boundary is $x = \cos t, y = \sin t$, then $df/dt = \mathbf{2 \sin t \cos t}$. It is zero at the points $\mathbf{t = 0, \pi/2, \pi, 3\pi/2}$. The maximum is at $(\mathbf{0, \pm 1})$ and the minimum is at $(\mathbf{\pm 1, 0})$. Inside the circle f has an absolute minimum at **(0,0)**.

To separate maximum from minimum from **saddle point**, compute the **second** derivatives at a **stationary** point. The tests for a minimum are $\mathbf{f_{xx} > 0}$ **and** $\mathbf{f_{xx} f_{yy} > f_{xy}^2}$. The tests for a maximum are $\mathbf{f_{xx} < 0}$ **and** $\mathbf{f_{xx} f_{yy} > f_{xy}^2}$. In case $ac < \mathbf{b^2}$ or $f_{xx} f_{yy} < \mathbf{f_{xy}^2}$, we have a **saddle point**. At all points these tests decide between concave up and **concave down** and "indefinite". For $f = 8x^2 - 6xy + y^2$, the origin is a **saddle point**. The signs of f at (1,0) and (1,3) are **+ and** $-$.

The Taylor series for $f(x, y)$ begins with the terms $\mathbf{f(0,0) + x f_x + y f_y + \frac{1}{2} x^2 f_{xx} + xy f_{xy} + \frac{1}{2} y^2 f_{yy}}$. The coefficient of $x^n y^m$ is $\mathbf{\partial^{n+m} f / \partial x^n \partial y^m (0,0)}$ **divided by n!m!** To find a stationary point numerically, use

Newton's method or steepest descent.

18 Volume $= xyz = xy(1-3x-2y) = xy - 3x^2 - 2xy^2; V_x = y - 6x - 2y^2$ and $V_y = x - 4xy$; at $(0, \frac{1}{2}, 0)$ and $(\frac{1}{3}, 0, 0)$ and (0,0,1) the volume is $V = 0$ (minimum); at $(\frac{1}{48}, \frac{12}{48}, \frac{21}{48})$ the volume is $V = \mathbf{\frac{7}{3072}}$ (*maximum*)

22 $\frac{\partial f}{\partial x} = 2x + 2$ and $\frac{\partial f}{\partial y} = 2y + 4$. (a) Stationary point $(-1, -2)$ yields $f_{\min} = \mathbf{-5}$. (b) On the boundary $y = 0$ the minimum of $x^2 + 2x$ is $\mathbf{-1}$ at $(-1, 0)$ (c) On the boundary $x \geq 0, y \geq 0$ the minimum is $\mathbf{0}$ at (0,0).

28 $d_1 = x, d_2 = d_3 = \sqrt{(1-x)^2 + 1}, \frac{d}{dx}(x + 2\sqrt{(1-x)^2+1}) = 1 + \frac{2(x-1)}{\sqrt{(1-x)^2+1}} = 0$ when $(1-x)^2 + 1 = 4(x-1)^2$ or $1 - x = \frac{1}{\sqrt{3}}$ or $x = \mathbf{1 - \frac{1}{\sqrt{3}}}$. From that point to (1,1) the line goes up 1 and across $\frac{1}{\sqrt{3}}$, a 60° angle with the horizontal that confirms three 120° angles.

34 From the point $C = (0, -\sqrt{3})$ the lines to $(-1, 0)$ and $(1, 0)$ make a 60° angle. C is the **center** of the circle $x^2 + (y - \sqrt{3})^2 = 4$ through those two points. From any point on that circle, the lines to $(-1, 0)$ and (1,0) make an angle of $2 \times 60° = 120°$. Theorem from geometry: angle from circle $= 2 \times$ angle from center.

44 $\frac{\partial^{n+m}}{\partial x^n \partial y^m}(xe^y) = xe^y$ for $n = 0$, e^y for $n = 1$, zero for $n > 1$. Taylor series $xe^y = \mathbf{x + xy + \frac{1}{2!}xy^2 + \frac{1}{3!}xy^3} + \cdots$

50 $f(x+h, y+k) \approx f(x,y) + h\frac{\partial f}{\partial x}(x,y) + k\frac{\partial f}{\partial y}(x,y) + \frac{h^2}{2}\frac{\partial^2 f}{\partial x^2}(x,y) + hk\frac{\partial^2 f}{\partial x \partial y}(x,y) + \frac{k^2}{2}\frac{\partial^2 f}{\partial y^2}(x,y)$

58 A house costs p, a yacht costs q : $\frac{d}{dx}f(x, \frac{k-px}{q}) = \frac{\partial f}{\partial x} + \frac{\partial f}{\partial y}(-\frac{p}{q}) = 0$ gives $-\frac{\partial f}{\partial x}/\frac{\partial f}{\partial y} = -\frac{p}{q}$.

13.7 Constraints and Lagrange Multipliers (page 519)

In reality, a constraint $g(x, y) = k$ is very common. The point (x, y) is restricted to this curve, when we are minimizing or maximizing $f(x, y)$. (Not to the inside of the curve, but right *on* the curve.) It is like looking for a maximum at a boundary point. The great difficulty is that we lose the equations $\frac{\partial f}{\partial x} = 0$ and $\frac{\partial f}{\partial y} = 0$.

The great success of Lagrange multipliers is to bring back the usual equations "x derivative equals zero" and "y derivative equals zero." But these are not f_x and f_y. We must account for the constraint $g(x, y) = k$. The idea that works is to subtract an unknown multiple λ times $g(x, y) - k$. Now set derivatives to zero:

$$\frac{\partial}{\partial x}[f(x,y) - \lambda(g(x,y) - k)] = 0 \quad \text{or} \quad \frac{\partial f}{\partial x} = \lambda\frac{\partial g}{\partial x}$$
$$\frac{\partial}{\partial y}[f(x,y) - \lambda(g(x,y) - k)] = 0 \quad \text{or} \quad \frac{\partial f}{\partial y} = \lambda\frac{\partial g}{\partial y}.$$

The text explains the reasoning that leads to these equations. Here we solve them for x, y, and λ. That locates the *constrained* maximum or minimum.

1. (This is Problem 13.7.6) Maximize $f(x, y) = x + y$ subject to $g(x, y) = x^{1/3}y^{2/3} = 1$. That is a special case of the Cobb-Douglas constraint: $x^c y^{1-c} = k$.

   - $f_x = \lambda g_x$ is $1 = \lambda(\frac{1}{3}x^{-2/3}y^{2/3})$ and $f_y = \lambda g_y$ is $1 = \lambda(\frac{2}{3}x^{1/3}y^{-1/3})$. The constraint is $1 = x^{1/3}y^{2/3}$.

   Square the second equation and multiply by the first to get $1 = (\frac{2\lambda}{3})^2(\frac{\lambda}{3})$ or $(\frac{\lambda}{3})^3 = \frac{1}{4}$ or $\frac{\lambda}{3} = 4^{-1/3}$. Then divide the constraint by the first equation to get $1 = \frac{3}{\lambda}x$ or $x = \frac{\lambda}{3} = 4^{-1/3}$. Divide the constraint by the second equation to get $1 = \frac{3}{2\lambda}y$ or $y = \frac{2\lambda}{3} = 2 \cdot 4^{-1/3}$. The constrained maximum is $f = x + y = 3 \cdot 4^{-1/3}$.

2. (This is Problem 13.7.22 and also Problem 13.7.8 with a twist. It gives the shortest distance to a plane.) Minimize $f(x, y, z) = x^2 + y^2 + z^2$ with the constraint $g(x, y, z) = ax + by + cz = d$.

- Now we have three variables x, y, z (also λ for the constraint). The method is the same:

$$f_x = \lambda f_x \text{ is } 2x = \lambda a \qquad f_y = \lambda g_y \text{ is } 2y = \lambda b \qquad f_z = \lambda g_z \text{ is } 2z = \lambda c.$$

Put $x = \frac{1}{2}\lambda a$ and $y = \frac{1}{2}\lambda b$ and $z = \frac{1}{2}\lambda c$ in the constraint to get $\frac{1}{2}\lambda(a^2 + b^2 + c^2) = d$. That yields λ.
The constrained minimum is $x^2 + y^2 + z^2 = (\frac{\lambda}{2})^2(a^2 + b^2 + c^2) = \frac{d^2}{a^2+b^2+c^2}$.
The shortest distance to the plane is the square root $\frac{|d|}{\sqrt{a^2+b^2+c^2}}$. This agrees with the formula $\frac{|d|}{|\mathbf{N}|}$ from Section 11.2, where the normal vector to the plane was $\mathbf{N} = a\,\mathbf{i} + b\,\mathbf{j} + c\,\mathbf{k}$.

The text explains how to handle *two* constraints $g(x, y, z) = k_1$ and $h(x, y, z) = k_2$. There are two Lagrange multipliers λ_1 and λ_2. The text also explains *inequality* constraints $g(x, y) \leq k$. The point (x, y) is either *on* the boundary where $g(x, y) = k$ or it is *inside* where $g(x, y) < k$. We are back to our old problem:

The minimum of $f(x, y)$ may be at a boundary point. Using Lagrange multipliers we find $\lambda > 0$.

The minimum of $f(x, y)$ may be at a stationary point. Using Lagrange multipliers we find $\lambda = 0$.

The second case has an inside minimum. The equation $f_x = \lambda g_x$ becomes $f_x = 0$. Similarly $f_y = 0$. Lagrange is giving us one unified way to handle stationary points (inside) and boundary points. Rough points are handled separately. Problems 15–18 develop part of the theory behind λ. I am most proud of including what calculus authors seldom attempt – ***the meaning of*** λ. It is the derivative of $f_{\min}$ with respect to k. Thus λ measures *the sensitivity of the answer to a change in the constraint.*

This section is not easy but it is really important. Remember it when you need it.

Read-throughs and selected even-numbered solutions :

A restriction $g(x, y) = k$ is called a **constraint.** The minimizing equations for $f(x, y)$ subject to $g = k$ are $\mathbf{\partial f/\partial x = \lambda \partial g/\partial x, \partial f/\partial y = \lambda \partial g/\partial y}$, and $\mathbf{g = k}$. The number λ is the Lagrange **multiplier.** Geometrically, grad f is **parallel** to grad g at the minimum. That is because the **level** curve $f = f_{\min}$ is **tangent** to the constraint curve $g = k$. The number λ turns out to be the derivative of $f_{\min}$ with respect to $\mathbf{k}$. The Lagrange function is $L = \mathbf{f(x, y) - \lambda(g(x, y) - k)}$ and the three equations for x, y, λ are $\mathbf{\partial L/\partial x = 0}$ **and** $\mathbf{\partial L/\partial y = 0}$ **and** $\mathbf{\partial L/\partial \lambda = 0}$.

To minimize $f = x^2 - y$ subject to $g = x - y = 0$, the three equations for x, y, λ are $\mathbf{2x = \lambda, -1 = -\lambda}$, $\mathbf{x - y = 0}$. The solution is $\mathbf{x = \frac{1}{2}, y = \frac{1}{2}, \lambda = 1}$. In this example the curve $f(x, y) = f_{\min} = -\frac{1}{4}$ is a **parabola** which is **tangent** to the line $g = 0$ at $(x_{\min}, y_{\min})$.

With two constraints $g(x, y, z) = k_1$ and $h(x, y, z) = k_2$ there are **two** multipliers λ_1 **and** λ_2. The five unknowns are $\mathbf{x, y, z, \lambda_1}$, **and** λ_2. The five equations are $\mathbf{f_x = \lambda_1 g_x + \lambda_2 h_x, f_y = \lambda_1 g_y + \lambda_2 h_x}$, $\mathbf{f_z = \lambda_1 g_z + \lambda_2 h_z, g = 0}$, **and** $\mathbf{h = 0}$. The level surface $f = f_{\min}$ is **tangent** to the curve where $g = k_1$ and $h = k_2$. Then grad f is **perpendicular** to this curve, and so are grad g and **grad h.** With nine variables and six constraints, there will be **six** multipliers and eventually **15** equations. If a constraint is an **inequality** $g \leq k$, then its multiplier must satisfy $\lambda \leq 0$ at a minimum.

2 $x^2 + y^2 = 1$ and $2xy = \lambda(2x)$ and $x^2 = \lambda(2y)$ yield $2\lambda^2 + \lambda^2 = 1$. Then $\lambda = \frac{1}{\sqrt{3}}$ gives $x_{\max} = \pm\frac{\sqrt{6}}{3}$, $y_{\max} = \frac{\sqrt{3}}{3}$, $\mathbf{f_{\max} = \frac{2\sqrt{3}}{9}}$. Also $\lambda = -\frac{1}{\sqrt{3}}$ gives $\mathbf{f_{\min} = -\frac{2\sqrt{3}}{9}}$.

18 $f = 2x + y = 1001$ at the point $x = 1000, y = -999$. The Lagrange equations are $2 = \lambda$ and $1 = \lambda$ (no solution). Linear functions with linear constraints generally have no maximum.

20 (a) $yz = \lambda, xz = \lambda, xy = \lambda$, and $x + y + z = k$ give $x = y = z = \frac{k}{3}$ and $\lambda = \frac{k^2}{9}$ (b) $\mathbf{V_{max}} = (\frac{\mathbf{k}}{\mathbf{3}})^{\mathbf{3}}$ so $\partial V_{\max}/\partial k = \mathbf{k^2/9}$ (which is λ!) (c) Approximate $\Delta V = \lambda$ times $\Delta k = \frac{108^2}{9}(111 - 108) = \mathbf{3888}\ \text{in}^3$. Exact $\Delta V = (\frac{111}{3})^3 - (\frac{108}{3})^3 = \mathbf{3677}\ \text{in}^3$.

26 Reasoning: By increasing k, *more* points satisfy the constraints. *More* points are available to minimize f. Therefore $f_{\min}$ goes **down**.

28 $\lambda = 0$ when $h > k$ (not $h = k$) at the minimum. Reasoning: An increase in k leaves the same minimum. Therefore $f_{\min}$ is **unchanged**. Therefore $\lambda = df_{\min}/dk$ is **zero**.

13 Chapter Review Problems

Graph Problems

G1 Draw the level curves of the function $f(x, y) = y - x$. Describe the surface $z = y - x$.

G2 Draw the level curves of $f(x, y) = \frac{y-1}{x-2}$. Label the curve through (3,3). Which points (x, y) are not on any level curve? The surface has an infinite crack like an asymptote.

Computing Problems

C1 Set up Newton's method to give two equations for Δx and Δy when the original equations are $y = x^5$ and $x = y^5$. Start from various points (x_0, y_0) to see which solutions Newton converges to. Compare the basins of attraction to Figure 13.3 and the front cover of this Guide.

Review Problems

R1 For $f(x, y) = x^n y^m$ find the partial derivatives $f_x, f_y, f_{xx}, f_{xy}, f_{yx}$, and f_{yy}.

R2 If $z(x, y)$ is defined implicitly by $F(x, y, z) = xy - yz + xz = 0$, find $\partial z/\partial x$ and $\partial z/\partial y$.

R3 Suppose z is a function of x/y. From $z = f(x/y)$, show that $x\frac{\partial z}{\partial x} + y\frac{\partial z}{\partial y} = 0$.

R4 Write down a formula for the linear approximation to $z = f(x, y)$ around the origin. If $f(x, y) = 9 + xy$ show that the linear approximation at (1,1) gives $f \approx 11$ while the correct value is 10.

R5 Find the gradient vector for the function $f(x, y) = xy^2$. How is the direction of the gradient at the point $x = 1, y = 2$ related to the level curve $xy^2 = 4$?

R6 Find the gradient vector in three dimensions for the function $F(x, y, z) = z - x^2y^2$. How is the direction of the gradient related to the surface $z = x^2y^2$?

R7 Give a chain rule for df/dt when $f = f(x, y, z)$ and x, y, z are all functions of t.

R8 Find the maximum value of $f(x, y) = x + 2y - x^2 + xy - 2y^2$.

R9 The minimum of $x^2 + y^2$ occurs on the boundary of the region R (not inside) for which regions?

R10 To minimize $x^2 + y^2$ on the line $x + 3y = k$, introduce a Lagrange multiplier λ and solve the three equations for x, y, λ. Check that the derivative $df_{\min}/dk$ equals λ.

Drill Problems

D1 If $z = \ln\sqrt{x^2+y^2}$ show that $x\frac{\partial z}{\partial x} + y\frac{\partial z}{\partial y} = 1$ and $z_{xx} + z_{yy} = 0$ except at ____.

D2 The equation of the tangent plane to $z = x^2 + y^3$ at (1,1,2) is ____.

D3 The normal vector to the surface $xyz^2 = 1$ at (1,1,1) is $\mathbf{N} =$ ____.

D4 The linear approximation to $x^2 + y^2$ near the basepoint (1,2) is ____.

D5 Find the directional derivative of $f(x,y) = xe^y$ at the point (2,2) in the 45° direction $y = x$. What is $\mathbf{u}$? Compare with the ordinary derivative of $f(x) = xe^x$ at $x = 2$.

D6 What is the steepest slope on the plane $z = x + 2y$? Which direction is steepest?

D7 From the chain rule for $f(x,y) = xy^2$ with $x = u + v$ and $y = uv$ compute $\frac{\partial f}{\partial u}$ at $u = 2, v = 3$. Check by taking the derivative of $(u+v)(uv)^2$.

D8 What equations do you solve to find stationary points of $f(x,y)$? What is the tangent plane at those points? How do you know from f_{xx}, f_{xy}, and f_{yy} whether you have a saddle point?

D9 Find two functions $f(x,y)$ that have $\partial f/\partial x = \partial f/\partial y$ at all points. Which is the steepest direction on the surface $z = f(x,y)$? Which is the level direction?

D10 If $x = r\cos\theta$ and $y = r\sin\theta$ compute the determinant $J = \begin{vmatrix} \partial x/\partial r & \partial x/\partial \theta \\ \partial y/\partial r & \partial y/\partial \theta \end{vmatrix}$

D11 If $r = \sqrt{x^2+y^2}$ and $\theta = \tan^{-1}\frac{y}{x}$ compute the determinant $J^* = \begin{vmatrix} \partial r/\partial x & \partial r/\partial y \\ \partial \theta/\partial x & \partial \theta/\partial y \end{vmatrix} = \frac{1}{J}$.

CHAPTER 14 MULTIPLE INTEGRALS

14.1 Double Integrals (page 526)

The most basic double integral has the form $\int\int_R dA$ or $\int\int_R dy\,dx$ or $\int\int_R dx\,dy$. It is the integral of 1 over the region R in the xy plane. The integral equals the ***area of*** R. When we write dA, we are not committed to xy coordinates. The coordinates could be r and θ (polar) or any other way of chopping R into small pieces. When we write $dy\,dx$, we are planning to chop R into vertical strips (width dx) and then chop each strip into very small pieces (height dy). The y integral assembles the pieces and the x integral assembles the strips.

Suppose R is the rectangle with $1 \le x \le 4$ and $2 \le y \le 7$. The side lengths are 3 and 5. The area is 15:

$$\int_{x=1}^{4}\int_{y=2}^{7} 1\,dy\,dx = \int_{x=1}^{4}(7-2)dx = [5x]_1^4 = 20 - 5 = 15$$

The inner integral gave $\int_2^7 1\,dy = [y]_2^7 = 7-2$. This is the height of the strips.

My first point is that this is nothing new. We have written $\int y\,dx$ for a long time, to give the area between a curve and the x axis. The height of the strips is y. *We have short-circuited the inner integral* of $1\,dy$.

Remember also the area between two curves. That is $\int(y_2 - y_1)dx$. Again we have already done the inner integral, between the lower curve y_1 and the upper curve y_2. The integral of $1\,dy$ was just $y_2 - y_1 =$ height of strip. We went directly to the outer integral – the x-integral that adds up the strips.

So what is new? First, the regions R get more complicated. The limits of integration are not as easy as $\int_2^7 dy$ or $\int_1^4 dx$. Second, we don't always integrate the function "1". In particular, double integrals often give ***volume***:

$$\int_1^4\int_2^7 f(x,y)\,dy\,dx \text{ is the volume between the surface } z = f(x,y) \text{ and the } xy \text{ plane.}$$

To be really truthful, volume starts as a triple integral. It is $\int\int\int 1\,dz\,dy\,dx$. The inner integral $\int 1\,dz$ gives z. The lower limit on z is 0, at the xy plane. The upper limit is $f(x,y)$, at the surface. So the inner integral $\int 1\,dz$ between these limits is $f(x,y)$. When we find volume from a *double* integral, we have short-circuited the z-integral $\int 1\,dz = f(x,y)$.

The second new step is to go beyond areas and volumes. We can compute masses and moments and averages of all kinds. The integration process is still $\int\int_R f(x,y)dy\,dx$, if we choose to do the y-integral first. In reality the main challenge of double integrals is to find the limits. You get better by doing examples. We borrow a few problems from Schaum's Outline and other sources, to display the steps for double integrals – and the difference between $\int\int f(x,y)dy\,dx$ and $\int\int f(x,y)dx\,dy$.

1. Evaluate the integral $\int_{x=0}^{1}\int_{y=0}^{x}(x+y)dy\,dx$. Then reverse the order to $\int\int(x+y)dx\,dy$.

   - The inner integral is $\int_{y=0}^{x}(x+y)dy = [xy + \frac{1}{2}y^2]_0^x = \frac{3}{2}x^2$. This is a function of x. The outer integral is a completely ordinary x-integral $\int_0^1 \frac{3}{2}x^2 dx = [\frac{1}{2}x^3]_0^1 = \frac{1}{2}$.
   - Reversing the order is simple for rectangles. But we don't have a rectangle. The inner integral goes from $y = 0$ on the x axis up to $y = x$. This top point is on the 45° line. *We have a triangle* (see figure). When we do the x-integral first, it starts at the 45° line and ends at $x = 1$. The inner x-limits can depend on y, they can't depend on x. The outer limits are *numbers* 0 and 1.

$$\begin{array}{llll} \textbf{\textit{inner}} & \int_y^1(x+y)dx & = [\frac{1}{2}x^2 + xy]_y^1 & = \frac{1}{2} + y - \frac{3}{2}y^2 \\ \textbf{\textit{outer}} & \int_0^1(\frac{1}{2} + y - \frac{3}{2}y^2)dy & = [\frac{1}{2}y + \frac{1}{2}y^2 - \frac{1}{2}y^3]_0^1 & = \frac{1}{2} + \frac{1}{2} - \frac{1}{2} = \frac{1}{2}. \end{array}$$

The answer $\frac{1}{2}$ is the same in either order. The work is different.

2. Evaluate $\int\int_R y^2 dA$ in both orders $dA = dy\,dx$ and $dA = dx\,dy$. The region R is bounded by $y = 2x$, $y = 5x$, and $x = 1$. Please draw your own figures – vertical strips in one, horizontal strips in the other.

   - The vertical strips run from $y = 2x$ up to $y = 5x$. Then x goes from 0 to 1:

$$\textbf{inner} \quad \int_{2x}^{5x} y^2 dy = [\frac{1}{3}y^3]_{2x}^{5x} = \frac{1}{3}(125x^3 - 8x^3) = 39x^3 \qquad \textbf{outer} \quad \int_0^1 39x^3 dx = \frac{39}{4}.$$

For the reverse order, the limits are not so simple. The figure shows why. In the lower part, horizontal strips go between the sloping lines. The inner integral is an x-integral so change $y = 5x$ and $y = 2x$ to $x = \frac{1}{5}y$ and $x = \frac{1}{2}y$. The outer integral in the lower part is from $y = 0$ to 2:

$$\textbf{inner} \quad \int_{y/5}^{y/2} y^2 dx = [y^2 x]_{y/5}^{y/2} = (\frac{1}{2} - \frac{1}{5})y^3 \qquad \textbf{outer} \quad \int_0^2 (\frac{1}{2} - \frac{1}{5})y^3 dy = (\frac{1}{2} - \frac{1}{5})\frac{2^4}{4}.$$

The upper part has horizontal strips from $x = \frac{1}{5}y$ to $x = 1$. The outer limits are $y = 2$ and $y = 5$:

$$\textbf{inner} \quad \int_{y/5}^{1} y^2 dx = [y^2 x]_{y/5}^{1} = y^2 - \frac{1}{5}y^3 \qquad \textbf{outer} \quad \int_2^5 (y^2 - \frac{1}{5}y^3) dy = [\frac{1}{3}y^3 - \frac{1}{20}y^4]_2^5 = \frac{125 - 8}{3} - \frac{625 - 16}{20}$$

Add the two parts, preferably by calculator, to get 9.75 which is $\frac{39}{4}$. *Same answer.*

3. Reverse the order of integration in $\int_0^2 \int_0^{x^2} (x + 2y) dy\,dx$. What *volume* does this equal?

   - The region is bounded by $y = 0$, $y = x^2$, and $x = 2$. When the x-integral goes first it starts at $x = \sqrt{y}$. It ends at $x = 2$, where the horizontal strip ends. Then the outer y-integral ends at $y = 4$:

$$\textit{The reversed order is} \quad \int_0^4 \int_{\sqrt{y}}^2 (x + 2y) dx\,dy. \qquad \textit{Don't reverse } x + 2y \textit{ into } y + 2x!$$

Read-throughs and selected even-numbered solutions :

The double integral $\int\int_R f(x, y) dA$ gives the volume between R and **the surface z = f(x,y)**. The base is first cut into small **squares** of area ΔA. The volume above the ith piece is approximately $\mathbf{f(x_i, y_i)\Delta A}$. The limit of the sum $\sum \mathbf{f(x_i, y_i)\Delta A}$ is the volume integral. Three properties of double integrals are $\int\int \mathbf{(f + g)dA} = \int\int \mathbf{fdA} + \int\int \mathbf{gdA}$ and $\int\int \mathbf{cfdA} = \mathbf{c} \int\int \mathbf{fdA}$ and $\int\int_{\mathbf{R}} \mathbf{f\,dA} = \int\int_{\mathbf{S}} \mathbf{f\,dA} + \int\int_{\mathbf{T}} \mathbf{f\,dA}$ **if R splits into S and T**.

If R is the rectangle $0 \le x \le 4, 4 \le y \le 6$, the integral $\int\int x\,dA$ can be computed two ways. One is $\int\int x\,dy\,dx$, when the inner integral is $\mathbf{xy]_4^6 = 2x}$. The outer integral gives $\mathbf{x^2]_0^4 = 16}$. When the x integral comes first it equals $\int x\,dx = \frac{1}{2}\mathbf{x^2}]_0^4 = 8$. Then the y integral equals $\mathbf{8y}]_4^6 = \mathbf{16}$. This is the volume between **the base rectangle and the plane z = x.**

The area R is $\iint \mathbf{1}dy\,dx$. When R is the triangle between $x=0, y=2x$, and $y=1$, the inner limits on y are **2x and 1.** This is the length of a **thin vertical** strip. The (outer) limits on x are **0 and** $\frac{1}{2}$. The area is $\frac{1}{4}$. In the opposite order, the (inner) limits on x are **0 and** $\frac{1}{2}\mathbf{y}$. Now the strip is **horizontal** and the outer integral is $\int_0^1 \frac{1}{2}\mathbf{y}\,\mathbf{dy} = \frac{1}{4}$. When the density is $\rho(x,y)$, the total mass in the region R is $\iint \rho\,\mathbf{dx\,dy}$. The moments are $M_y = \iint \rho\mathbf{x}\,\mathbf{dx\,dy}$ and $M_x = \iint \rho\mathbf{y}\,\mathbf{dx\,dy}$. The centroid has $\bar{x} = M_y/M$.

10 The area is all below the axis $y=0$, where horizontal strips cross from $x=y$ to $x=|y|$ (which is $-y$). Note that the y integral stops at $y=0$. Area $= \int_{-1}^0 \int_y^{-y} dx\,dy = \int_{-1}^0 -2y\,dy = [-y^2]_{-1}^0 = \mathbf{1}$.

16 The triangle in Problem 10 had sides $x=y, x=-y$, and $y=-1$. Now the vertical strips go from $y=-1$ up to $y=x$ on the left side: area $= \int_{-1}^0 \int_{-1}^x dy\,dx = \int_{-1}^0 (x+1)dx = \frac{1}{2}(x+1)^2]_{-1}^0 = \frac{1}{2}$. The strips go from -1 up to $y=-x$ on the right side: area $= \int_0^1 \int_{-1}^{-x} dy\,dx = \int_0^1(-x+1)dx = \frac{1}{2}$. Check: $\frac{1}{2}+\frac{1}{2}=1$.

24 The top of the triangle is (a,b). From $x=0$ to a the vertical strips lead to $\int_0^a \int_{dx/c}^{bx/a} dy\,dx = [\frac{bx^2}{2a} - \frac{dx^2}{2c}]_0^a = \frac{ba}{2} - \frac{da^2}{2c}$. From $x=a$ to c the strips go up to the third side: $\int_a^c \int_{dx/c}^{b+(x-a)(d-b)/(c-a)} dy\,dx = [bx + \frac{(x-a)^2(d-b)}{2(c-a)} - \frac{dx^2}{2c}]_a^c = b(c-a) + \frac{(c-a)(d-b)}{2} - \frac{dc}{2} + \frac{da^2}{2c}$. The sum is $\frac{ba}{2} + \frac{b(c-a)}{2} + \frac{d(c-a)}{2} - \frac{dc}{2} = \mathbf{\frac{bc-ad}{2}}$. This is half of a parallelogram.

26 $\int_0^b \int_0^a \frac{\partial f}{\partial x} dx\,dy = \int_0^b [f(a,y) - f(0,y)]dy$.

32 The height is $z = \frac{1-ax-by}{c}$. Integrate over the triangular base ($z=0$ gives the side $ax+by=1$) : volume $= \int_{x=0}^{1/a} \int_{y=0}^{(1-ax)/b} \frac{1-ax-by}{c} dy\,dx = \int_0^{1/a} \frac{1}{c}[y - axy - \frac{1}{2}by^2]_0^{(1-ax)/b} dx = \int_0^{1/a} \frac{1}{c}\frac{(1-ax)^2}{2b} dx = -\frac{(1-ax)^3}{6abc}]_0^{1/a} = \mathbf{\frac{1}{6abc}}$.

36 The area of the quarter-circle is $\frac{\pi}{4}$. The moment is zero around the axis $y=0$ (by symmetry): $\bar{\mathbf{x}} = \mathbf{0}$. The other moment, with a factor 2 that accounts for symmetry of left and right, is $2\int_0^{\sqrt{2}/2} \int_x^{\sqrt{1-x^2}} y\,dy\,dx = 2\int_0^1 (\frac{1-x^2}{2} - \frac{x^2}{2})dx = 2[\frac{x}{2} - \frac{x^3}{3}]_0^{\sqrt{2}/2} = \frac{\sqrt{2}}{3}$. Then $\bar{y} = \frac{\sqrt{2}/3}{\pi/4} = \mathbf{\frac{4\sqrt{2}}{3\pi}}$.

14.2 Change to Better Coordinates (page 534)

This title really means "better than xy coordinates." Mostly those are the best, but not always. For regions cut out by circles and rays from the origin, polar coordinates are better. This is the most common second choice, but ellipses and other shapes lead to third and fourth choices. So we concentrate on the change to $r\theta$, but we explain the rules for other changes too.

1. Compute the area between the circles $r=2$ and $r=3$ and between the rays $\theta=0$ and $\theta=\frac{\pi}{2}$.

   - The region is a quarter of a ring. When you draw it, you see that xy coordinates are terrible. Strips start and end in complicated ways. Polar coordinates are extremely easy:

$$\int_0^{\pi/2} \int_{r=2}^3 r\,dr\,d\theta = \int_0^{\pi/2} \frac{1}{2}(3^2-2^2)d\theta = \frac{1}{2}(3^2-2^2)\frac{\pi}{2} = \frac{5\pi}{4}.$$

Notice first that dA is not $dr\ d\theta$. It is $r\ dr\ d\theta$. The extra factor r gives this the dimension of (length)^2. The area of a small polar rectangle is $r\ dr\ d\theta$.

Notice second the result of the inner integral of $r\ dr$. It gives $\frac{1}{2}r^2$. This leaves the outer integral as our old formula $\int \frac{1}{2}r^2 d\theta$ from Chapter 9.

Notice third the result of *limits* on that inner integral. They give $\frac{1}{2}(3^2 - 2^2)$. This leaves the outer integral as our formula for ring areas and washer areas and areas between two polar curves $r = F_1(\theta)$ and $r = F_2(\theta)$. That area was and still is $\int \frac{1}{2}(F_2^2 - F_1^2)d\theta$. For our ring this is $\int \frac{1}{2}(3^2 - 2^2)d\theta$.

2. (14.2.4) Find the centroid $(\bar{x},\ \bar{y})$ of the pie-shaped wedge $0 \le r \le 1$, $\frac{\pi}{4} \le \theta \le \frac{3\pi}{4}$. The average height $\bar{y}$ is $\int\int y\ dA / \int\int dA$. This corresponds to moment around the x axis divided by total mass or area.

   - The area is $\int_{\pi/4}^{3\pi/4} \int_0^1 r\ dr\ d\theta = \int_{\pi/4}^{3\pi/4} \frac{1}{2}d\theta = \frac{1}{2}(\frac{\pi}{2})$. The integral $\iint y\ dA$ has $y = r\sin\theta$:

$$\int_{\pi/4}^{3\pi/4} \int_0^1 (r\ \sin\theta) r\ dr\ d\theta = \int_{\pi/4}^{3\pi/4} [\frac{1}{3}r^3]_0^1 \sin\theta\ d\theta = \frac{1}{3}[-\cos\theta]_{\pi/4}^{3\pi/4} = \frac{1}{3}\cdot 2\cdot \frac{\sqrt{2}}{2} = \frac{\sqrt{2}}{3}.$$

   Now divide to find the average $\bar{y} = (\frac{\sqrt{2}}{3})/(\frac{\pi}{4}) = \frac{4\sqrt{2}}{3\pi}$. This is the height of the centroid.

   Symmetry gives $\bar{x} = 0$. The region for negative x is the mirror image of the region for positive x. This answer zero also comes from integrating $x\ dy\ dx$ or $(r\ \cos\theta)(r\ dr\ d\theta)$. Integrating $\cos\theta d\theta$ gives $\sin\theta$. Since $\sin\frac{\pi}{4} = \sin\frac{3\pi}{4}$, the definite integral is zero.

The text explains the "stretching factor" for any coordinates. It is a 2 by 2 determinant J. Write the old coordinates in terms of the new ones, as in $x = r\cos\theta$ and $y = r\sin\theta$. For these polar coordinates the stretching factor is the r in $r\ dr\ d\theta$.

$$J = \begin{vmatrix} \partial x/\partial r & \partial x/\partial\theta \\ \partial y/\partial r & \partial y/\partial\theta \end{vmatrix} = \begin{vmatrix} \cos\theta & -r\sin\theta \\ \sin\theta & r\cos\theta \end{vmatrix} = r(\cos^2\theta + \sin^2\theta) = r.$$

3. Explain why $J = 1$ for the coordinte change $x = u\cos\alpha - v\sin\alpha$ and $y = u\sin\alpha + v\cos\alpha$.

   - This is a ***pure rotation***. The xy axes are at a 90° angle and the uv axes are also at a 90° angle (just rotated through the angle α). The area $dA = dx\ dy$ just rotates into $dA = du\ dv$. The factor is

$$J = \begin{vmatrix} \partial x/\partial u & \partial x/\partial v \\ \partial y/\partial u & \partial y/\partial v \end{vmatrix} = \begin{vmatrix} \cos\alpha & -\sin\alpha \\ \sin\alpha & \cos\alpha \end{vmatrix} = \cos^2\alpha + \sin^2\alpha = 1.$$

4. Show that $\int_0^\infty e^{-x^2}dx = \frac{1}{2}\sqrt{\pi}$. This is exact even if we can't do $\int_0^1 e^{-x^2}dx$!

   - This is half of Example 4 in the text. There we found the integral $\sqrt{\pi}$ from $-\infty$ to ∞. But e^{-x^2} is an even function – same value for x and $-x$. Therefore the integral from 0 to ∞ is $\frac{1}{2}\sqrt{\pi}$.

Read-throughs and selected even-numbered solutions :

We change variables to improve the **limits** of integration. The disk $x^2 + y^2 \le 9$ becomes the rectangle $0 \le r \le \mathbf{3}, 0 \le \theta \le \mathbf{2\pi}$. The inner limits of $\iint\ dy\ dx$ are $y = \pm\sqrt{\mathbf{9 - x^2}}$. In polar coordinates this area integral becomes $\iint \mathbf{r\ dr\ d\theta} = \mathbf{9\pi}$.

A polar rectangle has sides dr and $\mathbf{r\ d\theta}$. Two sides are not **straight** but the angles are still **90°**. The area between the circles $r = 1$ and $r = 3$ and the rays $\theta = 0$ and $\theta = \pi/4$ is $\mathbf{\frac{1}{8}(3^2 - 1^2)} = \mathbf{1}$. The integral $\iint x\ dy\ dx$

changes to $\iint \mathbf{r^2 \cos\theta \, dr \, d\theta}$. This is the **moment** around the **y** axis. Then $\bar{x}$ is the ratio $\mathbf{My/M}$. This is the x coordinate of the **centroid**, and it is the **average** value of x.

In a rotation through α, the point that reaches (u,v) starts at $x = u \cos\alpha - v\sin\alpha, y = \mathbf{u \sin\alpha + v \cos\alpha}$. A rectangle in the uv plane comes from a **rectangle** in xy. The areas are **equal** so the stretching factor is $J = \mathbf{1}$. This is the determinant of the matrix $\begin{bmatrix} \cos\alpha & -\sin\alpha \\ \sin\alpha & \cos\alpha \end{bmatrix}$. The moment of inertia $\iint x^2 dx\, dy$ changes to $\iint \mathbf{(u \cos\alpha - v \sin\alpha)^2} du\, dv$.

For single integrals dx changes to $\mathbf{(dx/du)}du$. For double integrals $dx\, dy$ changes to $J\, du\, dv$ with $J = \mathbf{\partial(x,y)/\partial(u,v)}$. The stretching factor J is the determinant of the 2 by 2 matrix $\begin{bmatrix} \partial x/\partial u & \partial x/\partial v \\ \partial y/\partial u & \partial y/\partial v \end{bmatrix}$. The functions $x(u,v)$ and $y(u,v)$ connect an xy region R to a uv region S, and $\iint_R dx\, dy = \iint_S \mathbf{J\, du\, dv}$ = area of $\mathbf{R}$. For polar coordinates $x = \mathbf{u \cos v}$ and $y = \mathbf{u \sin v}$ (**or** $\mathbf{r \sin\theta}$). For $x = u, y = u + 4v$ the 2 by 2 determinant is $J = \mathbf{4}$. A square in the uv plane comes from a **parallelogram** in xy. In the opposite direction the change has $u = x$ and $v = \frac{1}{4}(y - x)$ and a new $J = \mathbf{\frac{1}{4}}$. This J is constant because this change of variables is **linear.**

2 Area $= \int_{-\sqrt{2}/2}^{\sqrt{2}/2} \int_{|x|}^{\sqrt{1-x^2}} dy\, dx$ splits into two equal parts left and right of $x = 0 : 2\int_0^{\sqrt{2}/2} \int_x^{\sqrt{1-x^2}} dy\, dx =$ $2\int_0^{\sqrt{2}/2} (\sqrt{1-x^2} - x)dx = [x\sqrt{1-x^2} + \sin^{-1} x - x^2]_0^{\sqrt{2}/2} = \sin^{-1}\frac{\sqrt{2}}{2} = \frac{\pi}{4}$. The limits on $\iint dx\, dy$ are $\int_0^{\sqrt{2}/2} \int_{-y}^{y} \mathbf{dx\, dy}$ for the lower triangle plus $\int_{\sqrt{2}/2}^{1} \int_{-\sqrt{1-y^2}}^{\sqrt{1-y^2}} \mathbf{dx\, dy}$ for the circular top.

6 Area of wedge $= \frac{b}{2\pi}(\pi a^2)$. Divide $\int_0^b \int_0^a (r \cos\theta) r\, dr\, d\theta = \mathbf{\frac{a^3}{3} \sin b}$ by this area $\frac{ba^2}{2}$ to find $\bar{x} = \mathbf{\frac{2a}{3b}} \sin b$. (*Interesting limit:* $\bar{x} \to \frac{2}{3}a$ as the wedge angle b approaches zero: This is like the centroid of a triangle.) For $\bar{y}$ divide $\int_0^b \int_0^a (r \sin\theta) r\, dr\, d\theta = \mathbf{\frac{a^3}{3}(1 - \cos b)}$ by the area $\frac{ba^2}{2}$ to find $\bar{y} = \mathbf{\frac{2a}{3b}(1 - \cos b)}$.

12 $I_x = \int_{\pi/4}^{3\pi/4} \int_0^1 (r \sin\theta + 1)^2 r\, dr\, d\theta = \frac{1}{4}\int \sin^2\theta\, d\theta + \frac{2}{3}\int \sin\theta d\theta + \frac{1}{2}\int d\theta = [\frac{\theta}{8} - \frac{\sin 2\theta}{16} - \frac{2}{3}\cos\theta + \frac{\theta}{2}]_{\pi/4}^{3\pi/4} =$ $\mathbf{\frac{5\pi}{16} + \frac{2}{16} + \frac{4}{3}\frac{\sqrt{2}}{2}}$; $I_y = \iint (r \cos\theta)^2 r\, dr\, d\theta = \frac{\pi}{16} - \frac{1}{8}$ (as in Problem 11); $I_0 = I_x + I_y = \mathbf{\frac{3\pi}{8} + \frac{4}{3}\frac{\sqrt{2}}{2}}$.

24 Problem 18 has $J = \begin{vmatrix} 3 & 2 \\ 1 & 1 \end{vmatrix} = 1$. So the area of R is $1\times$ area of unit square $= \mathbf{1}$.

Problem 20 has $J = \begin{vmatrix} v & u \\ -2u & 2v \end{vmatrix} = 2(u^2 + v^2)$, and integration over the square gives area of $R =$ $\int_0^1 \int_0^1 2(u^2 + v^2) du\, dv = \mathbf{\frac{4}{3}}$. Check in x, y coordinates: area of $R = 2\int_0^1 (1 - x^2)dx = \frac{4}{3}$.

26 $\begin{vmatrix} \partial r/\partial x & \partial r/\partial y \\ \partial\theta/\partial x & \partial\theta/\partial y \end{vmatrix} = \begin{vmatrix} x/r & y/r \\ -y/r^2 & x/r^2 \end{vmatrix} = \frac{x^2+y^2}{r^3} = \frac{1}{r}$. As in equation 12, this new J is $\frac{1}{\text{old} J}$.

34 (a) **False** (forgot the stretching factor J) (b) **False** (x can be larger than x^2) (c) **False** (forgot to divide by the area) (d) **True** (odd function integrated over symmetric interval) (e) **False** (the straight-sided region is a trapezoid: angle from 0 to θ and radius from r_1 to r_2 yields area $\frac{1}{2}(r_2^2 - r_1^2)\sin\theta\cos\theta$).

14.3 Triple Integrals (page 540)

For a triple integral, the plane region R changes to a solid V. The basic integral is $\int\int\int dV = \int\int\int dz\, dy\, dx$.

This integral of "1" equals the volume of V. Similarly $\iiint x\,dV$ gives the moment. Divide by the volume for $\bar{x}$.

As always, the limits are the hardest part. The inner integral of dz is z. The limits depend on x and y (unless the top and bottom of the solid are flat). Then the middle integral is ***not*** $\int dy = y$. We are not integrating "1" any more, when we reach the second integral. We are integrating $z_{top} - z_{\text{bottom}}$ = function of x and y. The limits give y as a function of x. Then the outer integral is an ordinary x-integral (but it is not $\int 1\,dx$!).

1. Compute the triple integral $\int_0^1 \int_0^x \int_0^y dz\,dy\,dx$. What solid volume does this equal?

   - The inner integral is $\int_0^y dz = y$. The middle integral is $\int_0^x y\,dy = \frac{1}{2}x^2$. The outer integral is $\int_0^1 \frac{1}{2}x^2 dx = \frac{1}{6}$. The y-integral goes across to the line $y = x$ and the x-integral goes from 0 to 1.
   - In the xy plane this gives a triangle (between the x axis and the 45° line $y = x$). Then the z-integral goes up to the sloping plane $z = y$. I think we have a ***tetrahedron*** – a pyramid with a triangular base and three triangular sides. *Draw it.*

   Check: Volume of pyramid $= \frac{1}{3}$(base)(height) $= \frac{1}{3}(\frac{1}{2})(1) = \frac{1}{6}$. This is one of the six solids in Problem 14.3.3. It is quickly described by $0 \le z \le y \le x \le 1$. Can you see those limits in our triple integral?

2. Find the limits on $\int\int\int dz\,dy\,dx$ for the volume between the surfaces $x^2 + y^2 = 9$ and $x + z = 4$ and $z = 0$. Describe those surfaces and the region V inside them.

   - The inner integral is from $z = 0$ to $z = 4 - x$. (Key point: We just solved $x + z = 4$ to find z.) The middle integral is from $y = -\sqrt{9 - x^2}$ to $y = +\sqrt{9 - x^2}$. The outer integral is from $x = -3$ to $x = +3$.

   Where did -3 and 3 come from? That is the smallest possible x and the largest possible x, when we are inside the surface $x^2 + y^2 = 9$. This surface is a *circular cylinder*, a pipe around the z axis. It is chopped off by the horizontal plane $z = 0$ and the sloping plane $x + z = 4$. The triple integral turns out to give the volume 36π.

   If we change $x + z = 4$ to $x^2 + z^2 = 4$, we have a harder problem. The limits on z are $\pm\sqrt{4 - x^2}$. But now x can't be as large as 3. The solid is now an *intersection of cylinders*. I don't know its volume.

Read-throughs and selected even-numbered solutions :

Six important solid shapes are **a box, prism, cone, cylinder, tetrahedron, and sphere.** The integral $\iiint dx\,dy\,dz$ adds the volume $\mathbf{dx\,dy\,dz}$ of small **boxes.** For computation it becomes **three** single integrals. The inner integral $\int dx$ is the **length** of a line through the solid. The variables **y** and **z** are constant. The double integral $\iint dx\,dy$ is the **area** of a slice, with **z** held constant. Then the z integral adds up the volumes of **slices.**

If the solid region V is bounded by the planes $x = 0, y = 0, z = 0$, and $x + 2y + 3z = 1$, the limits on the inner x integral are **0 and 1 − 2y − 3z**. The limits on y are **0 and $\frac{1}{2}$(1 − 3z)**. The limits on z are **0 and $\frac{1}{3}$**. In the new variables $u = x, v = 2y, w = 3z$, the equation of the outer boundary is **u + v + w = 1.** The volume of the tetrahedron in uvw space is $\mathbf{\frac{1}{6}}$. From $dx = du$ and $dy = dv/2$ and $dz = \mathbf{dw/3}$, the volume of an xyz box is $dx\,dy\,dz = \mathbf{\frac{1}{6}}du\,dv\,dw$. So the volume of V is $\mathbf{\frac{1}{36}}$.

To find the average height $\bar{z}$ in V we compute $\iiint \mathbf{z\,dV}/\iiint \mathbf{dV}$. To find the total mass if the density is $\rho = e^z$ we compute the integral $\iiint \mathbf{e^z\,dx\,dy\,dz}$. To find the average density we compute $\iiint \mathbf{e^z dV}/\iiint \mathbf{dV}$. In the order $\iiint dz\,dx\,dy$ the limits on the inner integral can depend on **x and y**. The limits on the middle integral can depend on **y**. The outer limits for the ellipsoid $x^2 + 2y^2 + 3z^2 \le 8$ are $\mathbf{-2 \le y \le 2}$.

4 $\int_0^1 \int_0^z \int_0^y x\,dx\,dy\,dz = \int_0^1 \int_0^z \frac{y^2}{2}dy\,dz = \int_0^1 \frac{z^3}{6}dz = \frac{1}{24}$. Divide by the volume $\frac{1}{6}$ to find $\bar{\mathbf{x}} = \mathbf{\frac{1}{4}}$;

$\int_0^1 \int_0^z \int_0^y y\, dx\, dy\, dz = \int_0^1 \int_0^z y^2 dy\, dz = \int_0^1 \frac{z^3}{3} dz = \frac{1}{12}$ and $\bar{\mathbf{y}} = \frac{\mathbf{1}}{\mathbf{2}}$; by symmetry $\bar{\mathbf{z}} = \frac{\mathbf{3}}{\mathbf{4}}$.

14 Put dz last and stop at $z = 1 : \int_0^1 \int_0^{4-z} \int_0^{(4-y-z)/2} dx\, dy\, dz = \int_0^1 \int_0^{4-z} \frac{4-y-z}{2} dy\, dz =$ $\int_0^1 \frac{(4-z)^2}{4} dz = [-\frac{(4-z)^3}{12}]_0^1 = \frac{4^3-3^3}{12} = \frac{\mathbf{37}}{\mathbf{12}}$.

22 Change variables to $X = \frac{x}{a}, Y = \frac{y}{b}, Z = \frac{z}{c}$; then $dXdYdZ = \frac{dx\,dy\,dz}{abc}$. Volume $= \int\int\int abc\, dXdYdZ =$ $\frac{\mathbf{1}}{\mathbf{6}}\mathbf{abc}$. Centroid $(\bar{x}, \bar{y}, \bar{z}) = (a\bar{X}, b\bar{Y}, c\bar{Z}) = (\frac{\mathbf{a}}{\mathbf{4}}, \frac{\mathbf{b}}{\mathbf{4}}, \frac{\mathbf{c}}{\mathbf{4}})$. (Recall volume $\frac{1}{6}$ and centroid $(\frac{1}{4}, \frac{1}{4}, \frac{1}{4})$ of standard tetrahedron: this is Example 2.)

30 In one variable, the midpoint rule is correct for the functions 1 and x. In three variables it is correct for **1, x, y, z, xy, xz, yz, xyz.**

14.4 Cylindrical and Spherical Coordinates (page 547)

I notice in Schaum's Outline that very few triple integrals use $dx\ dy\ dz$. Most use cylindrical or spherical coordinates. The small pieces have volume $dV = r\ dr\ d\theta\ dz$ when they are wedges from a cylinder – the base is $r\ dr\ d\theta$ and the height is dz. The pieces of spheres have volume $dV = \rho^2 \sin\phi\ d\rho\ d\phi\ d\theta$. The reason for these coordinates is that many curved solids in practice have cylinders or spheres as boundary surfaces.

Notice that r is $\sqrt{x^2+y^2}$ and ρ is $\sqrt{x^2+y^2+z^2}$. Thus $r = 1$ is a cylinder and $\rho = 1$ is a sphere. For a cylinder on its side, you would still use $r\ dr\ d\theta$ but dy would replace dz. Just turn the whole system.

1. Find the volume inside the cylinder $x^2 + y^2 = 16$ (or $r = 4$), above $z = 0$ and below $z = y$.

   - The solid region is a *wedge*. It goes from $z = 0$ up to $z = y$. The base is half the disk of radius 4. It is not the whole disk, because when y is negative we can't go "up to $z = y$." We can't be above $z = 0$ and below $z = y$, unless y is positive – which puts the polar angle θ between 0 and π. The volume integral seems to be

$$\int_0^4 \int_0^{\sqrt{16-x^2}} \int_0^y dz\ dy\ dx \quad \text{or} \quad \int_0^\pi \int_0^4 \int_0^{r\sin\theta} r\ dz\ dr\ d\theta.$$

   The first gives $\int_0^4 \int_0^{\sqrt{16-x^2}} y\ dy\ dx = \int_0^4 \frac{1}{2}(16-x^2)dx = \frac{64}{3}$. The second is $\int_0^\pi \int_0^4 r^2 \sin\theta\ dr\ d\theta =$ $\int_0^\pi \frac{64}{3} \sin\theta\ d\theta = \frac{64}{3} \cdot 2$. ***Which is right?***

2. Find the average distance from the center of the unit ball $\rho \le 1$ to all other points of the ball.

   - We are looking for the average value of ρ, when ρ goes between 0 and 1. But the average is not $\frac{1}{2}$. There is more volume for large ρ than for small ρ. So the average $\bar{\rho}$ over the whole ball will be greater than $\frac{1}{2}$. The integral we want is

$$\bar{\rho} = \frac{1}{\text{volume}} \int\int\int \rho dV = \frac{1}{4\pi/3} \int_0^{2\pi} \int_0^\pi \int_0^1 \rho \cdot \rho^2 \sin\phi\ d\rho\ d\phi\ d\theta = \frac{3}{4\pi} \cdot \frac{1}{4} \cdot 2 \cdot 2\pi = \frac{3}{4}.$$

   The integration was quick because $\int \rho^3\ d\rho = \frac{1}{4}$ separates from $\int \sin\phi\ d\phi = 2$ and $\int d\theta = 2\pi$.

The same separation gives the volume of the unit sphere as $(\int \rho^2 d\rho = \frac{1}{3}) \times (\int \sin\phi\ d\phi = 2) \times (\int d\theta = 2\pi)$. The volume is $\frac{4\pi}{3}$. ***Notice that the angle ϕ from the North Pole has upper limit*** π (not 2π).

3. Find the centroid of the *upper half* of the unit ball. Symmetry gives $\bar{x} = \bar{y} = 0$. Compute $\bar{z}$.

- The volume is half of $\frac{4\pi}{3}$. The integral of $z\,dV$ (remembering $z = \rho\cos\phi$) is

$$\int_0^{2\pi}\int_0^{\pi/2}\int_0^1 \rho\cos\phi\cdot\rho^2\sin\phi\,d\rho\,d\phi\,d\theta = \int_0^1 \rho^3 d\rho\int_0^{\pi/2}\cos\phi\sin\phi\,d\phi\int_0^{2\pi}d\theta = \frac{1}{4}\cdot\frac{1}{2}\cdot 2\pi = \frac{\pi}{4}.$$

Divide by the volume to find the average $\bar{z} = \frac{\pi}{4}/\frac{2\pi}{3} = \frac{3}{8}$. The ball goes up to $z = 1$, but it is fatter at the bottom so the centroid is below $z = \frac{1}{2}$. This time ϕ stops at $\pi/2$, the Equator.

The text explains Newton's famous result for the gravitational attraction of a sphere. The sphere acts as if all its mass were concentrated at the center. Problem 26 gives the proof.

Read-throughs and selected even-numbered solutions :

The three **cylindrical** coordinates are $r\theta z$. The point at $x = y = z = 1$ has $r = \mathbf{\sqrt{2}}, \theta = \boldsymbol{\pi/4}, z = \mathbf{1}$. The volume integral is $\iiint \mathbf{r\,dr\,d\theta\,dz}$. The solid region $1 \le r \le 2, 0 \le \theta \le 2\pi, 0 \le z \le 4$ is a **hollow cylinder (a pipe)**. Its volume is $\mathbf{12\pi}$. From the r and θ integrals the area of a **ring (or washer)** equals $\mathbf{3\pi}$. From the z and θ integrals the area of *a* **shell** equals $\mathbf{2\pi rz}$. In $r\theta z$ coordinates **cylinders** are convenient, while **boxes** are not.

The three **spherical** coordinates are $\rho\phi\theta$. The point at $x = y = z = 1$ has $\rho = \mathbf{\sqrt{3}}, \phi = \mathbf{\cos^{-1} 1/\sqrt{3}}, \theta = \boldsymbol{\pi/4}$. The angle ϕ is measured from **the z axis.** θ is measured from **the x axis.** ρ is the distance to **the origin,** where r was the distance to **the z axis.** If $\rho\phi\theta$ are known then $\mathbf{x} = \boldsymbol{\rho}\,\mathbf{sin}\,\boldsymbol{\phi}\,\mathbf{cos}\,\boldsymbol{\theta}, \mathbf{y} = \boldsymbol{\rho}\,\mathbf{sin}\,\boldsymbol{\phi}\,\mathbf{sin}\,\boldsymbol{\theta}, \mathbf{z} = \boldsymbol{\rho}\,\mathbf{cos}\,\boldsymbol{\phi}$. The stretching factor J is a 3 by 3 **determinant** and volume is $\iiint \mathbf{r^2\sin\phi\,dr\,d\phi\,d\theta}$.

The solid region $1 \le \rho \le 2, 0 \le \phi \le \pi, 0 \le \theta \le 2\pi$ is a **hollow sphere.** Its volume is $\mathbf{4\pi(2^3 - 1^3)/3}$. From the ϕ and θ integrals the area of a **spherical shell** at radius ρ equals $\mathbf{4\pi\rho^2}$. Newton discovered that the outside gravitational attraction of a **sphere** is the same as for an equal mass located at **the center.**

6 $(x, y, z) = (\frac{3}{2}, \frac{\sqrt{3}}{2}, 1); (r, \theta, z) = (\sqrt{3}, \frac{\pi}{6}, 1)$

14 This is the volume of a half-cylinder (because of $0 \le \theta \le \pi$) : height π, radius π, volume $\mathbf{\frac{1}{2}\pi^4}$.

22 The curve $\rho = 1 - \cos\phi$ is a **cardioid** in the xz plane (like $r = 1 - \cos\theta$ in the xy plane). So we have a **cardioid of revolution.** Its volume is $\mathbf{\frac{8\pi}{3}}$ as in Problem 9.3.35.

26 *Newton's achievement* The cosine law (see hint) gives $\cos\alpha = \frac{D^2+q^2-\rho^2}{2qD}$. Then integrate $\frac{\cos\alpha}{q^2}$: $\iiint(\frac{D^2-\rho^2}{2q^3D} + \frac{1}{2qD})dV$. The second integral is $\frac{1}{2D}\iiint\frac{dV}{q} = \frac{4\pi R^3/3}{2D^2}$. The first integral over ϕ uses the same $u = D^2 - 2\rho D\cos\phi + \rho^2 = q^2$ as in the text: $\int_0^\pi \frac{\sin\phi d\phi}{q^3} = \int\frac{du/2\rho D}{u^{3/2}} = [\frac{-1}{\rho D u^{1/2}}]_{\phi=0}^{\phi=\pi} = \frac{1}{\rho D}(\frac{1}{D-\rho} - \frac{1}{D+\rho}) = \frac{2}{D(D^2-\rho^2)}$. The θ integral gives 2π and then the ρ integral is $\int_0^R 2\pi\frac{2}{D(D^2-\rho^2)}\frac{D^2-\rho^2}{2D}\rho^2 d\rho = \frac{4\pi R^3/3}{2D^2}$. The two integrals give $\frac{4\pi R^3/3}{D^2}$ as Newton hoped and expected.

30 $\iint q\,dA = 4\pi\rho^2 D + \frac{4\pi}{3}\frac{\rho^4}{D}$. Divide by $4\pi\rho^2$ to find $\bar{q} = \mathbf{D} + \mathbf{\frac{\rho^2}{3D}}$ for the shell. Then the integral over ρ gives $\iiint q\,dV = \frac{4\pi}{3}R^3 D + \frac{4\pi}{15}\frac{R^5}{D}$. Divide by the volume $\frac{4\pi}{3}R^3$ to find $\bar{q} = \mathbf{D} + \mathbf{\frac{R^2}{5D}}$ for the solid ball.

42 The ball comes to a stop at Australia and **returns** to its starting point. It continues to oscillate in harmonic motion $y = R\cos(\sqrt{c/m}\,t)$.

14 Chapter Review Problems

Review Problems

R1 Integrate $1 + x + y$ over the triangle R with corners (0,0) and (2,0) and (0,2).

R2 Integrate $\frac{x^2}{x^2+y^2}$ over the unit circle using polar coordinates.

R3 Show that $\int_1^2 \int_0^y x\sqrt{y^2 - x^2}\, dx\, dy = \frac{5}{4}$.

R4 Find the area A_n between the curves $y = x^{n+1}$ and $y = x^n$. The limits on x are 0 and 1. Draw A_1 and A_2 on the same graph. Explain why $A_1 + A_2 + A_3 + \cdots$ equals $\frac{1}{2}$.

R5 Convert $y = \sqrt{2x - x^2}$ to $r = 2\cos\theta$. Show that $\int_0^2 \int_0^{\sqrt{2x-x^2}} x\, dy\, dx = \frac{\pi}{2}$ using polar coordinates.

R6 The polar curve $r = 2\cos\theta$ is a unit circle. Find the average $\bar{r}$ for points inside. This is the average distance from points inside to the point (0,0) on the circle. (Answer: $\frac{1}{\text{area}} \iint r\, dA = \frac{32}{9\pi}$.)

R7 Sketch the region whose area is $\int_0^2 \int_{x^2}^{2x} dy\, dx$. Reverse the order of integration to $\iint dx\, dy$.

R8 Write six different triple integrals starting with $\iiint dx\, dy\, dz$ for the volume of the solid with $0 \le x \le 2y \le 4z \le 8$.

R9 Write six different triple integrals beginning with $\iiint r\, dr\, d\theta\, dz$ for the volume limited by $0 \le r \le z \le 1$. Describe this solid.

Drill Problems

D1 The point with cylindrical coordinates $(2\pi, 2\pi, 2\pi)$ has $x =$ ____, $y =$ ____, $z =$ ____.

D2 The point with spherical coordinates $(\frac{\pi}{2}, \frac{\pi}{2}, \frac{\pi}{2})$ has $x =$ ____, $y =$ ____, $z =$ ____.

D3 Compute $\int_0^\infty \frac{e^{-x}}{\sqrt{x}} dx$ by substituting u for $\sqrt{x}$.

D4 Find the area of the ellipse $\frac{x^2}{a^2} + \frac{y^2}{b^2} = 1$ by substituting $u = \frac{x}{a}$ and $v = \frac{y}{b}$.

D5 Find $(\bar{x}, \bar{y})$ for the infinite region under $y = e^{-x^2}$. Use page 535 and integration by parts.

D6 What integral gives the area between $x + y = 1$ and $r = 1$?

D7 Show that $\iint_R e^{x^2+y^2} dy\, dx = \frac{\pi}{2}(e - 1)$ when R is the upper half of the unit circle.

D8 If the xy axes are rotated by 30°, the point $(x, y) = (2, 4)$ has new coordinates $(u, v) =$ ____.

D9 In Problem D8 explain why $x^2 + y^2 = u^2 + v^2$. Also explain $dx\, dy = du\, dv$.

D10 ***True or false*** : The centroid of a region is inside that region.

CHAPTER 15 VECTOR CALCULUS

15.1 Vector Fields (page 554)

An ordinary function assigns a value $f(x)$ to each point x. A vector field assigns a ***vector*** $\mathbf{F}(x,y)$ to each point (x,y). Think of the vector as going out from the point (not out from the origin). The vector field is like a head of hair! We are placing a straight hair at every point. Depending on how the hair is cut and how it is combed, the vectors have different lengths and different directions.

The vector at each point (x,y) has two components. Its horizontal component is $M(x,y)$, its vertical component is $N(x,y)$. *The vector field is* $\mathbf{F}(x,y) = M(x,y)\mathbf{i} + N(x,y)\mathbf{j}$. Remember: A vector from every point.

1. Suppose all of the vectors $\mathbf{F}(x,y)$ have length 1, and their directions are *outward* (or *radial*). Find their components $M(x,y)$ and $N(x,y)$.

   - At a point like (3,0) on the x axis, the outward direction is the x direction. The vector of length 1 from that point is $\mathbf{F}(3,0) = \mathbf{i}$. This vector goes outward *from the point.* At (0,2) the outward vector is $\mathbf{F}(0,2) = \mathbf{j}$. At the point $(-2,0)$ it is $\mathbf{F}(-2,0) = -\mathbf{i}$. (The *minus* x direction is outward.) At every point (x,y), the outward direction is parallel to $x\mathbf{i} + y\mathbf{j}$. This is the "position vector" $\mathbf{R}(x,y)$.

   We want an outward spreading field of **unit** vectors. So divide the position vector $\mathbf{R}$ by its length:

   $$\mathbf{F}(x,y) = \frac{\mathbf{R}(x,y)}{|\mathbf{R}(x,y)|} = \frac{x\mathbf{i} + y\mathbf{j}}{\sqrt{x^2+y^2}}. \quad \text{This special vector field is called } \mathbf{u}_r.$$

   The letter $\mathbf{u}$ is for "unit," the subscript r is for "radial." No vector is assigned to the origin, because the outward direction there can't be decided. Thus $\mathbf{F}(0,0)$ is not defined (for this particular field). Then we don't have to divide by $r = \sqrt{x^2+y^2} = 0$ at the origin.

   To repeat: The field of outward unit vectors is $\mathbf{u}_r = \frac{R}{r}$. Another way to write it is $\mathbf{u}_r = \cos\theta\ \mathbf{i} + \sin\theta\ \mathbf{j}$. The components are $M(x,y) = \frac{x}{\sqrt{x^2+y^2}} = \cos\theta$ and $N(x,y) = \frac{y}{\sqrt{x^2+y^2}} = \sin\theta$.

2. Suppose again that all the vectors $\mathbf{F}(x,y)$ are unit vectors. But change their directions to be *perpendicular* to $\mathbf{u}_r$. The vector at (3,0) is $\mathbf{j}$ instead of $\mathbf{i}$. Find a formula for this "***unit spin field.***"

   - We want to take the vector $\mathbf{u}_r$, at each point (x,y) except the origin, and ***turn that vector by*** 90°. The turn is counterclockwise and the new vector is called $\mathbf{u}_\theta$. It is still a unit vector, and its dot product with $\mathbf{u}_r$ is zero. Here it is, written in two or three different ways:

   $$\mathbf{u}_\theta = -\sin\theta\ \mathbf{i} + \cos\theta\ \mathbf{j} \text{ is perpendicular to } \mathbf{u}_r = \cos\theta\ \mathbf{i} + \sin\theta\ \mathbf{j}$$
   $$\mathbf{u}_\theta = \frac{-y}{\sqrt{x^2+y^2}}\mathbf{i} + \frac{x}{\sqrt{x^2+y^2}}\mathbf{j} = \frac{1}{r}(-y\ \mathbf{i} + x\ \mathbf{j}) = \mathbf{S}/r.$$

   Where $\mathbf{R} = x\mathbf{i} + y\mathbf{j}$ was the position field (outward), $\mathbf{S} = -y\mathbf{i} + x\mathbf{j}$ is the ***spin field*** (around the origin). The lengths of $\mathbf{R}$ and $\mathbf{S}$ are both $r = \sqrt{x^2+y^2}$, increasing as we move outward. For unit vectors $\mathbf{u}_r$ and $\mathbf{u}_\theta$ we divide by this length.

   Two other important fields are $\mathbf{R}/r^2$ and $\mathbf{S}/r^2$. At each point one is still outward (parallel to $\mathbf{R}$) and the other is "turning" (parallel to $\mathbf{S}$). But now the lengths decrease as we go outward. The length of $\frac{\mathbf{R}}{r^2}$ is $\frac{r}{r^2}$ or $\frac{1}{r}$. This is closer to a typical men's haircut.

So far we have six vector fields: three radial fields $\mathbf{R}$ and $\mathbf{u}_r = \mathbf{R}/r$ and $\mathbf{R}/r^2$ and three spin fields $\mathbf{S}$ and $\mathbf{u}_\theta = \mathbf{S}/r$ and $\mathbf{S}/r^2$. The radial fields point along ***rays***, out from the origin. The spin fields are tangent to ***circles***, going around the origin. These rays and circles are the ***field lines*** or ***streamlines*** for these particular vector fields.

The field lines give the direction of the vector $\mathbf{F}(x, y)$ at each point. The length is not involved (that is why $\mathbf{S}$ and $\mathbf{S}/r$ and $\mathbf{S}/r^2$ all have the same field lines). The direction is tangent to the field line so the slope of that line is $dy/dx = N(x, y)/M(x, y)$.

3. Find the field lines (streamlines) for the vector field $\mathbf{F}(x, y) = x\,\mathbf{i} - y\,\mathbf{j}$.

   - Solve $dy/dx = N/M = -y/x$ by separating variables. We have $\frac{dy}{y} = -\frac{dx}{x}$. Integration gives

$$\ln y = -\ln x + C. \text{ Therefore } \ln x + \ln y = C \text{ or } \ln xy = C \text{ or } xy = c.$$

The field lines $xy = c$ are hyperbolas. The vectors $\mathbf{F} = x\mathbf{i} - y\mathbf{j}$ are tangent to those hyperbolas.

4. ***A function*** $f(x, y)$ ***produces*** a ***gradient field*** $\mathbf{F}$. Its components are $M = \frac{\partial f}{\partial x}$ and $N = \frac{\partial f}{\partial y}$. This field has the special symbol $\mathbf{F} = \nabla f$. Describe this gradient field for the particular function $f(x, y) = xy$.

   - The partial derivatives of $f(x, y) = xy$ are $\partial f/\partial x = y$ and $\partial f/\partial y = x$. Therefore the gradient field is $\nabla \mathbf{f} = y\mathbf{i} + x\mathbf{j}$. Not a radial field and not a spin field.

Remember that the gradient vector gives the direction in which $f(x, y)$ changes fastest. This is the "steepest direction." Tangent to the curve $f(x, y) = c$ there is no change in f. ***Perpendicular to the curve there is maximum change***. This is the gradient direction. So the gradient field $y\mathbf{i}+x\mathbf{j}$ of Problem 4 is perpendicular to the field $x\mathbf{i} - y\mathbf{j}$ of Problem 3.

In Problem 3, the field is tangent to the hyperbolas $xy = c$. In Problem 4, the field is perpendicular to those hyperbolas. The hyperbolas are called ***equipotential lines*** because the "potential" xy is "equal" (or constant) along those curves $xy = c$.

Read-throughs and selected even-numbered solutions :

A vector field assigns a **vector** to each point (x, y) or (x, y, z). In two dimensions $\mathbf{F}(x, y) = \mathbf{M(x, y)i+N(x, y)j}$. An example is the position field $\mathbf{R} = \mathbf{x\,i} + \mathbf{y\,j}(+\mathbf{z\,k})$. Its magnitude is $|\mathbf{R}| = \mathbf{r}$ and its direction is **out from the origin.** It is the gradient field for $f = \frac{1}{2}(\mathbf{x^2} + \mathbf{y^2})$. The level curves are **circles,** and they are **perpendicular** to the vectors $\mathbf{R}$.

Reversing this picture, the spin field is $\mathbf{S} = -\mathbf{y\,i} + \mathbf{x\,j}$. Its magnitude is $|\mathbf{S}| = \mathbf{r}$ and its direction is **around the origin.** It is not a gradient field, because no function has $\partial f/\partial x = -\mathbf{y}$ and $\partial f/\partial y = \mathbf{x}$. $\mathbf{S}$ is the velocity field for flow going **around the origin.** The streamlines or **field** lines or integral **curves** are **circles.** The flow field $\rho\mathbf{V}$ gives the rate at which **mass** is moved by the flow.

A gravity field from the origin is proportional to $\mathbf{F} = \mathbf{R}/\mathbf{r^3}$ which has $|\mathbf{F}| = \mathbf{1}/\mathbf{r^2}$. This is Newton's **inverse** square law. It is a gradient field, with potential $f = \mathbf{1}/\mathbf{r}$. The equipotential curves $f(x, y) = c$ are **circles.** They are **perpendicular** to the field lines which are **rays.** This illustrates that the **gradient** of a function $f(x, y)$ is **perpendicular** to its level curves.

The velocity field $y\,\mathbf{i} + x\,\mathbf{j}$ is the gradient of $f = \mathbf{xy}$. Its streamlines are **hyperbolas.** The slope dy/dx of a streamline equals the ratio $\mathbf{N}/\mathbf{M}$ of velocity components. The field is **tangent** to the streamlines. Drop a leaf onto the flow, and it goes along **a streamline.**

2 $x\mathbf{i} + \mathbf{j}$ is the gradient of $f(x,y) = \mathbf{\frac{1}{2}x^2 + y}$, which has **parabolas** $\frac{1}{2}x^2 + y = c$ as equipotentials (they open down). The streamlines solve $dy/dx = 1/x$ (this is N/M). So $y = \ln x + C$ gives the streamlines.

6 $x^2\mathbf{i} + y^2\mathbf{j}$ is the gradient of $f(x,y) = \mathbf{\frac{1}{3}(x^3 + y^3)}$, which has closed curves $x^3 + y^3 =$ constant as equipotentials. The streamlines solve $dy/dx = y^2/x^2$ or $dy/y^2 = dx/x^2$ or $\mathbf{y^{-1} = x^{-1}}$ + constant.

14 $\frac{\partial f}{\partial x} = \mathbf{2x - 2}$ and $\frac{\partial f}{\partial y} = \mathbf{2y}; \mathbf{F} = (2x-2)\mathbf{i} + 2y\mathbf{j}$ leads to circles $(x-1)^2 + y^2 = c$ around the center (1,0).

26 $f(x,y) = \mathbf{\frac{1}{2}\ln(x^2+y^2) = \ln\sqrt{x^2+y^2}}$. This comes from $\frac{\partial f}{\partial x} = \frac{x}{x^2+y^2}$ or $f = \int \frac{x\,dx}{x^2+y^2}$.

32 From the gradient of $y - x^2$, $\mathbf{F}$ must be $\mathbf{-2xi + j}$ (or this is $-\mathbf{F}$).

36 $\mathbf{F}$ is the gradient of $f = \mathbf{\frac{1}{2}ax^2 + bxy + \frac{1}{2}cy^2}$. The equipotentials are **ellipses** if $ac > b^2$ and **hyperbolas** if $ac < b^2$. (If $ac = b^2$ we get straight lines.)

15.2 Line Integrals (page 562)

The most common line integral is along the x axis. We have a function $y(x)$ and we integrate to find $\int y(x)dx$. Normally this is just called an integral, without the word "line." But now we have functions defined at every point in the xy plane, so we can integrate along curves. A better word for what is coming would be "curve" integral.

Think of a curved wire. The density of the wire is $\rho(x,y)$, possibly varying along the wire. Then the total mass of the wire is $\int \rho(x,y)ds$. This is a line integral or curve integral (or wire integral). Notice $ds = \sqrt{(dx)^2 + (dy)^2}$. We use dx for integrals along the x axis and dy up the y axis and ds for integrals along other lines and curves.

1. A circular wire of radius R has density $\rho(x,y) = x^2y^2$. How can you compute its mass $M = \int \rho\, ds$?

   - Describe the circle by $x = R\cos t$ and $y = R\sin t$. You are free to use θ instead of t. The point is that we need a parameter to describe the path and to compute ds:

   $$ds = \sqrt{(\frac{dx}{dt})^2 + (\frac{dy}{dt})^2}\, dt = \sqrt{R^2\sin^2 t + R^2\cos^2 t}\, dt = R\, dt.$$

   Then the mass integral is $M = \int x^2y^2\, ds = \int_{t=0}^{2\pi}(R^2\cos^2 t)(R^2\sin^2 t)R\, dt$. I won't integrate.

This chapter is about vector fields. But we integrate scalar functions (like the density ρ). So if we are given a vector $\mathbf{F}(x,y)$ at each point, we take its dot product with another vector – to get an ordinary scalar function to be integrated. Two dot products are by far the most important:

$$\begin{aligned} \mathbf{F}\cdot\mathbf{T}\, ds &= \text{component of } \mathbf{F} \textit{ tangent} \text{ to the curve} = M\,dx + N\,dy \\ \mathbf{F}\cdot\mathbf{n}\, ds &= \text{component of } \mathbf{F} \textit{ normal} \text{ to the curve} = M\,dy - N\,dx \end{aligned}$$

The unit tangent vector is $\mathbf{T} = \frac{d\mathbf{R}}{ds}$ in Chapter 12. Then $\mathbf{F}\cdot\mathbf{T}\,ds$ is $\mathbf{F}\cdot d\mathbf{R} = (M\,\mathbf{i} + N\,\mathbf{j})\cdot(dx\,\mathbf{i} + dy\,\mathbf{j})$. This dot product is $M\,dx + N\,dy$, which we integrate. Its integral is the ***work*** done by $\mathbf{F}$ along the curve. Work is force times distance, but the distance is measured parallel to the force. This is why the tangent component $\mathbf{F}\cdot\mathbf{T}$ goes into the work integral.

2. Compute the work by the force $\mathbf{F} = x\mathbf{i}$ around the unit circle $x = \cos t,\ y = \sin t$.

   - Work $= \int M\,dx + N\,dy = \int x\,dx + 0\,dy = \int_{t=0}^{2\pi}(\cos t)(-\sin t\,dt)$. This integral is zero!

3. Compute the work by $\mathbf{F} = x\mathbf{i}$ around a square: along $y = 0$, up $x = 1$, back along $y = 1$, back down $x = 0$.

   - Along the x axis, the direction is $\mathbf{T} = \mathbf{i}$ and $\mathbf{F} \cdot \mathbf{T} = x\mathbf{i} \cdot \mathbf{i} = x$. Work $= \int_0^1 x\,dx = \frac{1}{2}$.

   Up the line $x = 1$, the direction is $\mathbf{T} = \mathbf{j}$. Then $\mathbf{F} \cdot \mathbf{T} = x\mathbf{i} \cdot \mathbf{j} = 0$. No work.

   Back along $y = 1$ the direction is $\mathbf{T} = -\mathbf{i}$. Then $\mathbf{F} \cdot \mathbf{T} = -x$. The work is $\int \mathbf{F} \cdot \mathbf{T}ds = \int -x\,dx = -\frac{1}{2}$.

   Note! You might think the integral should be $\int_1^0(-x)dx = +\frac{1}{2}$. Wrong. Going left, ds is $-dx$.

   The work down the y axis is again zero. $\mathbf{F} = x\mathbf{i}$ is perpendicular to the movement $d\mathbf{R} = \mathbf{j}\,dy$. So $\mathbf{F} \cdot \mathbf{T} = 0$.

   Total work around square $= \frac{1}{2} + 0 - \frac{1}{2} + 0 =$ *zero.*

4. Does the field $\mathbf{F} = x\mathbf{i}$ do zero work around *every* closed path? If so, why?

   - *Yes*, the line integral $\int \mathbf{F} \cdot \mathbf{T}\,ds = \int x\,dx + 0\,dy$ is always zero around closed paths. The antiderivative is $f = \frac{1}{2}x^2$. ***When the start and end are the same point** P **the definite integral is*** $f(P) - f(P) = 0$.

   We used the word "antiderivative." From now on we will say "***potential function.***" This is a function $f(x, y)$ – if it exists – such that $df = M\,dx + N\,dy$:

   The potential function has $\frac{\partial f}{\partial x} = M$ and $\frac{\partial f}{\partial y} = N$. Then $\int_P^Q M\,dx + N\,dy = \int_P^Q df = f(Q) - f(P)$.

   The field $\mathbf{F}(x, y)$ is the ***gradient*** of the potential function $f(x, y)$. Our example has $f = \frac{1}{2}x^2$ and $\mathbf{F} = \nabla f = x\mathbf{i}$. Conclusion: Gradient fields are conservative. The work around a *closed* path is zero.

5. Does the field $\mathbf{F} = y\mathbf{i}$ do zero work around every closed path? If not, why not?

   - This is not a gradient field. There is no potential function that $\frac{\partial f}{\partial x} = M$ and $\frac{\partial f}{\partial y} = N$. We are asking for $\frac{\partial f}{\partial x} = y$ and $\frac{\partial f}{\partial y} = 0$ which is impossible. The work around the unit circle $x = \cos t,\ y = \sin t$ is

$$\int M\,dx + N\,dy = \int y\,dx = \int_{t=0}^{2\pi}(\sin t)(-\sin t\,dt) = -\pi. \text{ Not zero!}$$

   Important A gradient field has $M = \frac{\partial f}{\partial x}$ and $N = \frac{\partial f}{\partial y}$. Every function has equal mixed derivatives $\frac{\partial^2 f}{\partial x\,\partial y} = \frac{\partial^2 f}{\partial y\,\partial x}$. Therefore the gradient field has $\partial M/\partial y = \partial N/\partial x$. This is the quick test "**D**" for a gradient field. $\mathbf{F} = y\mathbf{i}$ fails this test as we expected, because $\partial M/\partial y = 1$ and $\partial N/\partial x = 0$.

Read-throughs and selected even-numbered solutions :

Work is the **integral** of $\mathbf{F} \cdot d\mathbf{R}$. Here $\mathbf{F}$ is the **force** and $\mathbf{R}$ is the **position**. The **dot** product finds the component of $\mathbf{F}$ in the direction of movement $d\mathbf{R} = dx\,\mathbf{i} + dy\,\mathbf{j}$. The straight path $(x, y) = (t, 2t)$ goes from **(0,0)** at $t = 0$ to **(1,2)** at $t = 1$ with $d\mathbf{R} = dt\,\mathbf{i} + \mathbf{2dt}\,\mathbf{j}$.

Another form of $d\mathbf{R}$ is $\mathbf{T}ds$, where $\mathbf{T}$ is the **unit tangent** vector to the path and the arc length has $ds = \sqrt{(\mathbf{dx/dt})^2 + (\mathbf{dy/dt})^2}$. For the path $(t, 2t)$, the unit vector $\mathbf{T}$ is $(\mathbf{i} + \mathbf{2j})/\sqrt{\mathbf{5}}$ and $ds = \sqrt{\mathbf{5}}dt$. For $\mathbf{F} = 3\mathbf{i} + \mathbf{j}$, $\mathbf{F} \cdot \mathbf{T}\,ds$ is still $\mathbf{5}dt$. This $\mathbf{F}$ is the gradient of $f = \mathbf{3x + y}$. The change in $f = 3x + y$ from (0,0) to (1,2) is **5**.

When $\mathbf{F} = \text{grad } f$, the dot product $\mathbf{F} \cdot d\mathbf{R}$ is $(\partial f/\partial x)dx + (\mathbf{\partial f/\partial y})\mathbf{dy} = df$. The work integral from P to Q is $\int df = \mathbf{f(Q) - f(P)}$. In this case the work depends on the **endpoints** but not on the **path.** Around a closed path the work is **zero.** The field is called **conservative.** $\mathbf{F} = (1+y)\,\mathbf{i} + x\,\mathbf{j}$ is the gradient of $f = \mathbf{x} + \mathbf{xy}$. The work from (0,0) to (1,2) is **3**, the change in potential.

For the spin field $\mathbf{S} = -\mathbf{y}\,\mathbf{i} + \mathbf{x}\,\mathbf{j}$, the work does depend on the path. The path $(x,y) = (3\cos t, 3\sin t)$ is a circle with $\mathbf{S}\cdot d\mathbf{R} = -\mathbf{y}\ \mathbf{dx} + \mathbf{x}\ \mathbf{dy} = \mathbf{9}\ \mathbf{dt}$. The work is $\mathbf{18\pi}$ around the complete circle. Formally $\int g(x,y)ds$ is the limit of the sum $\sum \mathbf{g}(\mathbf{x_i}, \mathbf{y_i})\Delta\mathbf{s_i}$.

The four equivalent properties of a conservative field $\mathbf{F} = M\mathbf{i} + N\mathbf{j}$ are **A: zero work around closed paths, B: work depends only on endpoints, C: gradient field, D:** $\partial M/\partial y = \partial N/\partial x$. Test **D** is passed by $\mathbf{F} = (y+1)\mathbf{i} + x\mathbf{j}$. The work $\int \mathbf{F}\cdot d\mathbf{R}$ around the circle $(\cos t, \sin t)$ is **zero.** The work on the upper semicircle equals the work on **the lower semicircle (clockwise)**. This field is the gradient of $f = \mathbf{x} + \mathbf{xy}$, so the work to $(-1,0)$ is $-\mathbf{1}$ **starting from (0,0).**

4 Around the square $0 \le x, y \le 3$, $\int_3^0 y\,dx = -9$ along the top (backwards) and $\int_0^3 -x\,dy = -9$ up the right side. All other integrals are zero: answer $-\mathbf{18}$. By Section 15.3 this integral is always $-2\times$ area.

8 Yes The field $x\mathbf{i}$ is the gradient of $f = \frac{1}{2}x^2$. Here $M = x$ and $N = 0$ so $\int_P^Q M dx + N dy = f(Q) - f(P)$. More directly: up and down movement has no effect on $\int x\,dx$.

10 Not much. Certainly the limit of $\Sigma(\Delta s)^2$ is zero.

14 $\frac{\partial N}{\partial x} = \frac{\partial M}{\partial y}$ and $\mathbf{F}$ is the gradient of $f = xe^y$. Then $\int \mathbf{F}\cdot d\mathbf{R} = f(Q) - f(P) = -\mathbf{1}$.

18 $\frac{\mathbf{R}}{r^n}$ has $M = \frac{x}{(x^2+y^2)^{n/2}}$ and $\frac{\partial M}{\partial y} = -xny(x^2+y^2)^{-(n/2)-1}$. This agrees with $\frac{\partial N}{\partial x}$ so $\frac{\mathbf{R}}{r^n}$ is a **gradient field for all n.** The potential is $\mathbf{f} = \frac{\mathbf{r^{2-n}}}{\mathbf{2-n}}$ or $\mathbf{f} = \ln \mathbf{r}$ when $n = 2$.

32 $\frac{\partial N}{\partial x} \ne \frac{\partial M}{\partial y}$ (not conservative): $\int x^2y\,dx + xy^2dy = \int_0^1 2t^3dt = \frac{\mathbf{1}}{\mathbf{2}}$ but $\int_0^1 t^2(t^2)dt + t(t^2)^2(2t\,dt) = \frac{\mathbf{17}}{\mathbf{35}}$.

34 The potential is $f = \frac{\mathbf{1}}{\mathbf{2}}\ln(\mathbf{x^2} + \mathbf{y^2} + \mathbf{1})$. Then $f(1,1) - f(0,0) = \frac{\mathbf{1}}{\mathbf{2}}\ln \mathbf{3}$.

15.3 Green's Theorem (page 571)

The last section studied line integrals of $\mathbf{F}\cdot\mathbf{T}ds$. This section connects them to *double integrals*. The work can be found by integrating around the curve (with ds) or *inside the curve* (with $dA = dx\,dy$). The connection is by Green's Theorem. The theorem is for integrals around ***closed curves***:

$$\int_C M\ dx + N\ dy = \iint_R \left(\frac{\partial N}{\partial x} - \frac{\partial M}{\partial y}\right) dx\ dy.$$

We see again that this is zero for gradient fields. Their test is $\frac{\partial N}{\partial x} = \frac{\partial M}{\partial y}$ so the double integral is immediately zero.

1. Compute the work integral $\int M\ dx + N\ dy = \int y\ dx$ for the force $\mathbf{F} = y\mathbf{i}$ around the unit circle.

   - Use Green's Theorem with $M = y$ and $N = 0$. The line integral equals the double integral of $\frac{\partial N}{\partial x} - \frac{\partial M}{\partial y} = -1$. Integrate -1 over a circle of area π to find the answer $-\pi$. This agrees with Question 5 in the previous section of the Guide. It also means that the true-false Problem 15.2.44c has answer "False."

     Special case If $M = -\frac{1}{2}y$ and $N = \frac{1}{2}x$ then $\partial N/\partial x = \frac{1}{2}$ and $-\partial M/\partial y = \frac{1}{2}$. Therefore Green's Theorem is $\frac{1}{2}\int_C -y\ dx + x\ dy = \iint_R(\frac{1}{2}+\frac{1}{2})dx\ dy =$ area of R.

2. Use that special case to find the area of a triangle with corners (0,0), (x_1, y_1), and (x_2, y_2).

- We have to integrate $\frac{1}{2}\int -y\,dx + x\,dy$ around the triangle to get the area. The first side has $x = x_1 t$ and $y = y_1 t$. As t goes from 0 to 1, the point (x, y) goes from (0,0) to (x_1, y_1). The integral is $\int(-y_1 t)(x_1 dt) + (x_1 t)(y_1 dt) = 0$. Similarly the line integral between (0,0) and (x_2, y_2) is zero.
The third side has $x = x_1 + t(x_2 - x_1)$ and $y = y_1 + t(y_2 - y_1)$. It goes from (x_1, y_1) to (x_2, y_2).

$$\frac{1}{2}\int -y\,dx + x\,dy = \text{(substitute } x \text{ and } y \text{ simplify)} = \frac{1}{2}\int_0^1 (x_1 y_2 - x_2 y_1)dt = \frac{1}{2}(x_1 y_2 - x_2 y_1).$$

This is the area of the triangle. It is half the parallelogram area $= \begin{vmatrix} x_1 & y_1 \\ x_2 & y_2 \end{vmatrix}$ in Chapter 11.

Green's Theorem also applies to the flux integral $\int \mathbf{F}\cdot\mathbf{n}\,ds$ ***around a closed curve*** C. Now we are integrating $M\,dy - N\,dx$. By changing letters in the first form (the work form) of Green's Theorem, we get the second form (the flux form):

$$\text{Flow through curve } = \int_C M\,dy - N\,dx = \iint_R \left(\frac{\partial M}{\partial x} + \frac{\partial N}{\partial y}\right) dx\,dy.$$

We are no longer especially interested in gradient fields (which give zero work). Now we are interested in source-free fields (which give zero flux). The new test is $\frac{\partial M}{\partial x} + \frac{\partial N}{\partial y} = 0$. This quantity is the ***divergence*** of the field $\mathbf{F} = M(x,y)\mathbf{i} + N(x,y)\mathbf{j}$. A source-free field has *zero divergence*.

3. Is the position field $\mathbf{F} = x\,\mathbf{i} + y\,\mathbf{j}$ source-free? If not, find the flux $\int \mathbf{F}\cdot\mathbf{n}\,ds$ going out of a unit square.

   - The divergence of this $\mathbf{F}$ is $\frac{\partial M}{\partial x} + \frac{\partial N}{\partial y} = \frac{\partial x}{\partial x} + \frac{\partial y}{\partial y} = 1 + 1 = 2$. The field is not source-free. The flux is not zero. Green's Theorem gives flux $= \iint 2\,dx\,dy = 2 \times$ area of region $= 2$, for a unit square.

4. Is the field $\mathbf{F} = x\,\mathbf{i} - y\,\mathbf{j}$ source-free? Is it also a gradient field?

   - *Yes* and *yes*. The field has $M = x$ and $N = -y$. It passes both tests. Test **D** for a gradient field is $\frac{\partial M}{\partial y} = \frac{\partial N}{\partial x}$ which is $0 = 0$. Test **H** for a source-free field is $\frac{\partial M}{\partial x} + \frac{\partial N}{\partial y} = 1 - 1 = 0$.

This gradient field has the potential function $f(x,y) = \frac{1}{2}(x^2 - y^2)$. This source-free field also has a ***stream function*** $g(x,y) = xy$. The stream function satisfies $\frac{\partial g}{\partial y} = M$ and $\frac{\partial g}{\partial x} = -N$. Then $g(x,y)$ is the antiderivative for the flux integral $\int M\,dy - N\,dx$. When it goes around a closed curve from P to P, the integral is $g(P) - g(P) = 0$. This is what we expect for source-free fields, with stream functions.

5. Show how the combination of "conservative" plus "source-free" leads to Laplace's equation for f (and g).

   - $f_{xx} + f_{yy} = M_x + N_y$ because $f_x = M$ and $f_y = N$. But source-free means $M_x + N_y = 0$.

Read-throughs and selected even-numbered solutions :

The work integral $\oint M\,dx + N\,dy$ equals the double integral $\iint(\mathbf{N_x} - \mathbf{M_y})\mathbf{dx\,dy}$ by **Green's** Theorem. For $\mathbf{F} = 3\mathbf{i} + 4\mathbf{j}$ the work is **zero.** For $\mathbf{F} = \mathbf{xj}$ and $-\mathbf{yi}$ the work equals the area of R. When $M = \partial f/\partial x$ and $N = \partial f/\partial y$, the double integral is zero because $\mathbf{f_{xy}} = \mathbf{f_{yx}}$. The line integral is zero because $\mathbf{f(Q) = f(P)}$ **when** $\mathbf{Q = P}$ **(closed curve)**. An example is $\mathbf{F} = \mathbf{y\,i + x\,j}$. The direction on C is **counterclockwise** around the outside and **clockwise** around the boundary of a hole. If R is broken into very simple pieces with crosscuts between them, the integrals of $\mathbf{M\,dx + N\,dy}$ cancel along the crosscuts.

Test **D** for gradient fields is $\partial\mathbf{M}/\partial\mathbf{y} = \partial\mathbf{N}/\partial\mathbf{x}$. A field that passes this test has $\oint \mathbf{F}\cdot d\mathbf{R} = \mathbf{0}$. There is a solution to $f_x = \mathbf{M}$ and $f_y = \mathbf{N}$. Then $df = M\,dx + N\,dy$ is an **exact** differential. The spin field $\mathbf{S}/r^2$ passes test

D except at $\mathbf{r} = \mathbf{0}$. Its potential $f = \theta$ increases by $\mathbf{2\pi}$ going around the origin. The integral $\iint (N_x - M_y)dx\,dy$ is not zero but $\mathbf{2\pi}$.

The flow form of Green's heorem is $\oint_{\mathbf{C}} \mathbf{M\,dy - N\,dx} = \iint_{\mathbf{R}}(\mathbf{M_x + N_y})\mathbf{dx\,dy}$. The normal vector in $\mathbf{F}\cdot\mathbf{n}ds$ points **out across C** and $|\mathbf{n}| = \mathbf{1}$ and $\mathbf{n}\,ds$ equals $dy\,\mathbf{i} - dx\,\mathbf{j}$. The divergence of $M\mathbf{i} + N\mathbf{j}$ is $\mathbf{M_x + N_y}$. For $\mathbf{F} = x\mathbf{i}$ the double integral is $\iint \mathbf{1\,dt} =$ **area**. There is a source. For $\mathbf{F} = y\,\mathbf{i}$ the divergence is **zero.** The divergence of $\mathbf{R}/r^2$ is zero except at $\mathbf{r} = \mathbf{0}$. This field has a **point** source.

A field with no source has properties **E = zero flux through C, F = equal flux across all paths from P to Q, G = existence of stream function, H=** zero divergence. The stream function g satisfies the equations $\partial \mathbf{g}/\partial \mathbf{y} = \mathbf{M}$ **and** $\partial \mathbf{g}/\partial \mathbf{x} = -\mathbf{N}$. Then $\partial M/\partial x + \partial N/\partial y = 0$ because $\partial^2 g/\partial x \partial y = \partial^{\mathbf{2}}\mathbf{g}/\partial \mathbf{y}\,\partial \mathbf{x}$. The example $\mathbf{F} = y\mathbf{i}$ has $g = \mathbf{\frac{1}{2}y^2}$. There is not a potential function. The example $\mathbf{F} = x\mathbf{i} - y\mathbf{j}$ has $g = \mathbf{xy}$ and also $f = \mathbf{\frac{1}{2}x^2 - \frac{1}{2}y^2}$. This f satisfies Laplace's equation $\mathbf{f_{xx} + f_{yy} = 0}$, because the field $\mathbf{F}$ is both **conservative and source-free.** The functions f and g are connected by the **Cauchy-Riemann** equations $\partial f/\partial x = \partial g/\partial y$ and $\partial \mathbf{f}/\partial \mathbf{y} = -\partial \mathbf{g}/\partial \mathbf{x}$.

4 $\oint y\,dx = \int_0^1 t(-dt) = -\mathbf{\frac{1}{2}}$; $M = y, N = 0, \iint(-1)dx\,dy = -$ area $= -\mathbf{\frac{1}{2}}$.

12 Let R be the square with base from a to b on the x axis. Set $\mathbf{F} = f(x)\mathbf{j}$ so $M = 0$ and $N = f(x)$. The line integral $\oint M dx + N dy$ is $\mathbf{(b - a)f(b)}$ up the right side **minus** $\mathbf{(b-a)f(a)}$ down the left side. The double integral is $\iint \frac{df}{dx}dx\,dy = \mathbf{(b-a)\int_a^b \frac{df}{dx}dx}$. Green's Theorem gives equality; cancel $b - a$.

16 $\oint \mathbf{F}\cdot\mathbf{n}ds = \int xy\,dy = \mathbf{\frac{1}{2}}$ up the right side of the square where $\mathbf{n} = \mathbf{i}$ (other sides give zero). Also $\int_0^1\int_0^1 (y+0)dx\,dy = \mathbf{\frac{1}{2}}$.

22 $\oint \mathbf{F}\cdot\mathbf{n}\,ds$ is the same through a square and a circle because the difference is $\iint(\frac{\partial M}{\partial x} + \frac{\partial N}{\partial y})dx\,dy = \iint$ div $\mathbf{F}dx\,dy = 0$ over the region in between.

30 $\frac{\partial M}{\partial x} + \frac{\partial N}{\partial y} = 3y^2 - 3y^2 = 0$. Solve $\frac{\partial g}{\partial y} = 3xy^2$ for $g = \mathbf{xy^3}$ and check $\frac{\partial g}{\partial x} = y^3$.

38 $g(Q) = \int_P^Q \mathbf{F}\cdot\mathbf{n}ds$ starting from $g(P) = 0$. Any two paths give the same integral because forward on one and back on the other gives $\oint \mathbf{F}\cdot\mathbf{n}\,ds = 0$, provided the tests $E - H$ for a stream function are passed.

15.4 Surface Integrals (page 581)

The length of a curve is $\int ds$. The area of a surface is $\iint dS$. Curves are described by functions $y = f(x)$ in single-variable calculus. Surfaces are described by functions $z = f(x, y)$ in multivariable calculus. When you have worked with ds, you see dS as the natural next step:

$$ds = \sqrt{1 + (\frac{dy}{dx})^2}\,dx \qquad \text{and} \qquad dS = \sqrt{1 + (\frac{\partial z}{\partial x})^2 + (\frac{\partial z}{\partial y})^2}\,dx\,dy$$

The basic step dx is along the x axis. The extra $(\frac{dy}{dx})^2$ in ds accounts for the extra length when the curve slopes up or down. Similarly $dx\,dy$ is the area dA down in the base plane. The extra $(\frac{\partial z}{\partial x})^2$ and $(\frac{\partial z}{\partial y})^2$ account for the extra area when the surface slopes up or down.

1. Find the length of the line $x+y=1$ cut off by the axes $y=0$ and $x=0$. The line segment goes from (1,0) to (0,1). Find the area of the plane $x+y+z=1$ cut off by the planes $z=0$ and $y=0$ and $x=0$. This is a triangle with corners at (1,0,0) and (0,1,0) and (0,0,1).

   - The line $y=1-x$ has $\frac{dy}{dx}=-1$. Therefore $ds=\sqrt{1+(-1)^2}\,dx=\sqrt{2}\,dx$. The integral goes from $x=0$ to $x=1$ along the base. The length is $\int_0^1\sqrt{2}\,dx=\sqrt{2}$. Check: The line from (1,0) to (0,1) certainly has length $\sqrt{2}$.
   - The plane $z=1-x-y$ has $\frac{\partial z}{\partial x}=-1$ and $\frac{\partial z}{\partial y}=-1$. Therefore $dS=\sqrt{1+(-1)^2+(-1)^2}\,dx\,dy=\sqrt{3}\,dx\,dy$. ***The integral is down in the xy plane!*** The equilateral triangle in the sloping plane is over a right triangle in the base plane. Look only at the xy coordinates of the three corners: (1,0) and (0,1) and (0,0). Those are the corners of the *projection* (the "shadow" down in the base). This shadow triangle has area $\frac{1}{2}$. The surface area above is:

$$\text{area of sloping plane}=\iint_{\text{shadow}}dS=\iint_{\text{base area}}\sqrt{3}\,dx\,dy=\sqrt{3}\cdot\frac{1}{2}.$$

   Check: The sloping triangle has sides of length $\sqrt{2}$. That is the distance between its corners (1,0,0) and (0,1,0) and (0,0,1). An equilateral triangle with sides $\sqrt{2}$ has area $\sqrt{3}/2$.

   All these problems have three steps : *Find dS. Find the shadow. Integrate dS over the shadow.*

2. Find the area on the plane $x+2y+z=4$ which lies inside the vertical cylinder $x^2+y^2=1$.

   - The plane $z=4-x-2y$ has $\frac{\partial z}{\partial x}=-1$ and $\frac{\partial z}{\partial y}=-2$. Therefore $dS=\sqrt{1+(-1)^2+(-2)^2}\,dx\,dy=\sqrt{6}\,dx\,dy$. The shadow in the base is the inside of the circle $x^2+y^2=1$. This unit circle has area π. So the surface area on the sloping plane above it is $\iint\sqrt{6}\,dx\,dy=\sqrt{6}\times$ area of shadow $=\sqrt{6}\,\pi$.

   The region on that sloping plane is an *ellipse*. This is automatic when a plane cuts through a circular cylinder. The area of an ellipse is πab, where a and b are the half-lengths of its axes. The axes of this ellipse are hard to find, so the new method that gave area $=\sqrt{6}\pi$ is definitely superior.

3. Find the surface area on the sphere $x^2+y^2+z^2=25$ between the horizontal planes $z=2$ and $z=4$.

   - The lower plane cuts the sphere in the circle $x^2+y^2+2^2=25$. This is $r^2=21$. The upper plane cuts the sphere in the circle $x^2+y^2+4^2=25$. This is $r^2=9$. *The shadow in the xy plane is the ring between $r=3$ and $r=\sqrt{21}$.*

   The spherical surface has $x^2+y^2+z^2=25$. Therefore $2x+2z\frac{\partial z}{\partial x}=0$ or $\frac{\partial z}{\partial x}=-\frac{x}{z}$. Similarly $\frac{\partial z}{\partial y}=-\frac{y}{z}$:

$$dS=\sqrt{1+(-\frac{x}{z})^2+(-\frac{y}{2})^2}\,dx\,dy=\frac{\sqrt{x^2+y^2+z^2}}{z}\,dx\,dy=\frac{5}{z}\,dx\,dy.$$

   Remember $z=\sqrt{25-x^2-y^2}=\sqrt{25-r^2}$. Integrate $\frac{5}{z}$ over the shadow (the ring) using r and θ :

$$\begin{aligned}\text{Surface area}&=\iint_{\text{ring}}\tfrac{5}{z}\,dx\,dy=\int_0^{2\pi}\int_3^{\sqrt{21}}\tfrac{5\,r\,dr\,d\theta}{\sqrt{25-r^2}}\\&=(-5)(2\pi)\sqrt{25-r^2}\big]_3^{\sqrt{21}}=-10\pi(2-4)=20\pi.\end{aligned}$$

Surface equations with parameters Up to now the surface equation has been $z = f(x,y)$. This is restrictive. Each point (x,y) in the base has only one point above it in the surface. A complete sphere is not allowed. We solved a similar problem for curves, by allowing a parameter: $x = \cos t$ and $y = \sin t$ gave a complete circle. For surfaces we need *two parameters* u and v. Instead of $x(t)$ we have $x(u,v)$. Similarly $y = y(u,v)$ and $z = z(u,v)$. As u and v go over some region R, the points (x,y,z) go over the surface S.

For a circle, the parameter t is really the angle θ. For a sphere, the parameters u and v are the angle ϕ down from the North Pole and the angle θ around the Equator. These are just spherical coordinates from Section 14.4: $x = \sin u \cos v$ and $y = \sin u \sin v$ and $z = \cos u$. In this case the region R is $0 \le u \le \pi$ and $0 \le v \le 2\pi$. Then the points (x,y,z) cover the surface S of the unit sphere $x^2 + y^2 + z^2 = 1$.

We still have to find dS! The general formula is equation (7) on page 575. For our x, y, z that equation gives $dS = \sin u \, du \, dv$. (In spherical coordinates you remember the volume element $dV = \rho^2 \sin\phi \, d\rho \, d\phi \, d\theta$. We are on the surface $\rho = 1$. And the letters ϕ and θ are changed to u and v.) This good formula for dS is typical of good coordinate systems – equation (7) is not as bad as it looks.

Integrate dS over the base region R in uv space to find the surface area above.

4. Find the surface area (known to be 4π) of the unit sphere $x^2 + y^2 + z^2 = 1$.

- Integrate dS over R to find $\int_0^{2\pi} \int_0^{\pi} \sin u \, du \, dv = (2\pi)(-\cos u)]_0^{\pi} = 4\pi$.

5. Recompute Question 3, the surface area on $x^2 + y^2 + z^2 = 25$ between the planes $z = 2$ and $z = 4$.

- This sphere has radius $\sqrt{25} = 5$. Multiply the points (x,y,z) on the unit sphere by 5:

$$x = 5 \sin u \cos v \text{ and } y = 5 \sin u \sin v \text{ and } z = 5 \cos u \text{ and } dS = 25 \sin u \, du \, dv.$$

Now find the region R. The angle v (or θ) goes around from 0 to 2π. Since $z = 5\cos u$ goes from 2 to 4, the angle u is between $\cos^{-1}\frac{2}{5}$ and $\cos^{-1}\frac{4}{5}$. Integrate dS over this region R and compare with 20π above:

$$\text{Surface area } = \iint_R 25 \sin u \, du \, dv = 25(2\pi)(-\cos u)] = 50\pi(\frac{4}{5} - \frac{2}{5}) = 20\pi.$$

6. Find the surface area of the cone $z = 1 - \sqrt{x^2 + y^2}$ above the base plane $z = 0$.

- *Method 1:* Compute $\frac{\partial z}{\partial x}$ and $\frac{\partial z}{\partial y}$ and dS. Integrate over the shadow, a circle in the base plane. (Set $z = 0$ to find the shadow boundary $x^2 + y^2 = 1$.) The integral takes 10 steps in Schaum's Outline. The answer is $\pi\sqrt{2}$.
- *Method 2:* Use parameters. Example 2 on page 575 gives $x = u \cos v$ and $y = u \sin v$ and $z = u$ and $dS = \sqrt{2} u \, du \, dv$. The cone has $0 \le u \le 1$ (since $0 \le z \le 1$). The angle v (alias θ) goes from 0 to 2π. This gives the parameter region R and we integrate in one step:

$$\text{cone area } = \int_0^{2\pi} \int_0^1 \sqrt{2} \, u \, du \, dv = (2\pi)(\sqrt{2})(\frac{1}{2}) = \pi\sqrt{2}.$$

The discussion of surface integrals ends with the calculation of ***flow through a surface***. We are given the flow field $\mathbf{F}(x,y,z)$ – a vector field with three components $M(x,y,z)$ and $N(x,y,z)$ and $P(x,y,z)$. The flow *through* a surface is $\iint \mathbf{F} \cdot \mathbf{n} \, dS$. where $\mathbf{n}$ is the unit normal vector to the surface.

For the surface $z = f(x, y)$, you would expect a big square root for dS. It is there, but it is cancelled by a square root in $\mathbf{n}$. We divide the usual normal vector $\mathbf{N} = -\frac{\partial f}{\partial x}\,\mathbf{i} - \frac{\partial f}{\partial y}\,\mathbf{j} + \mathbf{k}$ by its length to get the unit vector $\mathbf{n} = \mathbf{N}/|\mathbf{N}|$. That length is the square root that cancels. This leaves

$$\text{Flow through surface } = \iint_S \mathbf{F}\cdot\mathbf{n}\, dS = \iint_R \mathbf{F}\cdot\mathbf{N}\, dx\, dy = \iint_R \left(-M\frac{\partial f}{\partial x} - N\frac{\partial f}{\partial y} + P\right) dx\, dy.$$

The main job is to find the shadow region R in the xy plane and integrate. The "shadow" is the range of (x, y) down in the base, while (x, y, z) travels over the surface. With parameters u and v, the shadow region R is in the uv plane. It gives the range of parameters (u, v) as the point (x, y, z) travels over the surface S. This is not the easiest section in the book.

***Read-throughs and selected even-numbered solutions* :**

A small piece of the surface $z = f(x, y)$ is nearly **flat**. When we go across by dx, we go up by $(\partial \mathbf{z}/\partial \mathbf{x})\mathbf{dx}$. That movement is $\mathbf{A}dx$, where the vector $\mathbf{A}$ is $\mathbf{i} + \mathbf{dz/dx\ k}$. The other side of the piece is $\mathbf{B}dy$, where $\mathbf{B} = \mathbf{j} + (\partial \mathbf{z}/\partial \mathbf{y})\mathbf{k}$. The cross product $\mathbf{A}\times\mathbf{B}$ is $\mathbf{N} = -\partial \mathbf{z}/\partial \mathbf{x}\,\mathbf{i} - \partial \mathbf{z}/\partial \mathbf{y}\,\mathbf{j} + \mathbf{k}$. The area of the piece is $dS = |\mathbf{N}|dx\, dy$. For the surface $x = xy$, the vectors are $A = \iint \sqrt{\mathbf{1} + \mathbf{x^2} + \mathbf{y^2}}\mathbf{dx\, dy}$ and $N = -\mathbf{y}\,\mathbf{i} - \mathbf{x}\ + \mathbf{k}$. The area integral is $\iint dS = \mathbf{i} + \mathbf{y\ k}$.

With parameters u and v, a typical point on a 45° cone is $x = u\ \cos\ v, y = \mathbf{u\,sinv}, z = \mathbf{u}$. A change in u moves that point by $\mathbf{A\ du} = (\mathbf{cos\ v\ i} + \ \mathbf{sinv\ j} + \mathbf{k})du$. The change in v moves the point by $\mathbf{B}dv = (-\mathbf{u\ sinv\ i} + \mathbf{u\ cos\ v\ j})\mathbf{dv}$. The normal vector is $\mathbf{N} = \mathbf{A}\times\mathbf{B} = -\mathbf{u\ cos\ v\ i} -\ \mathbf{u\ sinv\ j} + \mathbf{u\ k}$. The area is $dS = \sqrt{\mathbf{2}}\ \mathbf{u}du\ dv$. In this example $\mathbf{A}\cdot\mathbf{B} - \mathbf{0}$ so the small piece is a **rectangle** and $dS = |\mathbf{A}||\mathbf{B}|du\ dv$.

For flux we need $\mathbf{n}dS$. The **unit normal** vector $\mathbf{n}$ is $\mathbf{N} = \mathbf{A}\times\mathbf{B}$ divided by $|\mathbf{N}|$. For a surface $z = f(x, y)$, the product $\mathbf{n}dS$ is the vector $\mathbf{N}\ dx\ dy$ (to memorize from table). The particular surface $z = xy$ has $\mathbf{n}dS = (-\mathbf{yi} - \mathbf{xj} + \mathbf{k})dx\ dy$. For $\mathbf{F} = x\mathbf{i} + y\mathbf{j} + z\mathbf{k}$ the flux through $z = xy$ is $\mathbf{F}\cdot\mathbf{n}dS = -\mathbf{xy}\ dx\ dy$.

On a 30° cone the points are $x = 2u\cos v, y = 2u\sin v, z = u$. The tangent vectors are $\mathbf{A} = \mathbf{2\ cosv\ i} + \mathbf{2\ sinv\ j} + \mathbf{k}$ and $\mathbf{B} = -\mathbf{2\ u\ sin\ v\ i} + \mathbf{2u\ cos\ v\ j}$. This cone has $\mathbf{n}dS = \mathbf{A}\times\mathbf{B}\ du\ dv = (-\mathbf{2u\ cos\ v\ i} - \mathbf{2\ u\ sinv\ j} + \mathbf{4\ u\ k})\mathbf{du\ dv}$. For $\mathbf{F} = x\ \mathbf{i} + y\ \mathbf{j} + z\ \mathbf{k}$, the flux element through the cone is $\mathbf{F}\cdot\mathbf{n}dS =$ **zero.** The reason for this answer is **that F is along the cone.** The reason we don't compute flux through a Möbius strip **is that N cannot be defined (the strip is not orientable).**

2 $\mathbf{N} = -2x\,\mathbf{i} - 2y\,\mathbf{j} + \mathbf{k}$ and $dS = \sqrt{\mathbf{1} + \mathbf{4x^2} + \mathbf{4y^2}}\mathbf{dx\ dy}$. Then $\iint dS = \int_0^{2\pi}\int_2^{\sqrt{8}} \sqrt{1+4r^2}\, r\, dr\, d\theta = \frac{\pi}{6}(\mathbf{33^{3/2}} - \mathbf{17^{3/2}})$.

8 $\mathbf{N} = -\frac{x\mathbf{i}}{r} - \frac{y\mathbf{i}}{r} + \mathbf{k}$ and $dS = \frac{x^2+y^2+r^2}{r^2}dx\, dy = \sqrt{\mathbf{2}}\ \mathbf{dx\ dy}$. Then area $= \int_0^{2\pi}\int_a^b \sqrt{2}r dr\ d\theta = \sqrt{\mathbf{2}}\pi(\mathbf{b^2} - \mathbf{a^2})$.

16 On the sphere $dS = \sin\phi\ d\phi\ d\theta$ and $g = x^2 + y^2 = \sin^2\phi$. Then $\int_0^{2\pi}\int_0^{\pi/2} \sin^3\phi\ d\phi\ d\theta = 2\pi(\frac{2}{3}) = \frac{\mathbf{4\pi}}{\mathbf{3}}$.

20 $\mathbf{A} = v\mathbf{i} + \mathbf{j} + \mathbf{k}, \mathbf{B} = u\mathbf{i} + \mathbf{j} - \mathbf{k}, \mathbf{N} = \mathbf{A}\times\mathbf{B} = -2\mathbf{i} + (u+v)\mathbf{j} + (v-u)\mathbf{k}, dS = \sqrt{4 + 2u^2 + 2v^2}du\ dv$.

24 $\iint \mathbf{F}\cdot\mathbf{n}dS = \int_0^{2\pi}\int_2^{\sqrt{8}} -r^3 dr\ d\theta = -24\pi$.

30 $\mathbf{A} = \cos\theta\,\mathbf{i} + \sin\theta\,\mathbf{j} - 2r\mathbf{k}, \mathbf{B} = -r\sin\theta\,\mathbf{i} + r\cos\theta\,\mathbf{j}, \mathbf{N} = \mathbf{A}\times\mathbf{B} = 2r^2\cos\theta\,\mathbf{i} + 2r^2\sin\theta\,\mathbf{j} + r\mathbf{k}$,
$\iint \mathbf{k}\cdot\mathbf{n}\ dS = \iint \mathbf{k}\cdot\mathbf{N}du\ dv = \int_0^{2\pi}\int_0^a r dr\ d\theta = \pi\mathbf{a^2}$ as in Example 12.

15.5 The Divergence Theorem (page 588)

This theorem says that the total source inside a volume V equals the total flow through its closed surface S. We need to know how to measure the *source* and how to measure the *flow out*. Both come from the vector field $\mathbf{F}(x,y,z) = M\mathbf{i} + N\mathbf{j} + P\mathbf{k}$ which assigns a flow vector to every point inside V and on S :

$$\text{Source } = \text{divergence of } \mathbf{F} = \nabla \cdot \mathbf{F} = \frac{\partial M}{\partial x} + \frac{\partial N}{\partial y} + \frac{\partial P}{\partial z}$$

$$\text{Flow out } = \text{normal component of } \mathbf{F} = \mathbf{F} \cdot \mathbf{n} = \mathbf{F} \cdot \mathbf{N}/|\mathbf{N}|$$

The balance between source and outward flow is the *Divergence Theorem*. It is Green's Theorem in 3 dimensions:

$$\iint_S \mathbf{F} \cdot \mathbf{n}\, dS = \iiint_V (\text{divergence of } \mathbf{F})\, dV = \iiint_V \left(\frac{\partial M}{\partial x} + \frac{\partial N}{\partial y} + \frac{\partial P}{\partial z}\right) dx\, dy\, dz.$$

An important special case is a "source-free" field. This means that the divergence is zero. Then the integrals are zero and the total outward flow is zero. There may be flow out through one part of S and flow in through another part – they must cancel when div $\mathbf{F} = 0$ inside V.

1. For the field $\mathbf{F} = y\mathbf{i} + x\mathbf{j}$ in the unit ball $x^2 + y^2 + z^2 \le 1$, compute both sides of the Divergence Theorem.

- The divergence is $\frac{\partial y}{\partial x} + \frac{\partial x}{\partial y} + \frac{\partial 0}{\partial z} = 0$. This source-free field has $\iiint \text{div } \mathbf{F}\, dV = 0$.
- The normal vector to the unit sphere is radially outward: $\mathbf{n} = x\mathbf{i} + y\mathbf{j} + z\mathbf{k}$. Its dot product with $\mathbf{F} = y\mathbf{i} + x\mathbf{j}$ gives $\mathbf{F} \cdot \mathbf{n} = 2xy$ for flow out through S. This is not zero, but *its integral is zero*. One proof is by symmetry: $2xy$ is equally positive and negative on the sphere. The direct proof is by integration (use spherical coordinates):

$$\iint 2xy\, dS = \int_0^{2\pi}\int_0^{\pi} 2(\sin\phi\ \cos\theta)(\sin\phi\ \sin\theta)\sin\phi\, d\phi\, d\theta = 0 \text{ because } \int_0^{2\pi} 2\cos\theta\ \sin\theta\, d\theta = 0.$$

To emphasize: The flow out of *any volume* V is zero because this field has divergence $= 0$. For strange shapes we can't do the surface integral. But the volume integral is still $\iiint 0\, dV$.

2. (This is Problem 15.5.8) Find the divergence of $\mathbf{F} = x^3\mathbf{i} + y^3\mathbf{j} + z^3\mathbf{k}$ and the flow out of the sphere $\rho = a$.

- Divergence $= \frac{\partial x^3}{\partial x} + \frac{\partial y^3}{\partial y} + \frac{\partial z^3}{\partial z} = 3x^2 + 3y^2 + 3z^2 = 3\rho^2$. The triple integral for the total source is

$$\int_0^{2\pi}\int_0^{\pi}\int_0^{a} 3\rho^2(\rho^2 \sin\phi\, d\rho\, d\phi\, d\theta) = 3(\frac{a}{5})^5(2)(2\pi) = 12\pi a^5/5.$$

- Flow out has $\mathbf{F} \cdot \mathbf{n} = (x^3, y^3, z^3) \cdot (\frac{x}{a}, \frac{y}{a}, \frac{z}{a}) = \frac{x^4+y^4+z^4}{a}$. The integral of $\frac{z^4}{a}$ over the sphere $\rho = a$ is

$$\int_0^{2\pi}\int_0^{\pi} \frac{1}{a}(a\cos\phi)^4(a^2\sin\phi\, d\phi\, d\theta) = -2\pi a^5\frac{\cos^5\phi}{5}\Big]_0^{\pi} = \frac{4\pi a^5}{5}.$$

By symmetry $\frac{x^4}{a}$ and $\frac{y^4}{a}$ have this same integral. So multiply by 3 to get the same $12\pi a^5/5$ as above.

The text assigns special importance to the vector field $\mathbf{F} = \mathbf{R}/\rho^3$. This is radially outward (remember $\mathbf{R} = x\mathbf{i} + y\mathbf{j} + z\mathbf{k}$). The length of $\mathbf{F}$ is $|\mathbf{R}|/\rho^3 = \rho/\rho^3 = 1/\rho^2$. This is the *inverse-square law* – the force of gravity from a point mass at the origin decreases like $1/\rho^2$.

The special feature of this radial field is to have zero divergence – except at one point. This is not typical of radial fields: $\mathbf{R}$ itself has divergence $\frac{\partial x}{\partial x} + \frac{\partial y}{\partial y} + \frac{\partial z}{\partial z} = 3$. But dividing $\mathbf{R}$ by ρ^3 gives a field with div $\mathbf{F} = 0$. (Physically: No divergence where there is no mass and no source of gravity.) The exceptional point is (0,0,0), where there *is* a mass. That point source is enough to produce 4π on both sides of the divergence theorem (provided S encloses the origin). The point source has strength 4π. The divergence of $\mathbf{F}$ is 4π times a "delta function."

The other topic in this section is the vector form of two familiar rules: the *product rule* for the derivative of $u(x)v(x)$ and the reverse of the product rule which is *integration by parts*. Now we are in 2 or 3 dimensions and v is a vector field $\mathbf{V}(x, y, z)$. The derivative is replaced by the divergence or the curl. We just use the old product rule on uM and uN and uP. Then collect terms:

$$\text{div}(u\mathbf{V}) = u \text{ div}\mathbf{V} + (\text{grad } u) \cdot \mathbf{V} \qquad \text{curl}(u\mathbf{V}) = u \text{ curl } \mathbf{V} + (\text{grad } u) \times \mathbf{V}.$$

Read-throughs and selected even-numbered solutions :

In words, the basic balance law is **flow in = flow out.** The flux of $\mathbf{F}$ through a surface S is the double integral $\iint \mathbf{F} \cdot \mathbf{n}dS$. The divergence of $M\mathbf{i} + N\mathbf{j} + P\mathbf{k}$ is $\mathbf{M_x + N_y + P_z}$. It measures the **source at the point.** The total source is the triple integral $\iiint$ div $\mathbf{F}$ $\mathbf{dV}$. That equals the flux by the **Divergence** Theorem.

For $\mathbf{F} = 5z\mathbf{k}$ the divergence is **5.** If V is a cube of side a then the triple integral equals $\mathbf{5a^3}$. The top surface where $z = a$ has $\mathbf{n} = \mathbf{k}$ and $\mathbf{F} \cdot \mathbf{n} = 5a$. The bottom and sides have $\mathbf{F} \cdot \mathbf{n} =$ **zero**. The integral $\iint \mathbf{F} \cdot \mathbf{n}dS\mathbf{5a^3}$.

The field $\mathbf{F} = \mathbf{R}/\rho^3$ has div $\mathbf{F} = 0$ except **at the origin.** $\iint \mathbf{F} \cdot \mathbf{n}dS$ equals $\mathbf{4\pi}$ over any surface around the origin. This illustrates Gauss's Law: **flux =** 4π **times source strength.** The field $\mathbf{F} = x\,\mathbf{i} + y\,\mathbf{j} - 2z\,\mathbf{k}$ has div $\mathbf{F} = \mathbf{0}$ and $\iint \mathbf{F} \cdot \mathbf{n}dS = \mathbf{0}$. For this $\mathbf{F}$, the flux out through a pyramid and in through its base are **equal.**

The symbol ∇ stands for $(\partial/\partial \mathbf{x})\mathbf{i} + (\partial/\partial \mathbf{y})\mathbf{j} + (\partial/\partial \mathbf{z})\mathbf{k}$. In this notation div $\mathbf{F}$ is $\nabla \cdot \mathbf{F}$. The gradient of f is $\nabla \mathbf{f}$. The divergence of grad f is $\nabla \cdot \nabla \mathbf{f}$ **or** $\nabla^2 \mathbf{f}$. The equation div grad $f = 0$ is **Laplace's** equation.

The divergence of a product is $\text{div}(u\mathbf{V}) = \mathbf{u\,div\,V + (grad\,u) \cdot V}$. Integration by parts in 3D is $\iiint u \text{ div } \mathbf{V} dx\,dy\,dz = -\iiint \mathbf{V} \cdot \text{grad}\,\mathbf{u\,dx\,dy\,dz} + \iint \mathbf{u\,V \cdot n\,dS}$. In two dimensions this becomes $\iint \mathbf{u(\partial M/\partial x + \partial N/\partial y)dx\,dy} = -\int \mathbf{(M\,\partial u/\partial x + N\,\partial u/\partial y)dx\,dy} + \int \mathbf{u\,V \cdot n\,ds}$. In one dimension it becomes **integration by parts.** For steady fluid flow the continuity equation is div $\rho\mathbf{V} = -\partial\rho/\partial \mathbf{t}$.

14 $\mathbf{R} \cdot \mathbf{n} = (x\mathbf{i} + y\mathbf{j} + z\mathbf{k}) \cdot \mathbf{i} = x = 1$ on one face of the box. On the five other faces $\mathbf{R} \cdot \mathbf{n} = 2, 3, 0, 0, 0$. The integral is $\int_0^3 \int_0^2 1dy\,dz + \int_0^3 \int_0^1 2dx\,dz + \int_0^2 \int_0^1 3dx\,dy = \mathbf{18}$. Also div $\mathbf{R} = 1 + 1 + 1 = 3$ and $\int_0^3 \int_0^2 \int_0^1 3dx\,dy\,dz = \mathbf{18}$.

18 grad $f \cdot \mathbf{n}$ is the **directional** derivative in the normal direction $\mathbf{n}$ (also written $\frac{\partial f}{\partial n}$). The Divergence Theorem gives $\iiint$ div (grad f) $dV = \iint$ grad $f \cdot \mathbf{n}dS = \iint \frac{\partial f}{\partial n} dS$. But we are given that div (grad f) $= f_{xx} + f_{yy} + f_{zz}$ is zero.

26 When the density ρ is constant (incompressible flow), the continuity equation becomes div $\mathbf{V} = 0$. If the flow is irrotational then $\mathbf{F} =$ grad f and the continuity equation is div (ρ grad f) $= -d\rho/dt$. If also $\rho =$ constant, then **div grad f = 0**: Laplace's equation for the "potential."

30 The boundary of a solid ball is a sphere. A sphere has no boundary. Similarly for a cube or a cylinder – the boundary is a closed surface and **that surface's boundary** is empty. This is a crucial fact in topology.

15.6 Stokes' Theorem and the Curl of F (page 595)

The curl of $\mathbf{F}$ measures the "spin". A spin field in the plane is $\mathbf{S} = -y\mathbf{i} + x\mathbf{j}$, with third component $P = 0$:

$$\text{curl } \mathbf{S} = (\frac{\partial P}{\partial y} - \frac{\partial N}{\partial z})\mathbf{i} + (\frac{\partial M}{\partial z} - \frac{\partial P}{\partial x})\mathbf{j} + (\frac{\partial N}{\partial x} - \frac{\partial M}{\partial y})\mathbf{k} = 0\mathbf{i} + 0\mathbf{j} + (1+1)\mathbf{k}.$$

The curl is $2\ \mathbf{k}$. It points along the spin axis (the z axis, perpendicular to the plane of spin). Its magnitude is 2 times the rotation rate. This special spin field $\mathbf{S}$ gives a rotation counterclockwise in the xy plane. It is counterclockwise because $\mathbf{S}$ points that way, and also because of the right hand rule: thumb in the direction of curl $\mathbf{F}$ and fingers "curled in the direction of spin." Put your right hand on a table with thumb upward along $\mathbf{k}$.

Spin fields can go around any axis vector $\mathbf{a}$. The field $\mathbf{S} = \mathbf{a} \times \mathbf{R}$ does that. Its curl is $2\ \mathbf{a}$ (after calculation). Other fields have a curl that changes direction from point to point. Some fields have no spin at all, so their curl is *zero. These are gradient fields!* This is a key fact:

$$\text{curl } \mathbf{F} = \mathbf{0} \text{ whenever } \mathbf{F} = \text{grad } f = \frac{\partial f}{\partial x}\mathbf{i} + \frac{\partial f}{\partial y}\mathbf{j} + \frac{\partial f}{\partial z}\mathbf{k}.$$

You should substitute those three partial derivatives for M, N, P and see that the curl formula gives zero. ***The curl of a gradient is zero.*** The quick test $\mathbf{D}$ for a gradient field is curl $\mathbf{F} = \mathbf{0}$.

The twin formula is that *the divergence of a curl is zero.* The quick test $\mathbf{H}$ for a source-free "curl field" is div $\mathbf{F} = 0$. A gradient field can be a gravity field or an electric field. A curl field can be a magnetic field.

1. Show that $\mathbf{F} = yz\ \mathbf{i} + xy\ \mathbf{j} + (xy + 2z)\ \mathbf{k}$ passes the quick test $\mathbf{D}$ for a gradient field (a conservative field). The test is curl $\mathbf{F} = \mathbf{0}$. Find the potential function f that this test guarantees: $\mathbf{F}$ equals the gradient of f.

   - The curl of this $\mathbf{F}$ is $(\frac{\partial P}{\partial y} - \frac{\partial N}{\partial z})\mathbf{i} + (\frac{\partial M}{\partial z} - \frac{\partial P}{\partial x})\mathbf{j} + (\frac{\partial N}{\partial x} - \frac{\partial M}{\partial y})\mathbf{k} = (x - x)\mathbf{i} + (y - y)\mathbf{j} + (z - z)\mathbf{k} = \mathbf{0}$. The field passes test $\mathbf{D}$. There must be a function whose partial derivatives are M, N, P : $\frac{\partial f}{\partial x} = yz$ and $\frac{\partial f}{\partial y} = xz$ and $\frac{\partial f}{\partial z} = xy + 2z$ lead to the potential function $f = xyz + z^2$.

We end with ***Stokes' Theorem.*** It is like the original Green's Theorem, where work around a plane curve C was equal to the double integral of $\frac{\partial N}{\partial x} - \frac{\partial M}{\partial y}$. Now the curve C can go out of the plane. The region inside is a curved surface – also not in a plane. The field $\mathbf{F}$ is three-dimensional – its component $P\mathbf{k}$ goes out of the plane. Stokes' Theorem has a line integral $\int \mathbf{F} \cdot d\mathbf{R}$ (work around C) equal to a surface integral:

$$\int_C M\,dx + N\,dy + P\,dz = \iint_S (\text{curl } \mathbf{F}) \cdot \mathbf{n}\,dS = \iint_{\text{base}} (\text{curl } \mathbf{F}) \cdot \mathbf{N}\,dx\,dy.$$

If the surface is $z = f(x, y)$ then its normal is $\mathbf{N} = -\frac{\partial z}{\partial x}\mathbf{i} - \frac{\partial z}{\partial y}\mathbf{j} + \mathbf{k}$ and dz is $\frac{\partial z}{\partial x}dx + \frac{\partial z}{\partial y}dy$. Substituting for $\mathbf{N}$ and dz reduces the 3-dimensional theorem of Stokes to the 2-dimensional theorem of Green.

2. (Compare 15.6.12) For $\mathbf{F} = \mathbf{i} \times \mathbf{R}$ compute both sides in Stokes' Theorem when C is the unit circle.

   - The cross product $\mathbf{F} = \mathbf{i} \times \mathbf{R}$ is $\mathbf{i} \times (x\mathbf{i} + y\mathbf{j} + z\mathbf{k}) = y\mathbf{k} - z\mathbf{j}$. This spin field has curl $\mathbf{F} = 2\mathbf{i}$. either substitute $M = 0, N = -z, P = y$ in the curl formula or remember that curl $(\mathbf{a} \times \mathbf{R}) = 2\mathbf{a}$. Here $\mathbf{a} = \mathbf{i}$.

The line integral of $\mathbf{F} = y\mathbf{k} - z\mathbf{j}$ around the unit circle is $\int 0\,dx - z\,dy + y\,dz$. All those are zero because $z = 0$ for the circle $x^2 + y^2 = 1$ in the xy plane. ***The left side of Stokes' Theorem is zero for this* F.**

The double integral of (curl $\mathbf{F}$)$\cdot\mathbf{n} = 2\mathbf{i}\cdot\mathbf{n}$ is certainly zero if the surface S is the flat disk inside the unit circle. The normal vector to that flat surface is $\mathbf{n} = \mathbf{k}$. Then $2\mathbf{i}\cdot\mathbf{n}$ is $2\mathbf{i}\cdot\mathbf{k} = 0$.

The double integral of $2\mathbf{i}\cdot\mathbf{n}$ is zero even if the surface S is not flat. S can be a mountain (always with the unit circle as its base boundary). The normal $\mathbf{n}$ out from the mountain can have an $\mathbf{i}$ component, so $2\mathbf{i}\cdot\mathbf{n}$ can be non-zero. But Stokes' Theorem says: The *integral* of $2\mathbf{i}\cdot\mathbf{n}$ over the mountain is zero. That is the theorem. The mountain integral must be zero because the baseline integral is zero.

Read-throughs and selected even-numbered solutions :

The curl of $M\mathbf{i} + N\mathbf{j} + P\mathbf{k}$ is the vector $(\mathbf{P_y} - \mathbf{N_z})\mathbf{i} + (\mathbf{M_z} - \mathbf{P_x})\mathbf{j} + (\mathbf{N_x} - \mathbf{M_y})\mathbf{k}$. It equals the 3 by 3 determinant $\begin{vmatrix} \mathbf{i} & \mathbf{j} & \mathbf{k} \\ \partial/\partial \mathbf{x} & \partial/\partial \mathbf{y} & \partial/\partial \mathbf{z} \\ \mathbf{M} & \mathbf{N} & \mathbf{P} \end{vmatrix}$ The curl of $x^2\mathbf{i} + z^2\mathbf{k}$ is **zero.** For $\mathbf{S} = y\mathbf{i} - (x+z)\mathbf{j} + y\mathbf{k}$ the curl is $\mathbf{2i - 2k}$. This $\mathbf{S}$ is a **spin** field $\mathbf{a}\times\mathbf{R} = \frac{1}{2}(\text{ curl }\mathbf{F})\times\mathbf{R}$, with axis vector $\mathbf{a} = \mathbf{i} - \mathbf{k}$. For any gradient field $f_x\mathbf{i} + f_y\mathbf{j} + f_z\mathbf{k}$ the curl is **zero.** That is the important identity curl grad f = **zero**. It is based on $f_{xy} = f_{yx}$ and $\mathbf{f_{xz} = f_{zx}}$ and $\mathbf{f_{yz} = f_{zy}}$. The twin identity is **div curl F = 0.**

The curl measures the **spin (or turning)** of a vector field. A paddlewheel in the field with its axis along $\mathbf{n}$ has turning speed $\mathbf{\frac{1}{2}n}\cdot$ **curl F**. The spin is greatest when $\mathbf{n}$ is in the direction of **curl F**. Then the angular velocity is $\mathbf{\frac{1}{2}}|$ **curl F**$|$. Stokes' Theorem is $\oint_C \mathbf{F}\cdot d\mathbf{R} = \iint_S (\text{curl }\mathbf{F})\cdot\mathbf{n}\,dS$. The curve C is the **boundary** of the surface S. This is **Green's** Theorem extended to **three** dimensions. Both sides are zero when **F** is a gradient field because **the curl is zero.**

The four properties of a conservative field are **A** : $\oint \mathbf{F}\cdot d\mathbf{R} = 0$ **and B** : $\int_P^Q \mathbf{F}\cdot d\mathbf{R}$ **depends only on P and Q and C : F is the gradient of a potential function f(x, y, z) and D : curl F = 0.**
The field $y^2z^2\mathbf{i} + 2xy^2z\mathbf{k}$ fails test **D.** This field is the gradient of **no f.** The work $\int \mathbf{F}\cdot d\mathbf{R}$ from (0,0,0) to (1,1,1) is $\frac{3}{5}$ **along the straight path x = y = z = t.** For every field **F**, $\iint$ **curl F** $\cdot\,\mathbf{n}dS$ is the same out through a pyramid and up through its base **because they have the same boundary, so** $\oint \mathbf{F}\cdot d\mathbf{R}$ **is the same.**

14 $\mathbf{F} = (x^2 + y^2)\mathbf{k}$ so curl $\mathbf{F} = 2(y\mathbf{i} - x\mathbf{j})$. (Surprise that this $\mathbf{F} = \mathbf{a}\times\mathbf{R}$ has curl $\mathbf{F} = 2\mathbf{a}$ even with nonconstant $\mathbf{a}$.) Then $\oint \mathbf{F}\cdot d\mathbf{R} = \iint$ curl $\mathbf{F}\cdot\mathbf{n}dS = 0$ since $\mathbf{n} = \mathbf{k}$ is perpendicular to curl $\mathbf{F}$.

18 If curl $\mathbf{F} = \mathbf{0}$ then $\mathbf{F}$ is the gradient of a potential: $\mathbf{F}$ = grad f. Then div $\mathbf{F} = 0$ is div grad $f = 0$ which is Laplace's equation.

24 Start with one field that has the required curl. (Can take $\mathbf{F} = \frac{1}{2}\mathbf{i}\times\mathbf{R} = -\frac{z}{2}\mathbf{j} + \frac{y}{2}\mathbf{k}$). Then add any $\mathbf{F}$ with curl zero (particular solution plus homogeneous solution as always). The fields with curl $\mathbf{F} = \mathbf{0}$ are gradient fields $\mathbf{F}$ = grad f, since curl grad = $\mathbf{0}$. Answer: $\mathbf{F} = \frac{1}{2}\mathbf{i}\times\mathbf{R} +$ any grad f.

26 $\mathbf{F} = y\mathbf{i} - x\mathbf{k}$ has curl $\mathbf{F} = \mathbf{j} - \mathbf{k}$. (a) Angular velocity $= \frac{1}{2}$ curl $\mathbf{F}\cdot\mathbf{n} = \frac{1}{2}$ if $\mathbf{n} = \mathbf{j}$.
(b) Angular velocity $= \frac{1}{2}|\text{curl}\mathbf{F}| = \frac{\sqrt{2}}{2}$ (c) Angular velocity = 0.

36 curl $\mathbf{F} = \begin{vmatrix} \mathbf{i} & \mathbf{j} & \mathbf{k} \\ \partial/\partial x & \partial/\partial y & \partial/\partial z \\ z & x & xyz \end{vmatrix} = \mathbf{i}(xz) + \mathbf{j}(1 - yz) + \mathbf{k}(1)$ and $\mathbf{n} = x\mathbf{i} + y\mathbf{j} + z\mathbf{k}$. So curl $\mathbf{F}\cdot\mathbf{n} = x^2z + y - y^2z + z$. By symmetry $\iint x^2z\,dS = \iint y^2z\,dS$ on the half sphere and $\iint y\,dS = 0$. This leaves $\iint z\,dS = \int_0^{2\pi}\int_0^{\pi/2}\cos\phi(\sin\phi\,d\phi\,d\theta) = \frac{1}{2}(2\pi) = \pi$.

38 (The expected method is trial and error) $\mathbf{F} = 5yz\mathbf{i} + 2xy\mathbf{k} +$ any grad f.

15 Chapter Review Problems

Review Problems

R1 For $f(x,y) = x^2 + y^2$ what is the gradient field $\mathbf{F} = \nabla f$? What is the unit field $\mathbf{u} = \mathbf{F}/|\mathbf{F}|$?

R2 Is $\mathbf{F}(x,y) = \cos x\ \mathbf{i} + \sin x\ \mathbf{j}$ a gradient field? Draw the vectors $\mathbf{F}$ (0,0), $\mathbf{F}(0,\pi), \mathbf{F}(\pi,0), \mathbf{F}(\pi,\pi)$.

R3 Is $\mathbf{F}(x,y) = \cos y\ \mathbf{i} + \sin x\ \mathbf{j}$ a gradient field or not? Draw $\mathbf{F}$(0,0), $\mathbf{F}(0,\pi), \mathbf{F}(\pi,0), \mathbf{F}(\pi,\pi)$.

R4 Is $\mathbf{F}(x,y) = \cos x\ \mathbf{i} + \sin y\ \mathbf{j}$ a gradient field? If so find a potential $f(x,y)$ whose gradient is $\mathbf{F}$.

R5 Integrate $-y\ dx + x\ dy$ around the unit circle $x = \cos t, y = \sin t$. Why twice the area?

R6 Integrate the gradient of $f(x,y) = x^3 + y^3$ around the unit circle.

R7 With Green's Theorem find an integral around C that gives the area inside C.

R8 Find the flux of $\mathbf{F} = x^2\mathbf{i}$ through the unit circle from both sides of Green's Theorem.

R9 What integral gives the area of the surface $z = f(x,y)$ above the square $|x| \le 1,\ |y| \le 1$?

R10 Describe the cylinder given by $x = 2\cos v, y = 2\sin v$, and $z = u$. Is (2, 2, 2) on the cylinder? What parameters u and v produce the point (0, 2, 4)?

Drill Problems

D1 Find the gradient $\mathbf{F}$ of $f(x,y,z) = x^3 + y^3$. Then find the divergence and curl of $\mathbf{F}$.

D2 What integral gives the surface area of $z = 1 - x^2 - y^2$ above the xy plane?

D3 What is the area on the sloping plane $z = x + y$ above a base area (or shadow area) A?

D4 Write down Green's Theorem for $\int M dx + N dy = work$ and $\int M dy - N dx = flux$. Write down the flux form with $M dy - N dx$.

D5 Write down the Divergence Theorem. Say in words what it balances.

D6 If $\mathbf{F} = y^2\ \mathbf{i} + (1+2x)y\ \mathbf{j}$ is the gradient of $f(x,y)$, find the potential function f.

D7 When is the area of a surface equal to the area of its shadow on $z = 0$? (Surface is $z = f(x,y)$.

D8 For $\mathbf{F} = 3x\mathbf{i} + 4y\mathbf{j} + 5\mathbf{k}$, the flux $\iint \mathbf{F} \cdot \mathbf{n} dS$ equals ___ times the volume inside S.

D9 What vector field is the curl of $\mathbf{F} = xyz\ \mathbf{i}$? Find the gradient of that curl.

D10 What vector field is the gradient of $f = xyz$? Find the curl of that gradient.